Modular Maths

for Edexcel A/AS Level

Pure Mathematics 2

Series Editor Alan Smith

John Sykes
Val Hanrahan
Roger Porkess
Peter Secker

P2

Hodder & Stoughton
A MEMBER OF THE HODDER HEADLINE GROUP

ACKNOWLEDGEMENTS

We are grateful to the following companies, institutions and individuals who have given permission to reproduce photographs in this book. Every effort has been made to trace and acknowledge ownership of copyright. The publishers will be glad to make suitable arrangements with any copyright holders whom it has not been possible to contact.

David Simson (page 11)

Orders: please contact Bookpoint Ltd, 130 Milton Park, Abingdon, Oxon OX14 4SB. Telephone: (44) 01235 827720, Fax: (44) 01235 400454. Lines are open from 9.00–6.00, Monday to Saturday, with a 24 hour message answering service. Email address: orders@bookpoint.co.uk

British Library Cataloguing in Publication Data
A catalogue record for this title is available from the The British Library

ISBN 0 340 77990X

First published 2000
Impression number 10 9 8 7 6 5 4 3 2
Year 2005 2004 2003 2002 2001

This work includes material adapted from the MEI Structured Mathematics Series

Typeset by Pantek Arts Ltd, Maidstone, Kent.
Printed in Great Britain for Hodder & Stoughton Educational, a division of Hodder Headline Plc, 338 Euston Road, London NW1 3BH by J.W. Arrowsmiths Ltd, Bristol.

Edexcel Advanced Mathematics

The Edexcel course is based on units in the four strands of Pure Mathematics, Mechanics, Statistics and Decision Mathematics. The first unit in each of these strands is designated AS; all others are A2.

The units may be aggregated as follows:

3 units	AS Mathematics
6 units	A Level Mathematics
9 units	A Level Mathematics + AS Further Mathematics
12 units	A Level Mathematics + A Level Further Mathematics

Note the other titles (such as Applied Mathematics, Statistics and so on) are available for certain combinations of units. Full details can be found in the Edexcel Specification booklet.

The six units required for an award in A Level Mathematics must comprise Pure Mathematics 1, 2 and 3, plus three units from the remaining (applications) strands, including at least one other A2 unit. The synoptic requirement means that a specified pair of units must be examined together in the final session at the end of the course. Again, full details are given in the Edexcel Specification booklet.

Examinations are offered by Edexcel twice a year, in January and in June. Certain units with low candidate numbers (such as Mechanics 6) are offered during the summer sitting only.

Eighteen of the twenty units are assessed by examination only. The exceptions are Statistics 3 and 6, which each contain a project worth 25% of the total marks for that unit.

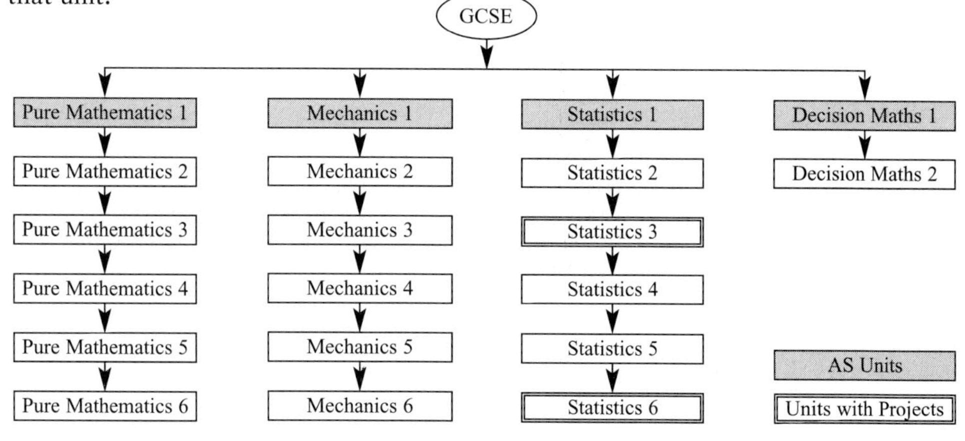

INTRODUCTION

This is the second book in a series written to support the Pure Mathematics units in the Edexcel Advanced Mathematics scheme. It has been adapted from the successful series written to support the MEI Structured Mathematics scheme, and has been substantially edited and rewritten to provide complete coverage of the new Edexcel Pure Mathematics 2 unit.

Throughout the series the emphasis is on understanding and applying a wide variety of mathematical skills, and examining conjectures using proof and counterexample.

There are eight chapters and an appendix in this book. A short introductory chapter on algebra is followed by a detailed treatment of functions. The sequences and series chapter builds on that in *Pure Mathematics 1* and goes on to develop the binomial series expansion. The chapter on trigonometry examines identities and standard results for double and half angles, and is followed by a short introduction to exponentials and logarithms.

The two calculus chapters, on differentiation and integration, extend the ideas met in *Pure Mathematics 1* to apply to a wider range of functions. The final chapter, on numerical methods, looks at ways of finding approximate solutions to unusual equations.

The book concludes with a short appendix on proof. At the time of writing the Edexcel specification explicity stated that proof will form the main content for one question in the Pure Mathematics 2 examination.

Throughout the book you will require the use of a calculator. Computer packages and graphics calculators may be helpful in clarifying new ideas, but you should remember that certain calculator restrictions may be enforced by the examining board. At the time of writing candidates were only allowed the use of a 'simple scientific calculator' for the Pure Mathematics 1 and Pure Mathematics 3 examinations.

I would like to thank the many people who have helped in the preparation and checking of material. Special thanks to Val Hanrahan, Roger Porkess and Peter Secker, who wrote the original MEI edition, and to Terry Heard for his helpful suggestions.

John Sykes

CONTENTS

Chapter one

Algebra and functions

She must know all the needs of a rational being.

Sir Owen Seaman

Review of algebraic fractions

If $f(x)$ and $g(x)$ are polynomials, the expression $\frac{f(x)}{g(x)}$ is an *algebraic fraction* or *rational function*. It may also be called a *rational expression*. There are many occasions in mathematics when a problem reduces to the manipulation of algebraic fractions, and the rules for this are exactly the same as those for numerical fractions.

Simplifying fractions

To simplify a fraction, you look for a factor common to both the numerator (top line) and the denominator (bottom line) and cancel by it.

For example, in arithmetic

$$\frac{15}{20} = \frac{5 \times 3}{5 \times 4} = \frac{3}{4}$$

and in algebra

$$\frac{6a}{9a^2} = \frac{2 \times 3 \times a}{3 \times 3 \times a \times a} = \frac{2}{3a}.$$

Note You must *factorise* both the numerator and denominator before cancelling, since it is only possible to cancel by a *common factor*. In some cases this involves putting brackets in. For example

$$\frac{2a + 4}{a^2 - 4} = \frac{2(a + 2)}{(a + 2)(a - 2)} = \frac{2}{(a - 2)}.$$

MULTIPLYING AND DIVIDING FRACTIONS

Multiplying fractions involves cancelling any factors common to the numerator and denominator. For example:

$$\frac{10a}{3b^2} \times \frac{9ab}{25} = \frac{6a^2}{5b}.$$

As with simplifying, it is often necessary to factorise any algebraic expressions first:

$$\frac{a^2 + 3a + 2}{9} \times \frac{12}{a + 1} = \frac{(a + 1)(a + 2)}{3 \times 3} \times \frac{3 \times 4}{(a + 1)}$$

$$= \frac{(a + 2)}{3} \times \frac{4}{1}.$$

$$= \frac{4(a + 2)}{3}.$$

Remember that when one fraction is divided by another, you change ÷ to × and invert the fraction which follows the ÷ symbol.

$$\frac{12}{x^2 - 1} \div \frac{4}{x + 1} = \frac{12}{(x + 1)(x - 1)} \times \frac{(x + 1)}{4}$$

$$= \frac{3}{(x - 1)}.$$

ADDITION AND SUBTRACTION OF FRACTIONS

To add or subtract two fractions they must be replaced by equivalent fractions, both of which have the same denominator.

For example:

$$\frac{2}{3} + \frac{1}{4} = \frac{8}{12} + \frac{3}{12} = \frac{11}{12}.$$

Similarly, in algebra:

$$\frac{2x}{3} + \frac{x}{4} = \frac{8x}{12} + \frac{3x}{12} = \frac{11x}{12}$$

and $$\frac{2}{3x} + \frac{1}{4x} = \frac{8}{12x} + \frac{3}{12x} = \frac{11}{12x}.$$

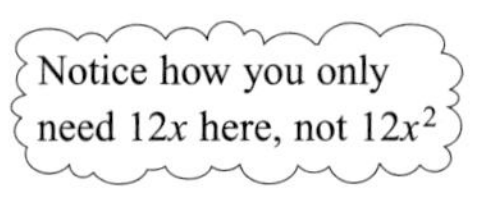

You must take particular care when the subtraction of fractions introduces a sign change. For example:

$$\frac{4x - 3}{6} - \frac{2x + 1}{4} = \frac{2(4x - 3) - 3(2x + 1)}{12}$$

$$= \frac{8x - 6 - 6x - 3}{12}$$

$$= \frac{2x - 9}{12}.$$

Note

In addition and subtraction, the new denominator is the *lowest common multiple* of the original denominators. When two denominators have no *common factor*, this product gives the new denominator. For example

$$\frac{2}{y+3} + \frac{3}{y-2} = \frac{2(y-2) + 3(y+3)}{(y+3)(y-2)}$$

$$= \frac{2y - 4 + 3y + 9}{(y+3)(y-2)}$$

$$= \frac{5y + 5}{(y+3)(y-2)}$$

$$= \frac{5(y+1)}{(y+3)(y-2)}.$$

It may be necessary to factorise denominators in order to identify common factors, as shown here:

$$\frac{2b}{a^2 - b^2} - \frac{3}{a+b} = \frac{2b}{(a+b)(a-b)} - \frac{3}{(a+b)}$$

$$= \frac{2b - 3(a-b)}{(a+b)(a-b)}$$

$$= \frac{5b - 3a}{(a+b)(a-b)}.$$

$(a + b)$ is a common factor

PROOF

You have to be able to show how to obtain a given result. The methods used were studied in *Pure Mathematics 1*. If one expression is equivalent to another the ≡ symbol is used. The following two examples illustrate the process. In each case we start with the left-hand side (LHS) of the equation and derive the right-hand side (RHS).

EXAMPLE 1.1

Prove that $\dfrac{x^2-1}{x+1} \times \dfrac{x}{x-1} \equiv x.$

Solution

$$\begin{aligned}\text{LHS} &= \frac{x^2-1}{x+1} \times \frac{x}{x-1}\\ &= \frac{(x+1)(x-1)}{x+1} \times \frac{x}{x-1}\\ &= x\\ &= \text{RHS.}\end{aligned}$$

Hence the result is proven.

EXAMPLE 1.2

Prove that $\dfrac{(x+1)(x+2)}{(x-3)(x+4)} - \dfrac{2}{x-3} \equiv \dfrac{(x+3)(x-2)}{(x-3)(x+4)}$

Solution

$$\begin{aligned}\text{LHS} &= \frac{(x+1)(x+2)}{(x-3)(x+4)} - \frac{2}{x-3}\\ &= \frac{x^2+3x+2-2(x+4)}{(x-3)(x+4)}\\ &= \frac{x^2+x-6}{(x-3)(x+4)}\\ &= \frac{(x+3)(x-2)}{(x-3)(x+4)}\\ &= \text{RHS.}\end{aligned}$$

Hence the result is proven.

Exercise 1A

Simplify the expressions in questions 1–10.

1 $\dfrac{6a}{b} \times \dfrac{a}{9b^2}$

2 $\dfrac{5xy}{3} \div 15xy^2$

3 $\dfrac{x^2-9}{x^2-9x+18}$

4 $\dfrac{5x-1}{x+3} \times \dfrac{x^2+6x+9}{5x^2+4x-1}$

5 $\dfrac{4x^2 - 25}{4x^2 + 20x + 25}$

6 $\dfrac{a^2 + a - 12}{5} \times \dfrac{3}{4a - 12}$

7 $\dfrac{4x^2 - 9}{x^2 + 2x + 1} \div \dfrac{2x - 3}{x^2 + x}$

8 $\dfrac{2p + 4}{5} \div (p^2 - 4)$

9 $\dfrac{a^2 - b^2}{2a^2 + ab - b^2}$

10 $\dfrac{x^2 + 8x + 16}{x^2 + 6x + 9} \times \dfrac{x^2 + 2x - 3}{x^2 + 4x}$

In questions 11–24 write each of the expressions as a single fraction in its simplest form.

11 $\dfrac{1}{4x} + \dfrac{1}{5x}$

12 $\dfrac{x}{3} - \dfrac{(x + 1)}{4}$

13 $\dfrac{a}{a + 1} + \dfrac{1}{a - 1}$

14 $\dfrac{2}{x - 3} + \dfrac{3}{x - 2}$

15 $\dfrac{x}{x^2 - 4} - \dfrac{1}{x + 2}$

16 $\dfrac{p^2}{p^2 - 1} - \dfrac{p^2}{p^2 + 1}$

17 $\dfrac{2}{a + 1} - \dfrac{a}{a^2 + 1}$

18 $\dfrac{2y}{(y + 2)^2} - \dfrac{4}{y + 4}$

19 $x + \dfrac{1}{x + 1}$

20 $\dfrac{2}{b^2 + 2b + 1} - \dfrac{3}{b + 1}$

21 $\dfrac{2}{3(x - 1)} + \dfrac{3}{2(x + 1)}$

22 $\dfrac{6}{5(x + 2)} - \dfrac{2x}{(x + 2)^2}$

23 $\dfrac{2}{a + 2} - \dfrac{a - 2}{2a^2 + a - 6}$

24 $\dfrac{1}{x - 2} + \dfrac{1}{x} + \dfrac{1}{x + 2}$

In questions 25–30 prove the given results.

25 $\dfrac{x^2 + 2x + 1}{x^2 + 3x + 2} \equiv \dfrac{x + 1}{x + 2}$

26 $\dfrac{1}{x + 1} - \dfrac{1}{x + 2} \equiv \dfrac{1}{(x + 1)(x + 2)}$

27 $\dfrac{(x + 1)(x + 3)}{(x - 1)(x - 3)} \div \dfrac{x^2 - 9}{x^2 - 1} \equiv \dfrac{(x + 1)^2}{(x - 3)^2}$

28 $\dfrac{3(x + 2)}{(x + 1)(x - 4)} - \dfrac{3}{x - 4} \equiv \dfrac{3}{(x + 1)(x - 4)}$

29 $\dfrac{x^2 + 8x + 12}{x^2 + 7x + 12} \times \dfrac{x^2 + 6x + 8}{x^2 + 9x + 18} \equiv \left(\dfrac{x + 2}{x + 3}\right)^2$

30 $\dfrac{(x + 1)(x + 4)}{(x - 2)(x + 3)} + \dfrac{2}{x - 2} \equiv \dfrac{(x + 2)(x + 5)}{(x - 2)(x + 3)}$

EQUATIONS INVOLVING ALGEBRAIC FRACTIONS

The easiest way to solve an equation involving fractions is usually to multiply both sides by an expression which will cancel out the fractions.

EXAMPLE 1.3

Solve $\frac{x}{3} + \frac{2x}{5} = 4$.

Solution Multiplying by 15 (the lowest common multiple of 3 and 5) gives

$$15 \times \frac{x}{3} + 15 \times \frac{2x}{5} = 15 \times 4$$

Notice that all three terms must be multiplied by 15

$$\Rightarrow \quad 5x + 6x = 60$$

$$\Rightarrow \quad 11x = 60$$

$$\Rightarrow \quad x = \frac{60}{11}.$$

A similar method applies when the denominator is algebraic.

EXAMPLE 1.4

Solve $\frac{5}{x} - \frac{4}{x+1} = 1$.

Solution Multiplying by $x(x + 1)$ (the least common multiple of x and $x + 1$) gives

$$\frac{5x(x+1)}{x} - \frac{4x(x+1)}{x+1} = x(x+1)$$

$$\Rightarrow \quad 5(x+1) - 4x = x(x+1)$$

$$\Rightarrow \quad 5x + 5 - 4x = x^2 + x$$

$$\Rightarrow \quad x^2 = 5$$

$$\Rightarrow \quad x = \pm\sqrt{5}.$$

In Example 1.4, the lowest common multiple of the denominators is their product, but this is not always the case

EXAMPLE 1.5

Solve $\dfrac{1}{(x-3)(x-1)} + \dfrac{1}{x(x-1)} = -\dfrac{1}{x(x-3)}$.

Solution Here you only need to multiply by $x(x-3)(x-1)$ to eliminate all the fractions. This gives

$$\frac{x(x-3)(x-1)}{(x-3)(x-1)} + \frac{x(x-3)(x-1)}{x(x-1)} = \frac{-x(x-3)(x-1)}{x(x-3)}$$

$$\Rightarrow x + (x-3) = -(x-1)$$

$$\Rightarrow \quad 2x - 3 = -x + 1$$

$$\Rightarrow \quad 3x = 4$$

$$\Rightarrow \quad x = \tfrac{4}{3}.$$

Fractional algebraic equations arise in a number of situations, including, as in the following example, problems connecting distance, speed and time. The relationship $\text{time} = \dfrac{\text{distance}}{\text{speed}}$ is useful here.

EXAMPLE 1.6

Each day I travel 10 km from home to work. One day, because of road works, my average speed was 5 km h^{-1} slower than usual, and my journey took an extra 10 minutes.

Taking x km h^{-1} as my usual speed:

(a) write down an expression in x which represents my usual time in hours;

(b) write down an expression in x which represents my time when I travel 5 km h^{-1} slower than usual;

(c) use these expressions to form an equation in x and solve it.

(d) How long did my journey usually take?

Solution **(a)** $\text{Time} = \dfrac{\text{distance}}{\text{speed}} \Rightarrow \text{usual time} = \dfrac{10}{x}$.

(b) I now travel at $(x - 5)$ km h^{-1}, so the longer time $= \dfrac{10}{x-5}$.

(c) The difference in these times is 10 minutes, or $\frac{1}{6}$ hour, so

$$\frac{10}{x-5} - \frac{10}{x} = \frac{1}{6}.$$

Multiplying by $6x(x-5)$ gives

$$\frac{60x(x-5)}{(x-5)} - \frac{60x(x-5)}{x} = \frac{6x(x-5)}{6}$$

$$\Rightarrow \quad 60x - 60(x-5) = x(x-5)$$

$$\Rightarrow \quad 60x - 60x + 300 = x^2 - 5x$$

$$\Rightarrow \quad x^2 - 5x - 300 = 0$$

$$\Rightarrow \quad (x-20)(x+15) = 0$$

$$\Rightarrow \quad x = 20 \quad \text{or} \quad x = -15.$$

(d) Reject $x = -15$, since x km h^{-1} is a speed.

Usual speed = 20 km h^{-1}.

Usual time = $\frac{10}{x}$ hours = 30 minutes.

EXERCISE 1B

1 Solve the following equations.

(a) $\frac{2x}{7} - \frac{x}{4} = 3$ **(b)** $\frac{5}{4x} + \frac{3}{2x} = \frac{11}{16}$

(c) $\frac{2}{x} - \frac{5}{2x-1} = 0$ **(d)** $x - 3 = \frac{x+2}{x-2}$

(e) $\frac{1}{x} + x + 1 = \frac{13}{3}$ **(f)** $\frac{2x}{x+1} - \frac{1}{x-1} = 1$

(g) $\frac{x}{x-1} - \frac{x-1}{x} = 2$

2 I have £6 to spend on crisps for a party. When I get to the shop I find that the price has been reduced by 1 pence per packet, and I can buy one packet more than I expected. Taking x pence as the original cost of a packet of crisps:

(a) write down an expression in x which represents the number of packets that I expected to buy;

(b) write down an expression in x which represents the number of packets bought at the reduced price;

(c) form an equation in x and solve it to find the original cost.

3 The distance from Manchester to Oxford is 270 km. One day, roadworks on the M6 meant that my average speed was 10 km h^{-1} less than I had anticipated, and so I arrived 18 minutes later than planned. Taking x km h^{-1} as the anticipated average speed:

(a) write down an expression in x for the anticipated and actual times of the journey;

(b) form an equation in x and solve it;

(c) find the time of my arrival in Oxford if I left home at 10 am.

4 Each time somebody leaves the firm of Honeys, he or she is taken out for a meal by the rest of the staff. On one such occasion the bill came to £272, and each member of staff remaining with the firm paid an extra £1 to cover the cost of the meal for the one who was leaving. Taking £x as the cost of the meal, write down an equation in x and solve it.

How many staff were left working for Honeys?

5 A Swiss Roll cake is 21 cm long. When I cut it into slices, I can get two extra slices if I reduce the thickness of each slice by $\frac{1}{4}$ cm. Taking x as the number of thicker slices, write down an equation in x and solve it.

6 Two electrical resistances may be connected in series or in parallel. In series, the equivalent single resistance is the sum of the two resistances, but in parallel, the two resistances R_1 and R_2 are equivalent to a single resistance R where

$$\frac{1}{R_1} + \frac{1}{R_2} = \frac{1}{R}.$$

(a) Find the single resistance which is equivalent to resistances of 3 and 4 ohms connected in parallel.

(b) What resistance must be added in parallel to a resistance of 6 ohms to give a resistance of 2.4 ohms?

(c) What is the effect of connecting two equal resistances in parallel?

EXERCISE 1C Examination-style questions

1 Express $x^3 + 1$ as the product of a linear and a quadratic factor.
Hence simplify

$$\frac{x^3 + 1}{x^2 + 3x + 2}.$$

2 Solve

$$1 + \frac{6}{x} = \frac{6}{x - 1}.$$

3 Express as a single fraction in its simplest form

$$\frac{x + 1}{x^2 - 4} \times \frac{x + 2}{x^2 - 1}.$$

4 Express as a single fraction in its simplest form

$$\frac{2(x+1)(x-2)}{(x-5)(x+4)} - \frac{x-1}{x-5}.$$

5 Prove that

$$\frac{5}{x+3} - \frac{10}{(x+3)^2} \equiv \frac{5(x+1)}{(x+3)^2}.$$

6 Express as a single fraction

$$\frac{x+3}{x^2-6x+5} + \frac{2}{x-1}.$$

7 Solve

$$\frac{2}{x} = 2 + \frac{3}{x+1}.$$

8 Simplify

$$\frac{x^3-1}{x^2-2x+1}.$$

9 Show that

$$\frac{r+1}{r+2} - \frac{r}{r+1} \equiv \frac{1}{(r+1)(r+2)}.$$

[Edexcel]

10 Express as a fraction in its simplest form

$$\frac{5(x-3)(x+1)}{(x-12)(x+3)} - \frac{3(x+1)}{x-12}.$$

[Edexcel]

KEY POINTS

1 A rational expression or rational function is a fraction of the form

$$\frac{\mathrm{f}(x)}{\mathrm{g}(x)}$$

where $\mathrm{f}(x)$ and $\mathrm{g}(x)$ are functions of x.

2 When simplifying fractions look for common factors to cancel. For example

$$\frac{2x+3}{2x^2+x-3} = \frac{2x+3}{(2x+3)(x-1)} = \frac{1}{x-1}.$$

3 When adding or subtracting fractions look for common denominators. For example

$$\frac{1}{x^2-1} + \frac{1}{x+1} = \frac{1+(x-1)}{(x+1)(x-1)} = \frac{x}{(x+1)(x-1)}$$

Chapter two

Functions

Still glides the stream and shall forever glide;
The form remains, the function never dies.

William Wordsworth

Mathematical mappings

Why fly to Geneva in January?

Several people arriving at Geneva airport from London were asked the main purpose of their visit. Their answers were recorded:

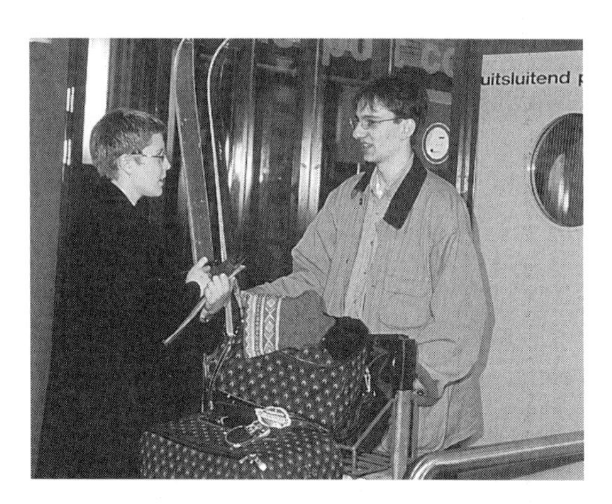

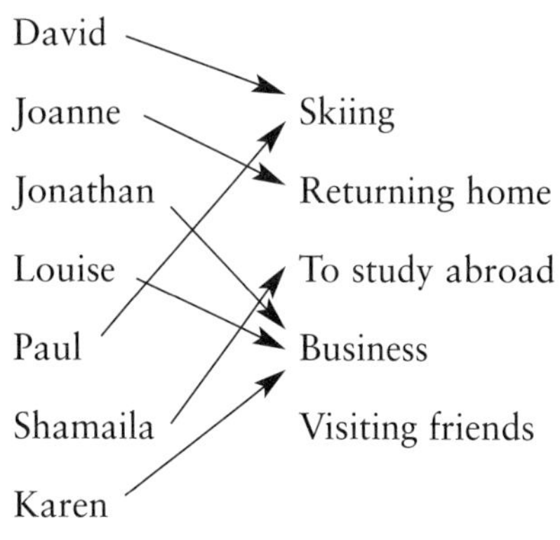

This is an example of a *mapping*.

A mapping is any rule which associates two sets of items. In this example, each of the names on the left is an *object*, or *input*, and each of the reasons on the right is an *image*, or *output*.

For a mapping to make sense or to have any practical application, the inputs and outputs must each form a natural collection or set. The set of possible inputs (in this case, all of the people who flew to Geneva from London in January) is called the *domain* of the mapping. The set of possible outputs (in this case, the set of all possible reasons for flying to Geneva including 'visiting friends') is called the *co-domain* of the mapping.

The seven people questioned in this example gave a set of four reasons, or outputs. These form the *range* of the mapping for this particular set of inputs. The range of any mapping forms part or all of its co-domain.

Notice that Jonathan, Louise and Karen are all visiting Geneva on business: each person gave only one reason for the trip, but the same reason was given by several people. This mapping is said to be many-to-one. A mapping can also be one-to-one, one-to-many, or many-to-many. The relationship between the people and their UK passport numbers will be one-to-one. The relationship between the people and their items of luggage is likely to be one-to-many, and that between the people and the countries they have visited in the last 10 years will be many-to-many.

Mappings expressed using algebra

In mathematics, many (but not all) mappings can be expressed using algebra. Here are some examples of mathematical mappings.

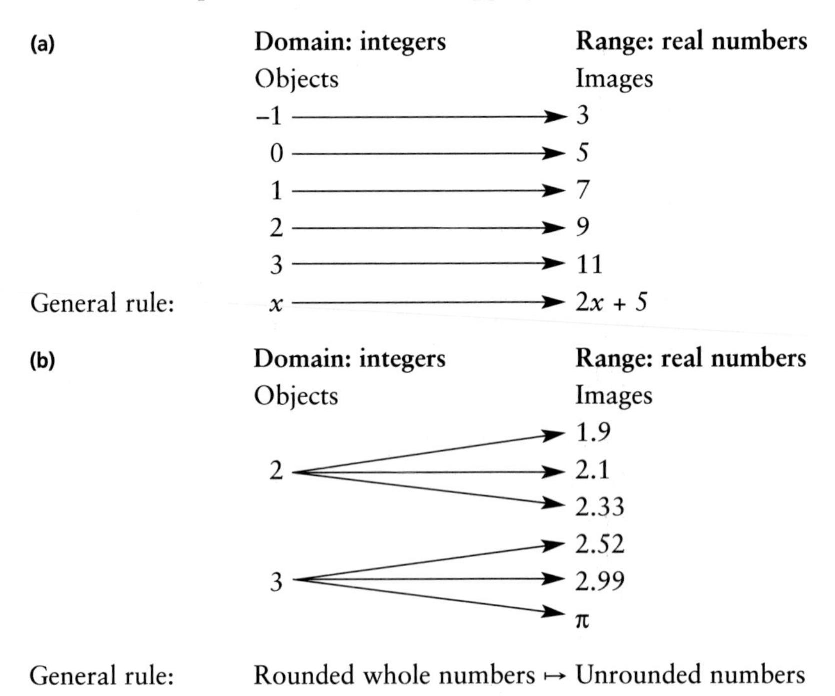

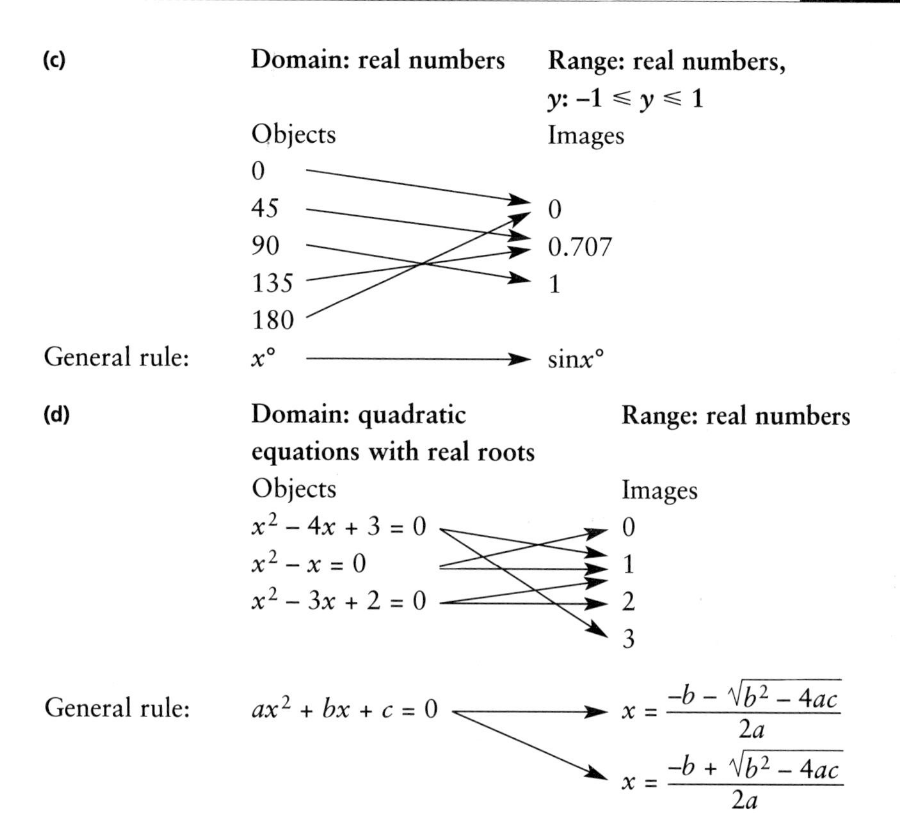

FUNCTIONS

Mappings which are one-to-one or many-to-one are of particular importance, since in these cases there is only one possible image for any object. Mappings of these types are called *functions*. For example, $x \mapsto x^2$ and $x \mapsto \cos x°$ are both functions, because in each case for any value of x there is only one possible answer. The mapping of rounded whole numbers onto unrounded numbers is not a function, since, for example, the rounded number 5 could mean any number between 4.5 and 5.5.

There are several different but equivalent ways of writing down a function. For example, the function which maps x onto x^2 can be written in any of the following ways.

- $y = x^2$
- $\mathrm{f}(x) = x^2$
- $\mathrm{f}: x \mapsto x^2$

Read this as 'f maps x onto x^2'

It is often helpful to represent a function graphically, as in the following example, which also illustrates the importance of knowing the domain.

EXAMPLE 2.1

Sketch the graph of $y = 3x + 2$ when the domain of x is:

(a) $x \in \mathbb{R}$

(b) $x \in \mathbb{R}^+$ (i.e. positive real numbers)

(c) $x \in \mathbb{N}$,

where $\mathbb{R}$ is the set of all real numbers, $\mathbb{R}^+$ is the set of positive real numbers and $\mathbb{N}$ is the set of natural numbers $\{1, 2, 3 \ldots\}$.

Solution

(a) When the domain is $\mathbb{R}$, all values of y are possible. The range is therefore $\mathbb{R}$, also.

(b) When x is restricted to positive values, all the values of y are greater than 2, so the range is $y > 2$.

(c) In this case the range is the set of points $\{2, 5, 8, \ldots\}$. These are clearly all of the form $3x + 2$ where x is a natural number $(1, 2, 3, \ldots)$. This set can be written neatly as $\{3x + 2 : x \in \mathbb{N}\}$.

The open circle shows that (0, 2) is not part of the line

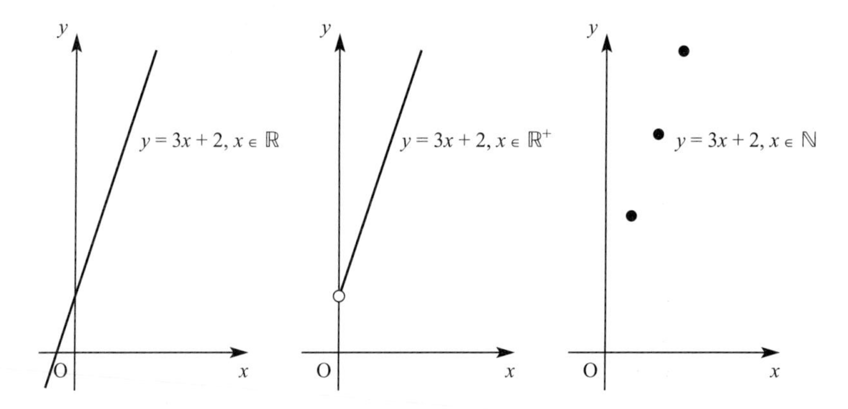

FIGURE 2.1

When you draw the graph of a mapping, the x coordinate of each point is an input value, the y coordinate is the corresponding output value. The table below shows this for the mapping $x \mapsto x^2$, or $y = x^2$, and figure 2.2 shows the resulting points on a graph.

Input (x)	Output (y)	Point plotted
–2	4	(–2, 4)
–1	1	(–1, 1)
0	0	(0, 0)
1	1	(1, 1)
2	4	(2, 4)

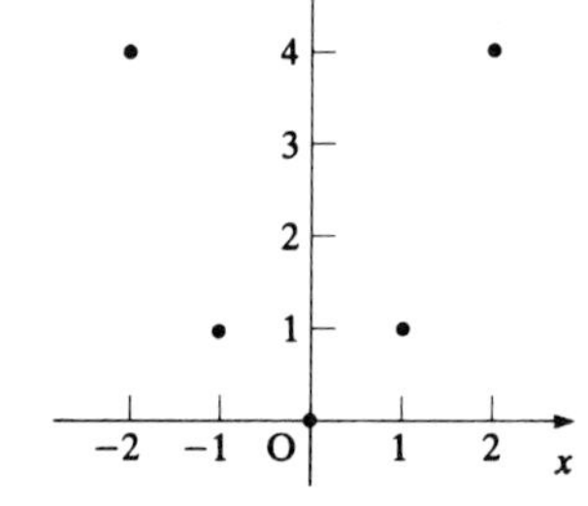

FIGURE 2.2

If the mapping is a function, there is one and only one value of y for every value of x in the domain. Consequently the graph of a function is a simple curve or line going from left to right, with no doubling back.

Figure 2.3 illustrates some different types of mapping. The graphs in (a) and (b) illustrate functions, those in (c) and (d) do not.

(a) *One-to-one*

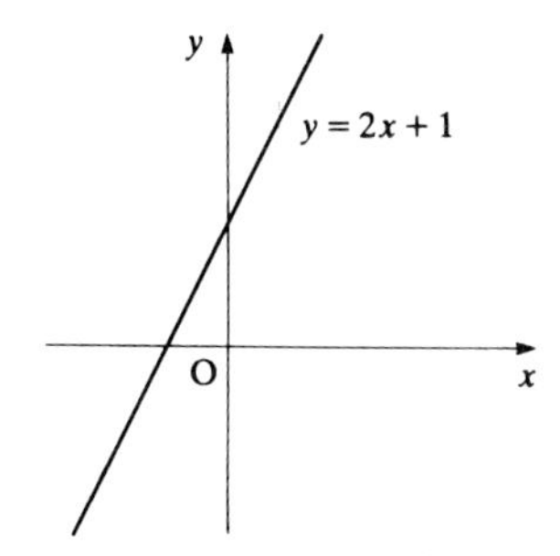

(b) *Many-to-one*

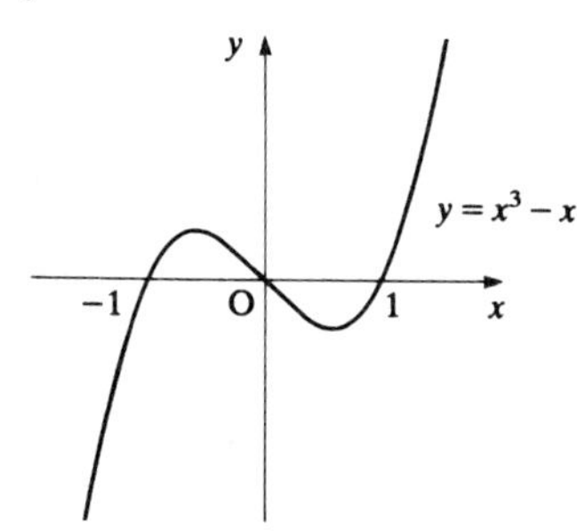

(c) *One-to-many*

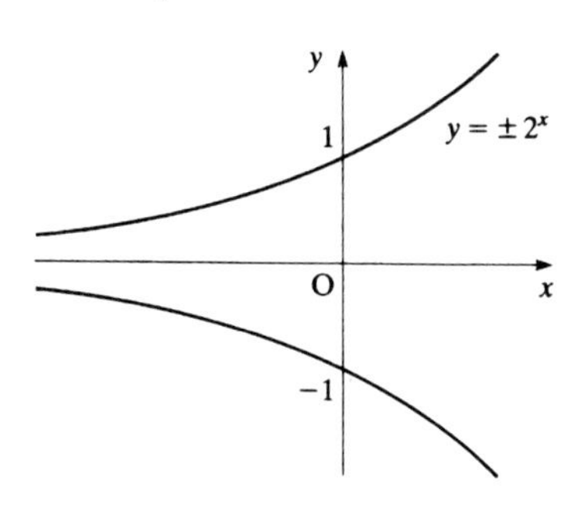

(d) *Many-to-many*

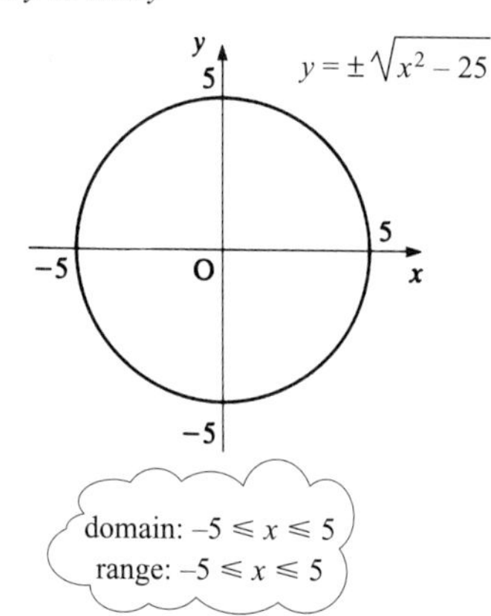

domain: $-5 \leqslant x \leqslant 5$
range: $-5 \leqslant x \leqslant 5$

FIGURE 2.3

EXAMPLE 2.2

Sketch the graph of $y = \mathrm{f}(x)$ where $\mathrm{f}(x) = \frac{1}{x}$, $x > 0$. State the type of mapping and whether or not $y = \mathrm{f}(x)$ is a function. State the range.

Solution The graph is as shown in figure 2.4.

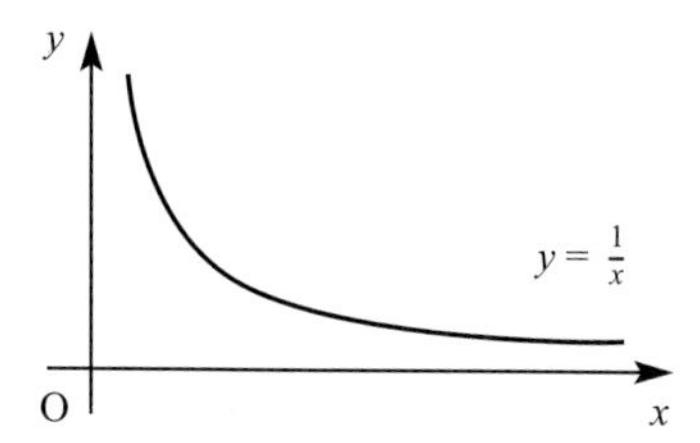

FIGURE 2.4

The mapping is one-to-one since each value of x maps onto a unique value of y.
Since the mapping is one-to-one, $f(x)$ is a function.
The range is $y > 0$.

SUMMARY

You have seen that mappings are:

- one-to-one if each one value of x maps onto exactly one value of y;
- many-to-one if two, or more, values of x map onto exactly one value of y;
- one-to-many if each one value of x maps onto two, or more, values of y;
- many-to many if two, or more, values of x map onto two, or more, values of y.

A mapping is said to be a *function* if it is either one-to-one or many-to-one.

The *domain* of the function is the set of x values that are allowed and the *range* is the corresponding set of y values.

EXERCISE 2A

1 Describe each of the following mappings as either one-to-one, many-to-one, one-to-many or many-to-many, and say whether it represents a function.

(a)

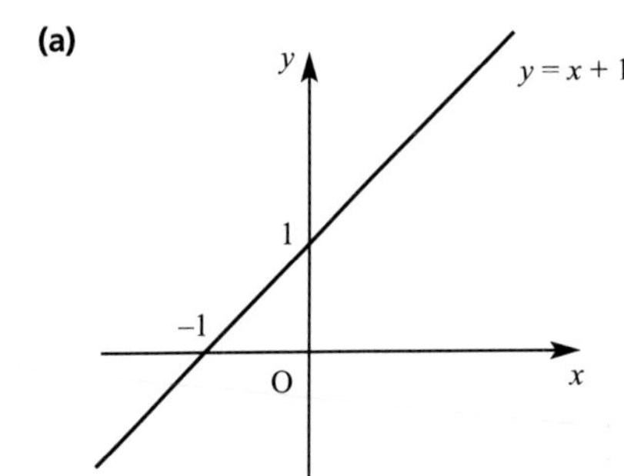

(b)

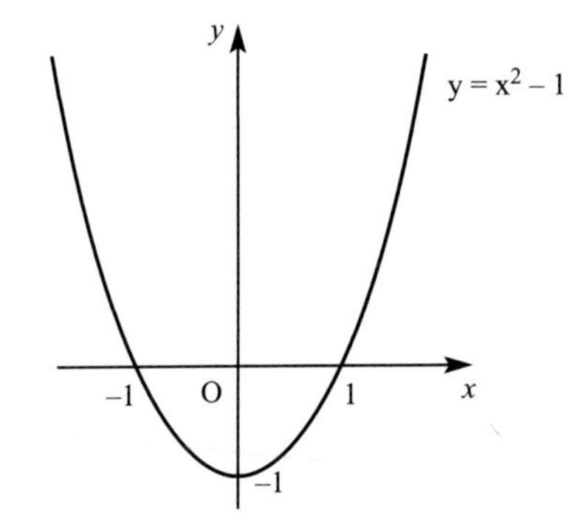

(c)

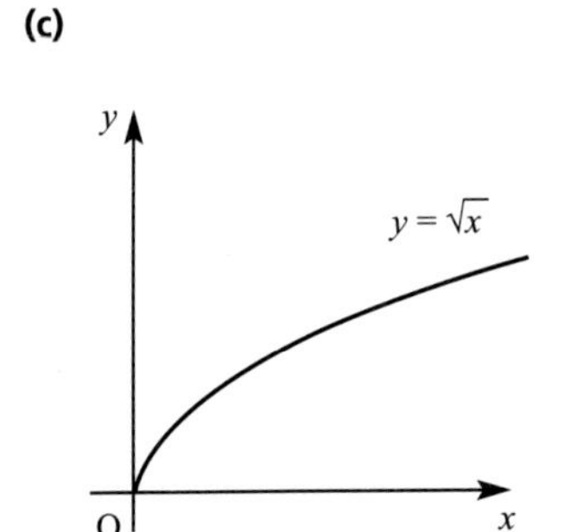

(d)

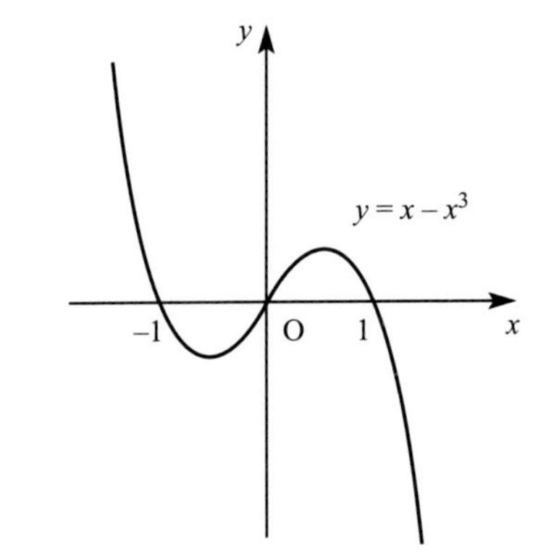

(e)

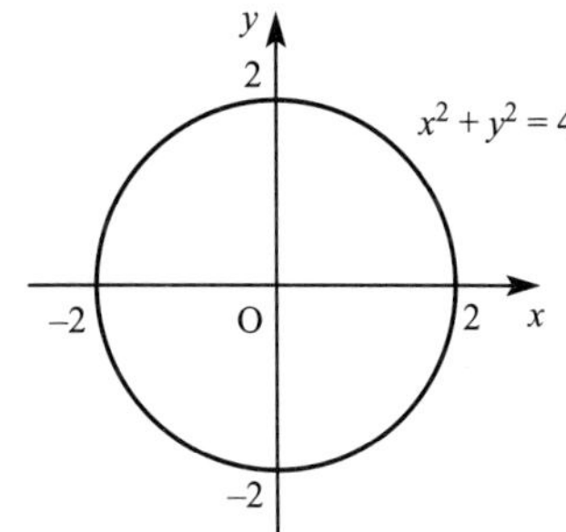

(f)

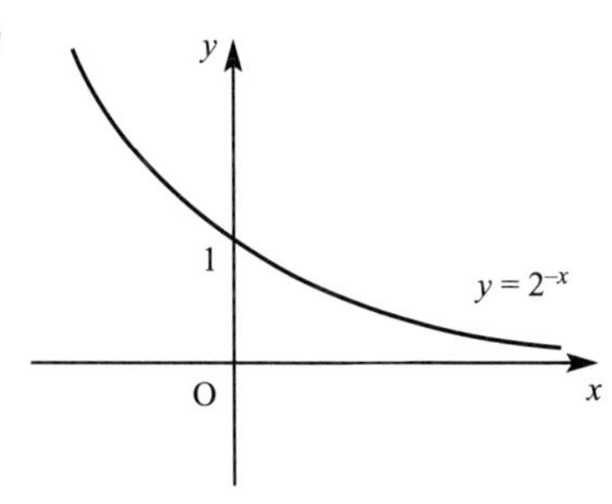

(g)

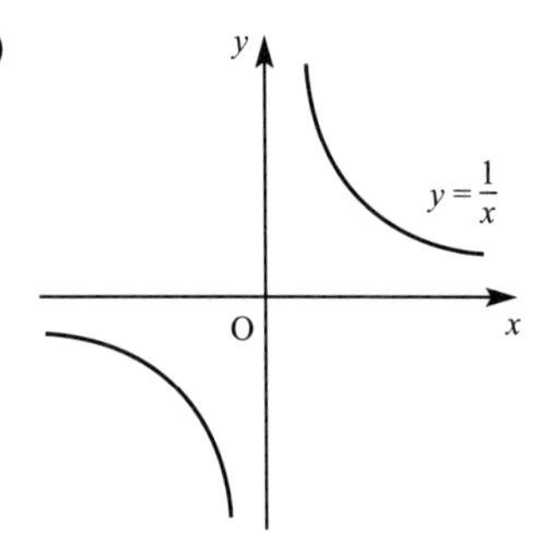

(h)

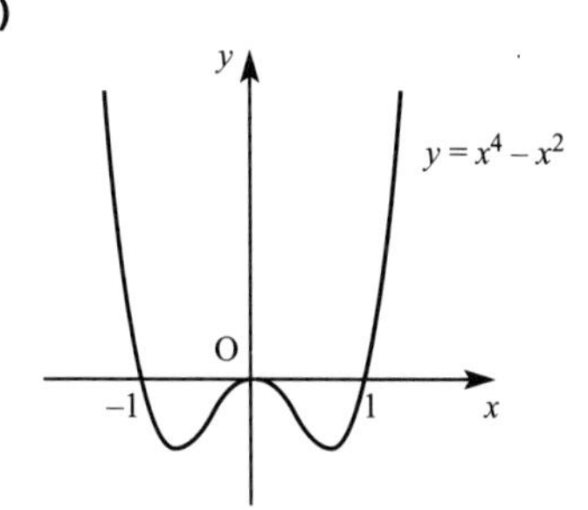

2 Draw the following graphs and state whether the mapping is either one-to-one, many-to-one, one-to-many or many-to-many. For each mapping that is a function state the domain and range.

(a) $y = 2x + 3$

(b) $y = \frac{1}{x^2}$

(c) $y = \pm\sqrt{x}$

(d) $y = 1 - \frac{1}{2}x$

(e) $x^2 + y^2 = 9$

(f) $y = \frac{1}{x}$

(g) $y = x^2 - 4$

(h) $y = (x - 1)x(x + 1)$

(i) $y = 3^x$

(j) $\frac{x^2}{4} + y^2 = 1$

3 (a) A function is defined by $f(x) = 2x - 5$. Write down the values of

(i) f(0) **(ii)** f(7) **(iii)** f(–3).

(b) A function is defined by g:(polygons) $\longmapsto$ (number of sides). Find

(i) g(triangle) **(ii)** g(pentagon) **(iii)** g(decagon).

(c) The function t maps Celsius temperatures onto Fahrenheit temperatures. It is defined by t: $C \longmapsto \frac{9C}{5} + 32$. Find

(i) t(0) **(ii)** t(28) **(iii)** t(–10)

(iv) the value of C when t(C) = C.

4 Find the range of each of the following functions. (You may find it helpful to draw the graph first.)

(a) $f(x) = 2 - 3x$ $\quad x \geqslant 0$

(b) $f(\theta) = \sin\theta$ $\quad 0° \leqslant \theta \leqslant 180°$

(c) $y = x^2 + 2$ $\quad x \in \{0, 1, 2, 3, 4\}$

(d) $y = \tan\theta$ $\quad 0° < \theta < 90°$

(e) $f : x \longmapsto 3x - 5$ $\quad x \in \mathbb{R}$

(f) $f : x \longmapsto 2^x$ $\quad x \in \{-1, 0, 1, 2\}$

(g) $y = \cos x$ $\quad -\frac{\pi}{2} \leqslant x \leqslant \frac{\pi}{2}$

(h) $f : \theta \longmapsto \dfrac{1}{\cos\theta}$ $\quad \theta \in \mathbb{R}$

(i) $f(x) = \dfrac{1}{1 + x^2}$ $\quad x \in \mathbb{R}$

(j) $f(x) = \sqrt{x - 3} + 3$ $\quad x \geqslant 3$

5 The mapping f is defined by $\quad f(x) = x^2 \quad 0 \leqslant x \leqslant 3$
$f(x) = 3x \quad 3 \leqslant x \leqslant 10.$

The mapping g is defined by $\quad g(x) = x^2 \quad 0 \leqslant x \leqslant 2$
$g(x) = 3x \quad 2 \leqslant x \leqslant 10.$

Explain why f is a function and g is not.

COMPOSITE FUNCTIONS

It is possible to combine functions in several different ways, and you have already met some of these. For example, if $f(x) = x^2$ and $g(x) = 2x$, then you could write:

$$f(x) + g(x) = x^2 + 2x.$$

In this example, two functions are added.

Similarly if $f(x) = x$ and $g(x) = \sin x$, then:

$$f(x).g(x) = x\sin x.$$

In this example, two functions are multiplied.

Sometimes you need to apply one function and then apply another to the answer. You are then creating a *composite function* or a *function of a function.*

EXAMPLE 2.3

A new mother is bathing her baby for the first time. She takes the temperature of the bath water with a thermometer which reads in Celsius, but then has to convert the temperature to degrees Fahrenheit to apply the rule that her own mother taught her:

At one o five
He'll cook alive
But ninety four
Is rather raw.

Write down the two functions that are involved, and apply them to readings of:
(a) 30°C **(b)** 38°C **(c)** 45°C.

Solution The first function converts the Celsius temperature C into a Fahrenheit temperature, *F*:

$$F = \frac{9C}{5} + 32.$$

The second function maps Fahrenheit temperatures onto the state of the bath:

$$\begin{array}{rl} F \leqslant 94 & \text{Too cold} \\ 94 \leqslant F \leqslant 105 & \text{All right} \\ F \geqslant 105 & \text{Too hot} \end{array}$$

This gives

(a) $30°\text{C} \longmapsto 86°\text{F} \longmapsto$ too cold
(b) $38°\text{C} \longmapsto 100.4°\text{F} \longmapsto$ all right
(c) $45°\text{C} \longmapsto 113°\text{C} \longmapsto$ too hot.

In this case the composite function would be (to the nearest degree):

$$\begin{array}{rl} C \leqslant 34°\text{C} & \text{too cold} \\ 35°\text{C} \leqslant C \leqslant 40°\text{C} & \text{all right} \\ C \leqslant 41°\text{C} & \text{too hot.} \end{array}$$

In algebraic terms, a composite function is constructed as

Input $x \stackrel{\text{f}}{\longmapsto}$ Output $\text{f}(x)$

Input $\text{f}(x) \stackrel{\text{g}}{\longmapsto}$ Output $\text{g}[\text{f}(x)]$ or $\text{gf}(x)$.

Read this as 'g of f of x'

Thus the composite function $\text{gf}(x)$ should be performed from right to left: start with x then apply f and then g.

NOTATION

To apply f twice in succession you would write $\text{f}^2(x)$, not $\text{ff}(x)$. Similarly $\text{g}^3(x)$ means three applications of g. In order to apply a function repeatedly the set of values of the range must be contained within the set of values for the domain.

ORDER OF FUNCTIONS

If f is the rule 'square the input value' and g is the rule 'add 1', then gf is given by

$$x \xmapsto[\text{square}]{f} x^2 \xmapsto[\text{add 1}]{g} x^2 + 1.$$

So $\quad \mathrm{gf}(x) = x^2 + 1.$

Notice that gf(x) is not the same as fg(x), since for fg(x) you must apply g first. In the example above, this would give:

$$x \xmapsto[\text{add 1}]{g} (x + 1) \xmapsto[\text{square}]{f} (x + 1)^2$$

and so $\mathrm{fg}(x) = (x + 1)^2$.

Clearly this is *not* the same result.

EXAMPLE 2.4

Given that $\mathrm{f}(x) = 2x$, $\mathrm{g}(x) = x^2$, and $\mathrm{h}(x) = \frac{1}{x}$, find:

(a) fg(x) **(b)** gf(x) **(c)** gh(x)
(d) $\mathrm{f}^2(x)$ **(e)** fgh(x) **(f)** hfg(x)

Solution

(a) $\mathrm{fg}(x) = \mathrm{f}[\mathrm{g}(x)] = \mathrm{f}(x^2) = 2x^2.$

(b) $\mathrm{gf}(x) = \mathrm{g}[\mathrm{f}(x)] = \mathrm{g}(2x) = (2x)^2 = 4x^2.$

(c) $\mathrm{gh}(x) = \mathrm{g}[\mathrm{h}(x)] = \mathrm{g}(\frac{1}{x}) = \frac{1}{x^2}.$

(d) $\mathrm{f}^2(x) = \mathrm{f}[\mathrm{f}(x)] = \mathrm{f}(2x) = 2(2x) = 4x.$

(e) $\mathrm{fgh}(x) = \mathrm{f}[\mathrm{gh}(x)] = \mathrm{f}\left(\frac{1}{x^2}\right)$ using (c) $= \frac{2}{x^2}.$

(f) $\mathrm{hfg}(x) = \mathrm{h}[\mathrm{fg}(x)] = \mathrm{h}(2x^2)$ using (a) $= \frac{1}{2x^2}.$

EXAMPLE 2.5

Functions f and g are defined by

$$\mathrm{f}: x \mapsto x - 2, \quad x \in \mathbb{R}$$
$$\mathrm{g}: x \mapsto x^2+1, \quad x \in \mathbb{R}.$$

(a) Find gf in terms of x stating its domain and range.
(b) Solve $\mathrm{fg}(x) = 15$.

Solution **(a)** gf means $g(f(x))$, i.e. substitute $f(x)$ into g,

so

$$gf(x) = (x - 2)^2 + 1$$
$$= x^2 - 4x + 5$$

Minimum value of gf(x) is when $x = 2$

with domain $x \in \mathbb{R}$ and range $y \geqslant 1$.

(b) To solve $fg(x) = 15$ first find the function $fg(x)$.

$$fg(x) = (x^2 + 1) - 2 = x^2 - 1$$

so

$$x^2 - 1 = 15$$
$$x^2 = 16$$
$$x = \pm 4.$$

EXERCISE 2B

1 The functions f, g and h are defined by $f(x) = x^3$, $g(x) = 2x$ and $h(x) = x + 2$. Find each of the following in terms of x.

(a) fg **(b)** gf **(c)** fh
(d) hf **(e)** fgh **(f)** ghf
(g) g^2 **(h)** $(fh)^2$ **(i)** h^2

2 The functions f and g are defined by

$f: x \mapsto x + 4, \quad x \in \mathbb{R}$
$g: x \mapsto x^2 - 2, \quad x \in \mathbb{R}.$

(a) Find gf in terms of x stating its domain and range.
(b) Solve $fg(x) = 27$.

3 The functions f and g are defined by

$f: x \mapsto x + 1, \quad x \in \mathbb{R}$
$g: x \mapsto \sqrt{x}, \quad x > 0.$

(a) Find $gf(x)$ stating its domain and range.
(b) Solve $gf(x) = 3$.
(c) Solve $fg(x) = 3$.

4 The functions f and g are defined by

$f: x \mapsto 2x + 3, \quad x \in \mathbb{R}$
$g: x \mapsto \dfrac{1}{x - 1}, \quad x \neq 1.$

(a) Find the composite function gf in its simplest form, stating its domain.
(b) Find the values of x for which

$$fg(x) = gf(x)$$

giving your answers to 3 significant figures.

5 The functions f, g and h are defined by

$f: x \mapsto 2x + 3, \quad x \in \mathbb{R}$

$g: x \mapsto \dfrac{3}{2x}, \quad x \neq 0$

$h: x \mapsto \dfrac{x-3}{2}, \quad x \in \mathbb{R}.$

(a) Find the composite function gh in terms of x.
(b) Find the composite function fgh in terms of x.
(c) Which pair of functions form a composite function equal to x?

6 Express the following functions in terms of f and g where

$f: x \mapsto \sqrt{x}$ and $g: x \mapsto x + 4$.

(a) $x \mapsto \sqrt{x+4}$ **(b)** $x \mapsto x + 8$
(c) $x \mapsto \sqrt{x+8}$ **(d)** $x \mapsto \sqrt{x} + 4$

7 The functions f and g are defined as

$f: x \mapsto x - 2, \quad x > 0$

$g: x \mapsto x^2 + 3 \quad x > 0.$

(a) Find, in terms of x, the composite function gf.
(b) Sketch $y = \text{gf}(x)$ for the appropriate domain.
(c) State the range of gf.
(d) Describe the mapping that takes x onto $\text{gf}(x)$.
(e) Is gf a function?

8 The functions f and g are defined by

$f: x \mapsto x^2 + 2x + 3 \quad x \in \mathbb{R}$

$g: x \mapsto px + q \quad x \in \mathbb{R}.$

(a) Given that $\text{fg}(1) = 6$ show that

$$p^2 + 2pq + q^2 + 2p + 2q - 3 = 0.$$

(b) Given that $\text{gf}(-2) = 7$ find another expression in terms of p and q.

Solve the simultaneous equations in (a) and (b) to find p and q.

9 The function f is defined for all real values of x by $f(x) = (x + 3)^{\frac{1}{3}}$.

(a) Find f(24).
(b) Find $f^2(122)$.
(c) Find $f^3(-1334)$.
(d) Find x if $f(x) = 4$.

10 The functions f and g are defined by

$f: x \mapsto 2x^2 + 2x + 3 \quad x \in \mathbb{R}$

$g: x \mapsto x - 1 \quad x \in \mathbb{R}.$

(a) Find fg as a function in terms of x.
(b) By the method of completing the square express fg in the form

$$p(x + q)^2 + r$$

where the values of the constants p, q, and r are to be stated.

(c) State the range of fg.

INVERSE FUNCTIONS

Look at the mapping $x \mapsto x + 2$ with domain and range the set of integers.

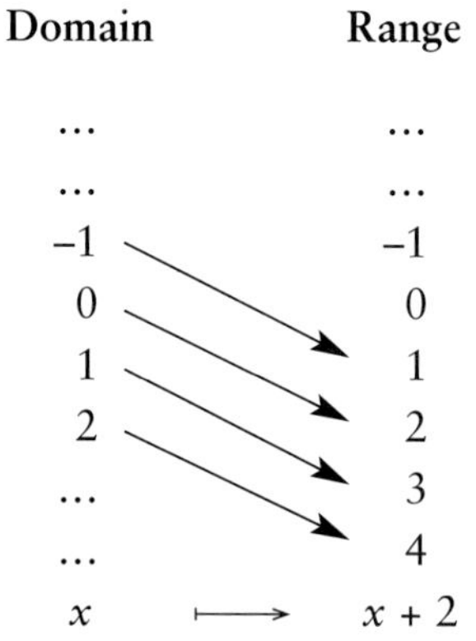

Domain		Range
...		...
...		...
–1		–1
0		0
1		1
2		2
...		3
...		4
x	$\longmapsto$	$x + 2$

The mapping is clearly a function, since for every input there is one and only one output, the number that is two greater than that input.

This mapping can also be seen in reverse. In that case, each number maps onto the number two less than itself: $x \mapsto x - 2$. The reverse mapping is also a function because for any input there is one and only one output. The reverse mapping is called the *inverse function*, f^{-1}.

Function: $\quad f : x \mapsto x + 2 \quad x \in \mathbb{Z}$ where $\mathbb{Z} = \{0, \pm1, \pm2, \pm3, \ldots\}$.

Inverse function: $\quad f^{-1} : x \mapsto x - 2 \quad x \in \mathbb{Z}$.

For a mapping to be a function which also has an inverse function, every object in the domain must have one and only one image in the range, and vice versa. This can only be the case if the mapping is one-to-one.

So the condition for a function f to have an inverse function is that, over the given domain and range, f represents a one-to-one mapping. This is a common situation, and many inverse functions are self-evident as in the following examples, for all of which the domain and range are the real numbers:

$f : x \mapsto x - 1$; $\quad f^{-1} : x \mapsto x + 1$

$g : x \mapsto 2x$; $\quad g^{-1} : x \mapsto \frac{1}{2}x$

$h : x \mapsto x^3$; $\quad h^{-1} : x \mapsto \sqrt[3]{x}$.

You can decide whether an algebraic mapping is a function, and whether it has an inverse function, by looking at its graph. The curve or line representing a one-to-one mapping does not double back on itself, has no turning points and covers the full domain and range. Figure 2.5 illustrates the functions given above.

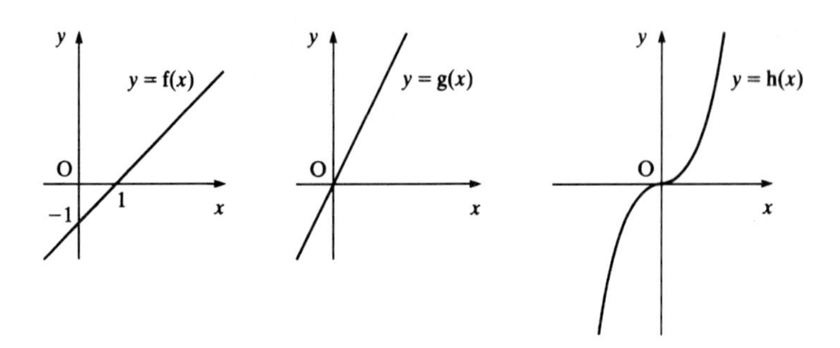

FIGURE 2.5

Now look at $f(x) = x^2$ for $x \in \mathbb{R}$ (figure 2.6). You can see that there are two distinct input values giving the same output: for example f(2) = f(–2) = 4. When you want to reverse the effect of the function, you have a mapping which for a single input of 4 gives two outputs, –2 and +2. Such a mapping is not a function.

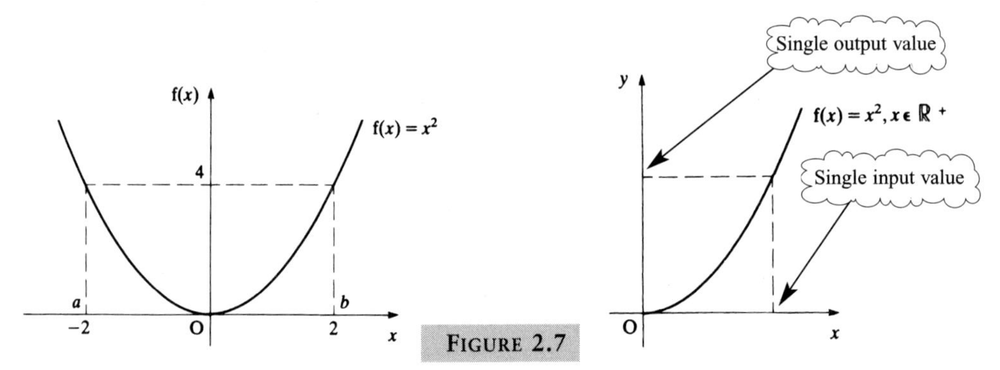

FIGURE 2.6

FIGURE 2.7

If the domain of $f(x) = x^2$ is restricted to $\mathbb{R}^+$ (the set of positive real numbers), you have the situation shown in figure 2.7. This shows that the function which is now defined is one-to-one. The inverse function is given by $f^{-1}(x) = \sqrt{x}$, since the sign $\sqrt{\ }$ means 'the positive square root of'.

It is often helpful to restrict the domain of a function so that its inverse is also a function. When you use the inv sin (i.e. $\sin^{-1}$ or arcsin) key on your calculator the answer is restricted to the range –90° to 90°, and is described as the *principal value.* Although there are infinitely many roots of the equation $\sin x = 0.5$ (…–330°, –210°, 30°, 150°, …), only one of these, 30°, lies in the restricted range and this is the value your calculator will give you.

THE GRAPH OF A FUNCTION AND ITS INVERSE

You have just looked at the graph of $f(x) = x^2$ for $x \in \mathbb{R}^+$. The reverse process of x^2 will be $\pm\sqrt{x}$. But for the function to be one-to-one the positive root only is taken. Drawing $y = x^2$ and $y = \sqrt{x}$ on the same diagram gives figure 2.8

Drawing $h(x) = x^3$ and $h^{-1}(x) = \sqrt[3]{x}$ on the same diagram gives figure 2.9.

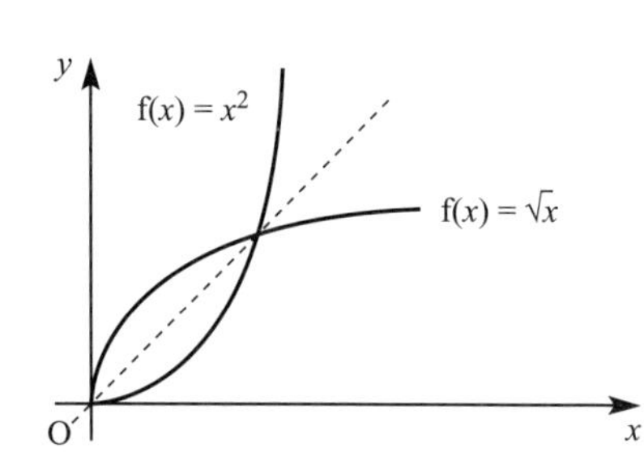

FIGURE 2.8

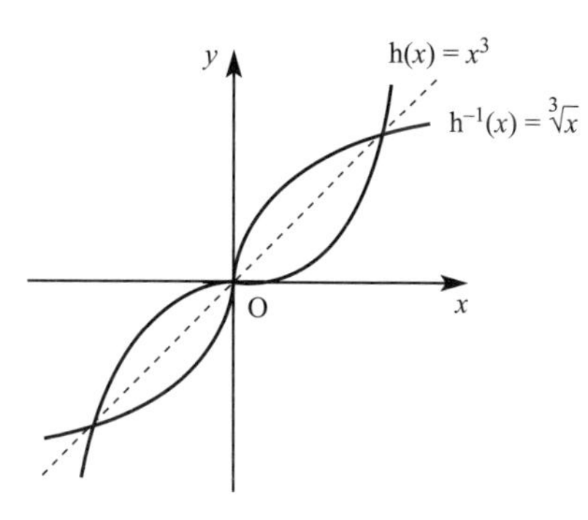

FIGURE 2.9

You have probably realised by now that the graph of the inverse function is the same shape as that of the function, but reflected in the line $y = x$. To see why this is so, think of a function $f(x)$ mapping a onto b; (a, b) is clearly a point on the graph of $f(x)$. The inverse function $f^{-1}(x)$, maps b onto a and so (b, a) is a point on the graph of $f^{-1}(x)$.

The point (b, a) is the reflection of the point (a, b) in the line $y = x$. This is shown for a number of points in figure 2.10.

This result can be used to obtain a sketch of the inverse function without having to find its equation, provided that the sketch of the original function uses the same scale on both axes.

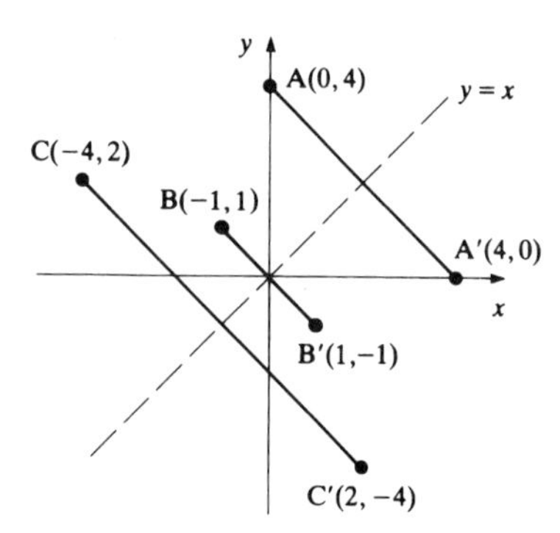

FIGURE 2.10

FINDING THE ALGEBRAIC FORM OF THE INVERSE FUNCTION

To find the algebraic form of the inverse of a function $f(x)$, you should start by changing notation and writing it in the form $y = \dots$.

Since the graph of the inverse function is the reflection of the graph of the original function in the line $y = x$, it follows that you may find its equation by interchanging y and x in the equation of the original function. You will then need to make y the subject of your new equation. This procedure is illustrated in Example 2.6.

EXAMPLE 2.6

Find $f^{-1}(x)$ when $f(x) = 2x + 1$.

Solution

The function f(x) is given by $y = 2x + 1$

Interchanging x and y gives $x = 2y + 1$

Re-arranging to make y the subject: $y = \frac{x-1}{2}$

So $f^{-1}(x) = \frac{x-1}{2}$.

Sometimes the domain of the function f will not include the whole of $\mathbb{R}$. When any real numbers are excluded from the domain of f, it follows that they will be excluded from the range of f^{-1}, and vice versa:

domain of f ≡ range of f^{-1}
range of f ≡ domain of f^{-1}.

EXAMPLE 2.7

Find $f^{-1}(x)$ when $f(x) = 2x - 3$ and the domain of f is $x \geqslant 4$.

Solution

	Domain	**Range**
Function: $y = 2x - 3$	$x \geqslant 4$	$y \geqslant 5$
Inverse function: $x = 2y - 3$	$x \geqslant 5$	$y \geqslant 4$

Rearranging the inverse function to make y the subject, $y = \frac{x+3}{2}$.

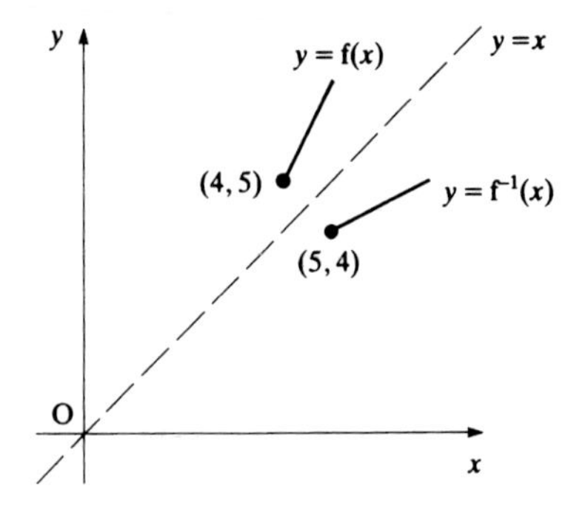

FIGURE 2.11

The full definition of the inverse function is therefore

$$f^{-1}(x) = \frac{x+3}{2} \text{ for } x \geqslant 5.$$

You can see in figure 2.11 that the inverse function is the reflection of a restricted part of the line $y = f(x)$ in the line $y = x$.

EXAMPLE 2.8

(a) Find $f^{-1}(x)$ when $f(x) = x^2 + 2,\ x \geqslant 0$.

(b) Find **(i)** f(7) **(ii)** $f^{-1}f(7)$.
What do you notice?

Solution **(a)**

	Domain	Range
Function: $y = x^2 + 2$	$x \geqslant 0$	$y \geqslant 2$
Inverse function: $x = y^2 + 2$	$x \geqslant 2$	$y \geqslant 0$

Rearranging the inverse function to make y its subject:

$$y^2 = x - 2.$$

This gives $y = \pm\sqrt{x-2}$, but since we know the range of the inverse function to be $y \geqslant 0$ we can write:

$$y = +\sqrt{x-2} \quad \text{or just} \quad y = \sqrt{x-2}.$$

The full definition of the inverse function is therefore:

$$f^{-1}(x) = \sqrt{x-2} \text{ for } x \geqslant 2.$$

The function and its inverse function are shown in figure 2.12.

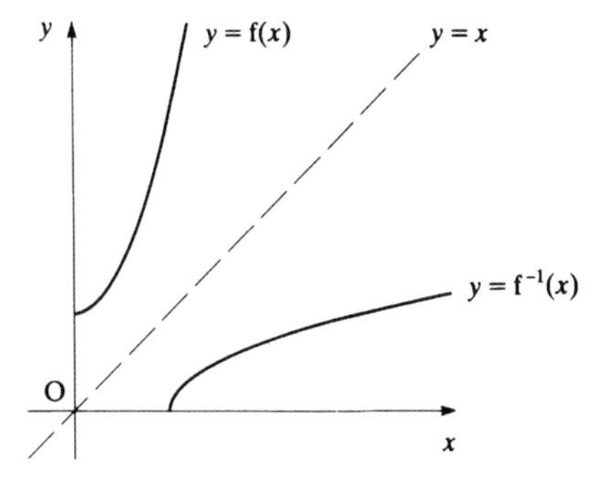

FIGURE 2.12

(b) **(i)** $f(7) = 7^2 + 2 = 51$
(ii) $f^{-1}f(7) = \sqrt{51-2} = 7$

Notice that applying the function followed by its inverse has brought us back to the original input value.

Note Part (b) of Example 2.8 illustrates an important general result. For any function $f(x)$ with an inverse $f^{-1}(x)$, $f^{-1}f(x) = x$. Similarly $ff^{-1}(x) = x$. The effects of a function and its inverse can be thought of as cancelling each other out.

EXERCISE 2C

1 Find the inverses of the following functions:

(a) $f(x) = 2x + 7$ **(b)** $f(x) = 4 - x$

(c) $f(x) = \dfrac{4}{2 - x}$ **(d)** $f(x) = x^2 - 3 \quad x \geqslant 0.$

2 The function f is defined by $f(x) = (x - 2)^2 + 3$ for $x \geqslant 2$.

(a) Sketch the graph of $f(x)$.

(b) On the same axes, sketch the graph of $f^{-1}(x)$ without finding its equation.

3 The functions f, g and h are defined by

$$f(x) = \frac{3}{x - 4} \qquad g(x) = x^2 \qquad h(x) = \sqrt{2 - x}.$$

(a) For each function, state any real values of x for which it is not defined.

(b) Find the inverse functions f^{-1} and h^{-1}.

(c) Explain why g^{-1} does not exist when the domain of g is $\mathbb{R}$.

(d) Suggest a suitable domain for g so that g^{-1} does exist.

(e) Is the domain for the composite function fg the same as for the composite function gf? Give reasons for your answer.

4 A function f is defined by:

$$f: x \mapsto \tfrac{1}{x} \qquad x \in \mathbb{R}, x \neq 0.$$

Find:

(a) $f^2(x)$ **(b)** $f^3(x)$ **(c)** $f^{-1}(x)$ **(d)** $f^{999}(x)$.

5 **(a)** Show that $x^2 + 4x + 7 = (x + 2)^2 + a$, where a is to be determined.

(b) Sketch the graph of $y = x^2 + 4x + 7$, giving the equation of its axis of symmetry and the coordinates of its vertex.

The function f is defined by $f : x \mapsto x^2 + 4x + 7$ and has as its domain the set of all real numbers.

(c) Find the range of f.

(d) Explain, with reference to your sketch, why f has no inverse with its given domain. Suggest a domain for f for which it has an inverse.

[MEI]

6 The function f is defined by: $f: x \mapsto 4x^3 + 3 \quad x \in \mathbb{R}$.

Give the corresponding definition of f^{-1}.

State the relationship between the graphs of f and f^{-1}.

[OCR]

7 The functions f and g are defined by

$f: x \mapsto 4x - 2, \quad x \in \mathbb{R}$

$g: x \mapsto x^2, \quad x \geqslant 0.$

(a) Find, in terms of x, the functions f^{-1} and g^{-1}.

(b) On the same diagram sketch $y = f^{-1}(x)$ and $y = g^{-1}(x)$.

(c) How many solutions are there to the equation $f^{-1}(x) = g^{-1}(x)$?

(d) Find the exact values of the solutions to $f^{-1}(x) = g^{-1}(x)$.

8 The function f is given by

$f : x \mapsto x^2 - 6x + 10, \quad x \in \mathbb{R}.$

(a) Express f in the form of $(x + a)^2 + b$, stating the values of a and b.

(b) State the coordinate of the minimum turning point on the graph of $y = f(x)$.

(c) Sketch the graph of $y = f(x)$.

(d) State a domain for which $y = f(x)$ is a one-to-one fuction. For this domain state the range.

(e) For your domain sketch the inverse function, f^{-1}, and describe its geometrical relationship with $y = f(x)$.

9 A function is defined by

$f : x \mapsto 4 - x^2, \quad x \geqslant 0.$

(a) Sketch the graph of $y = f(x)$.

(b) State the range of f.

(c) Find f^{-1} in terms of x and state its domain.

(d) Solve $f^{-1}(x) = 3$.

10 The functions f and g are defined as follows:

$f : x \mapsto x + 2 \quad x \in \mathbb{R}$

$g : x \mapsto x^2 + 1 \quad x > 0.$

(a) Show that $gf(x) = 0$ has no real roots.

(b) State the domain of g^{-1}.

(c) Find, in terms of x, an expression for $g^{-1}(x)$.

(d) Sketch, on a single diagram, the graph of $y = g(x)$ and $y = g^{-1}(x)$.

GRAPHS OF FUNCTIONS

Several of the curves with which you are familiar have symmetry of one form or another. For example

- the curve of any quadratic in x has a line of symmetry parallel to the y axis;
- the curve of $y = \cos x$ has the y axis as a line of symmetry;
- the curves of $y = \sin x$ and $y = \tan x$ have rotational symmetry of order 2 about the origin;
- all the trigonometrical graphs have a repeating pattern (translational symmetry).

In this section you will be looking at particular types of symmetry.

EVEN FUNCTIONS

A function is *even* if its graph has the y axis as a line of symmetry. This is true for all three of the functions in figure 2.13.

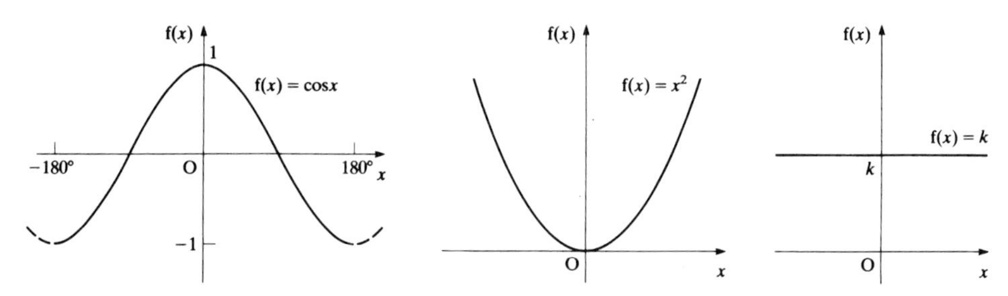

FIGURE 2.13

Reflecting a curve $y = f(x)$ in the y axis gives the curve $y = f(-x)$, so a curve which has the y axis as a line of symmetry satisfies the condition:

$$f(-x) = f(x).$$

This relationship can be used to check whether a function is even, without drawing its graph.

EXAMPLE 2.9

Show that the function $f(x) = x^4 - 2x^2 + 3$ is an even function.

Solution

$$\begin{aligned} f(-x) &= (-x)^4 - 2(-x)^2 + 3 \\ &= x^4 - 2x^2 + 3 \\ &= f(x), \text{ so the function is even.} \end{aligned}$$

Note In general, if $f(x)$ is any polynomial function containing only even powers of x, and possibly a constant term, then $f(x)$ is an even function.

ODD FUNCTIONS

A function whose curve has rotational symmetry of order 2 about the origin, like the curves shown in figure 2.14, is called an *odd function*.

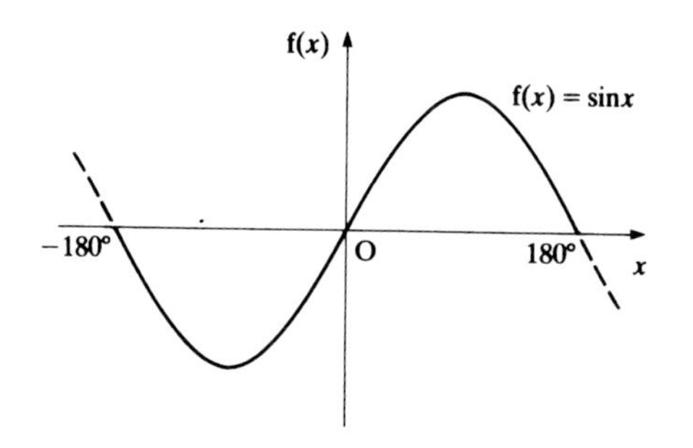

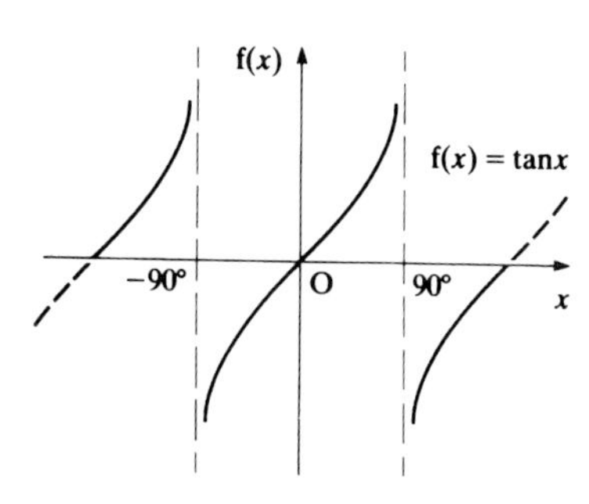

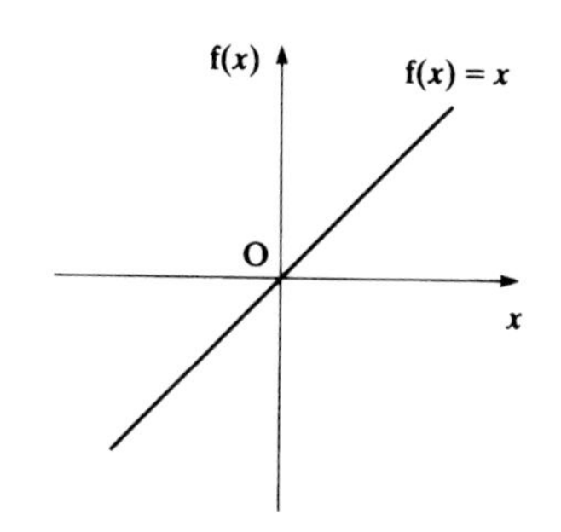

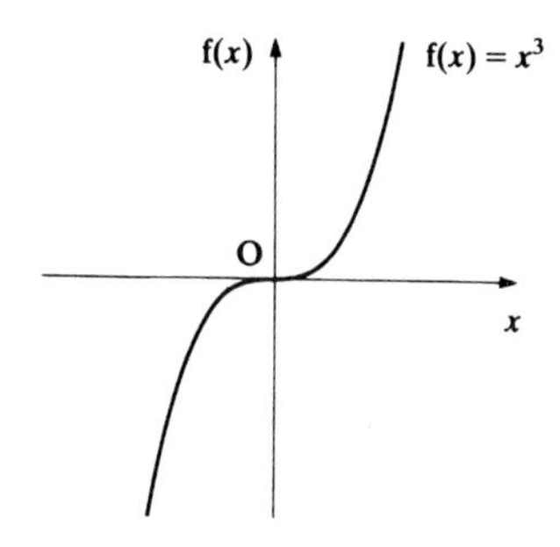

FIGURE 2.14

In all of these, the left-hand side of the graph is obtained from the right-hand side by rotating it through 180° around the origin.

In such cases, the condition for f(x) to be odd is

$$f(-x) = -f(x).$$

EXAMPLE 2.10

Show that the function $f(x) = 3x^5 - 2x^3 + x$ is an odd function.

Solution

$$\begin{aligned} f(-x) &= 3(-x)^5 - 2(-x)^3 + (-x) \\ &= -3x^5 + 2x^3 - x \\ &= -(3x^5 - 2x^3 + x) \\ &= -f(x), \end{aligned}$$

so the function is an odd function.

Note

Any polynomial function $f(x)$ containing only odd powers of x is an odd function.

Not all functions can be classified as even or odd – in fact the majority are neither.

EXAMPLE 2.11

For each of the graphs in figure 2.15, say whether the function is odd, even or neither.

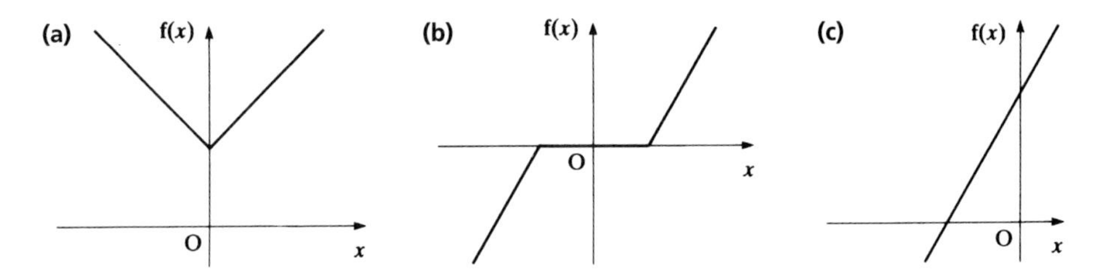

FIGURE 2.15

Solution

(a) The graph is symmetrical about the y axis, therefore the function is even.

(b) A rotation of 180° about the origin leaves the graph unchanged, therefore the function is odd.

(c) The graph is changed by a rotation of 180° about the origin, and the y axis is not a line of symmetry, therefore the function is neither odd nor even.

PERIODIC FUNCTIONS

A *periodic function* is one whose graph has a repeating pattern, just as a periodic sequence is a sequence which repeats itself at regular intervals. You have already met the most common periodic functions – the trigonometrical functions such as $f(x) = \sin x$ (figure 2.16).

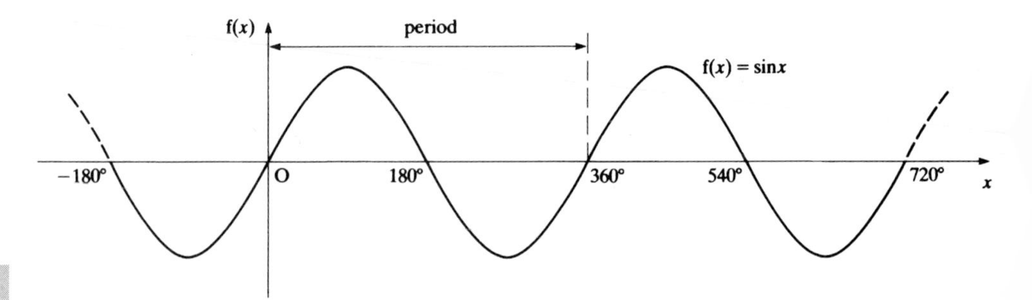

FIGURE 2.16

A periodic function $f(x)$ is such that there is some value of k for which

$$f(x + k) = f(x) \text{ for all values of } x.$$

The smallest value of k for which this is true is called the *period* of the function.

The functions $f(x) = \sin x$ and $f(x) = \cos x$ both have a period of 360° (or 2π radians), and $f(x) = \tan x$ has a period of 180° (or π radians).

EXAMPLE 2.12

(a) Sketch the curve of the function $f(x) = 3\sin(2x - 30°)$.

(b) State the period of this function.

Solution **(a)**

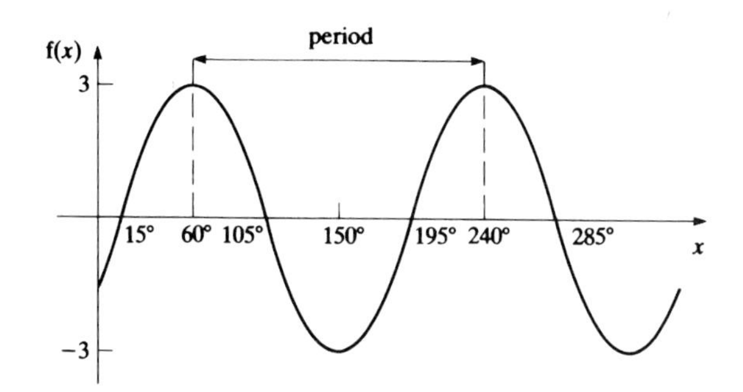

FIGURE 2.17

(b) Period = 180°

You can draw the graph of a periodic function if you know its behaviour over one period.

EXAMPLE 2.13

The function f(x) is periodic with period 2. Given that

$$f(x) = x^2 \qquad 0 \leqslant x < 1$$
$$f(x) = 2 - x \qquad 1 \leqslant x < 2,$$

sketch the graph of f(x) for $-2 \leqslant x < 4$.

Solution

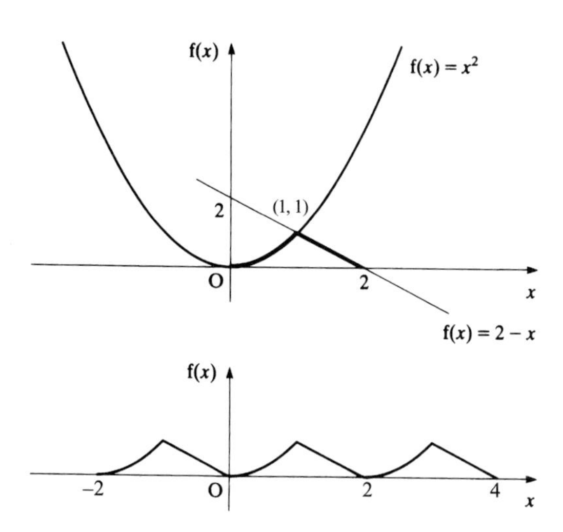

FIGURE 2.18

The first part of figure 2.18 shows the parts of the line and the curve which define f(x). These parts span an interval of length 2 (the period of the function) and thus form the basic repeating pattern. The second diagram shows this pattern repeated three times in the interval $-2 \leqslant x < 4$.

EXERCISE 2D

1 For each of the following curves, say whether the function is odd, even or neither.

(a)

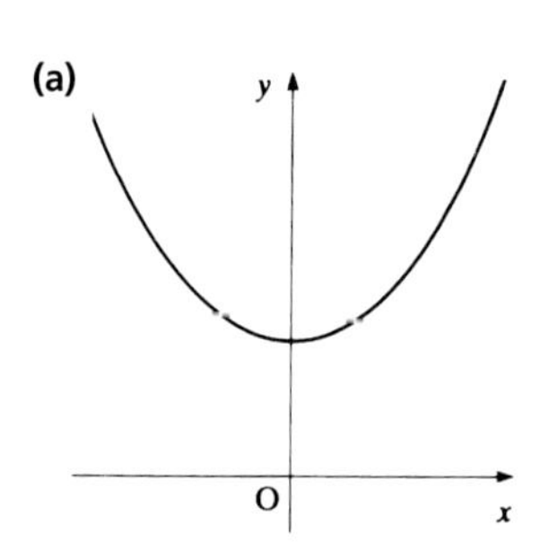

(b)

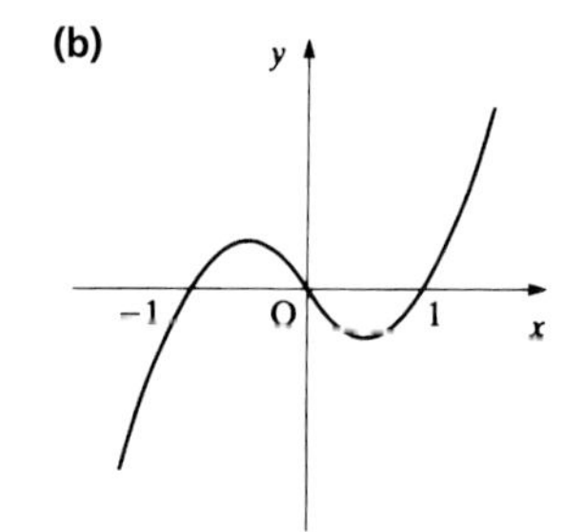

(c)

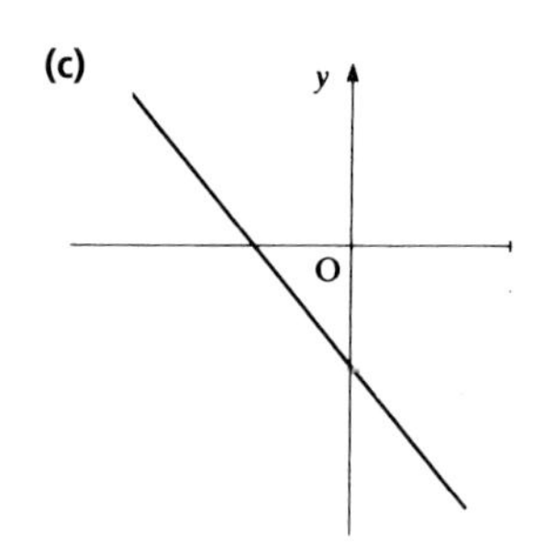

(d)

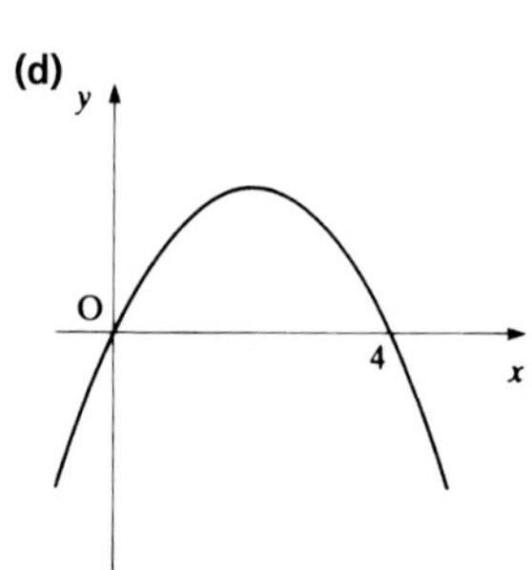

(e)

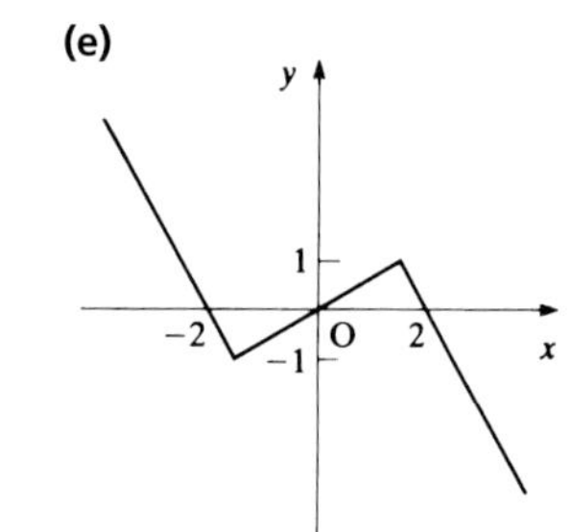

(f)

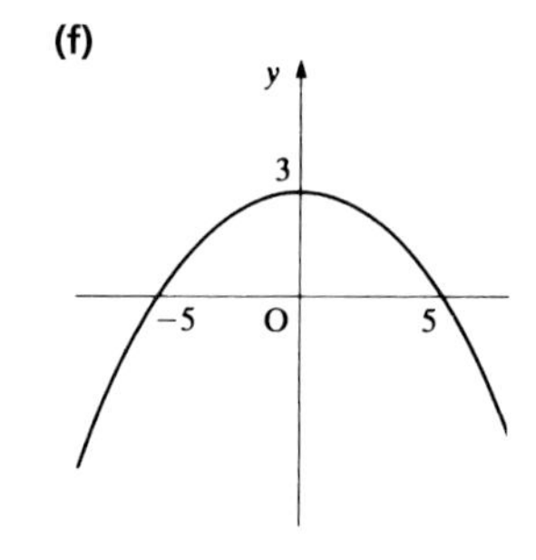

2 For each of the following functions, say whether it is odd, even, periodic, or any combination of these. For any function that is periodic, find its period.

(a) $f(x) = 2 - x^2$ **(b)** $f(x) = \sin 3x$

(c) $f(x) = x^2 + 2x - 3$ **(d)** $f(x) = 2x^3 - 3x$

(e) $f(x) = \sin x + \cos x$ **(f)** $f(x) = \sin x \cos x$

3 **(a)** Sketch the function $f(x) = \sin 2x$ for $0° \leqslant x \leqslant 360°$ and hence state its period.

(b) Say how the period of this function is related to the period of $\sin x$.

(c) What are the periods of the following functions?

(i) $f(x) = \sin 4x$ **(ii)** $f(x) = \sin 3x$ **(iii)** $f(x) = \sin \frac{x}{2}$

4 The function f is even, periodic with period 2, and for $0 \leqslant x \leqslant 1$, $f(x) = x$. Sketch the graph of $f(x)$ for $-4 \leqslant x \leqslant 4$.

5 On separate diagrams sketch

(a) $y = 2x^2$

(b) $y = -x^3$

(c) $y = \pm\sqrt{x}$.

For each diagram state whether the function is even, odd or neither.

FUNCTIONS OF THE TYPE $y = kx^n$

You are already familiar with a number of the functions of this type. For example, $y = x^2$ and $y = x^3$ have been drawn earlier in this chapter. Their inverses, provided the function is defined over an appropriate domain, are also known.

A number of the common functions of the type $y = x^n$ are shown in figure 2.19.

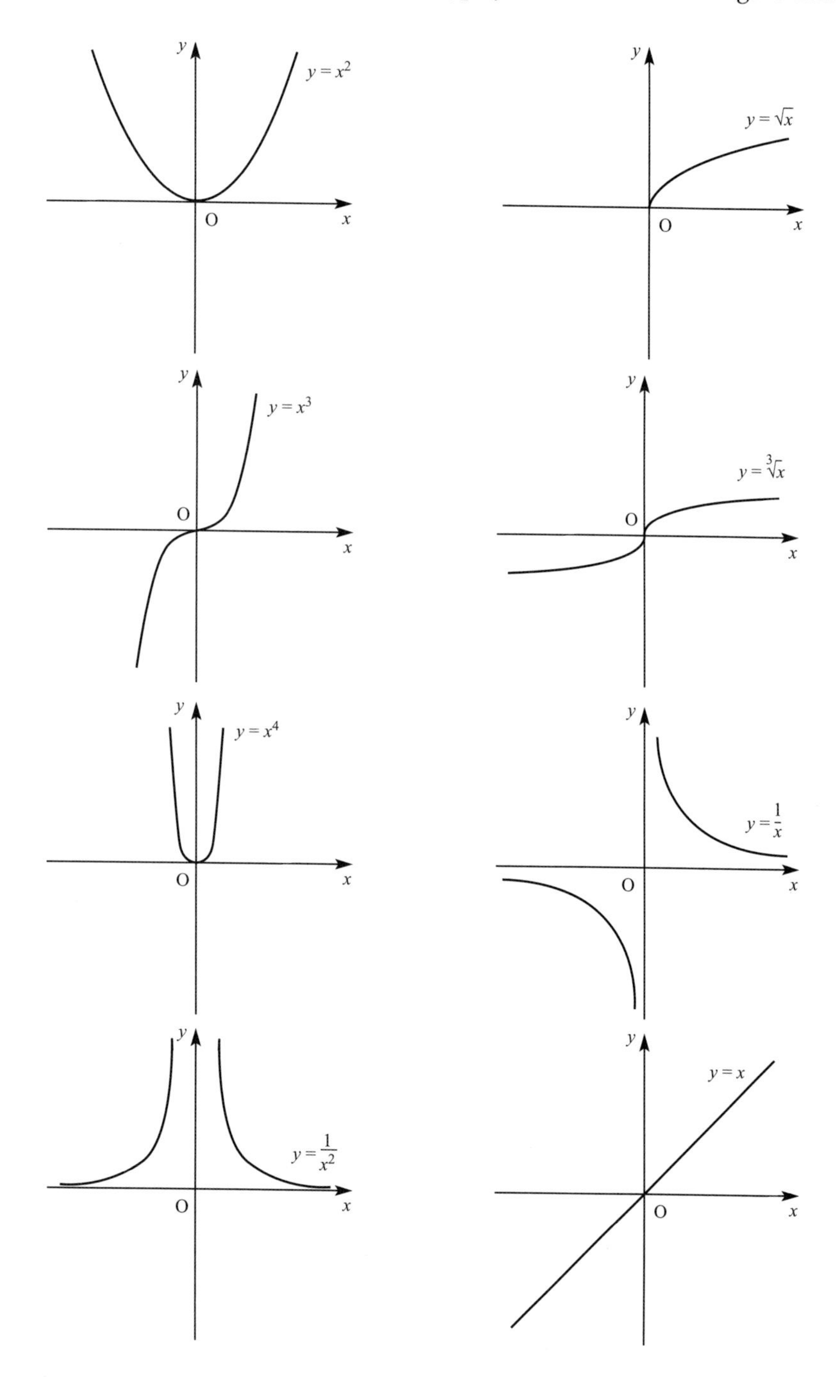

FIGURE 2.19

In all of the diagrams in figure 2.19 $k = 1$.

If k takes a value other than 1 then the magnitude of k gives the scale factor of a stretch that is parallel to the y axis with the x axis invariant. Figure 2.20 illustrates this.

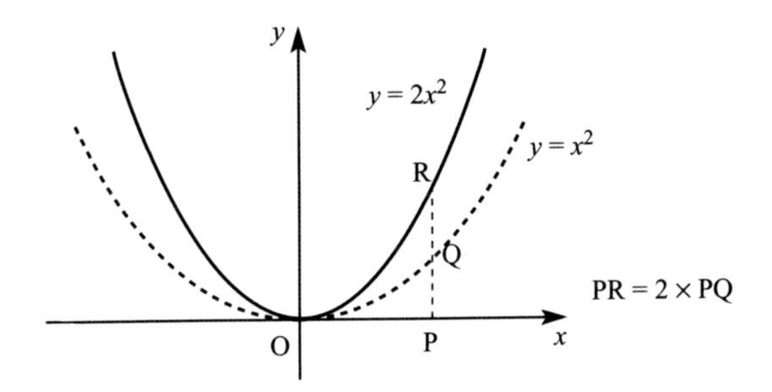

FIGURE 2.20

If $k<0$, i.e. k is negative, then the graph is reflected in the x axis. Figure 2.21 illustrates this.

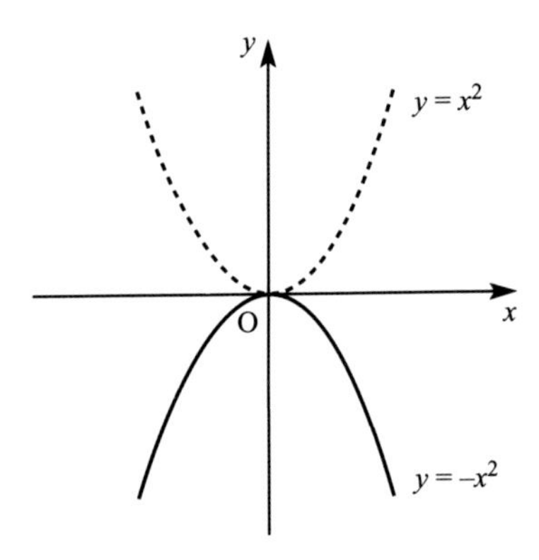

FIGURE 2.21

EXERCISE 2E

On the same diagram sketch the following pairs of graphs.

1 $y = x^3$ and $y = 2x^3$.

2 $y = f(x)$ and $y = f^{-1}(x)$ where $f(x) = \sqrt{x}$ and $x \geqslant 0$.

3 $y = \dfrac{1}{x}$ and $y = \dfrac{1}{2x}$.

4 $y = f(x)$ and $y = f^{-1}(x)$ where $f(x) = \sqrt[3]{x}$.

5 $y = \dfrac{1}{x^2}$ and $y = -\dfrac{1}{x^2}$.

6 $y = x^2$ and $y = -2x^2$.

7 $y = f(x)$ and $y = f^{-1}(x)$ where $f(x) = -\sqrt{x}$.

8 $y = x^4$ and $y = -\frac{1}{2}x^4$.

9 $y = f(x)$ and $y = f^{-1}(x)$ where $f(x) = -\sqrt[3]{x}$.

10 $y = \dfrac{1}{x}$ and $y = -\dfrac{2}{x}$.

THE MODULUS FUNCTION

THE GRAPH OF $y = |f(x)|$

The modulus of a number is its absolute value and it disregards the sign of the number. The modulus of x is labelled $|x|$.

So $\quad |5| = 5$

$\quad\quad |-5| = 5.$

The graph of $y = |x|$ has all y values positive and equal to the magnitude of the value of x. This is illustrated in figure 2.22.

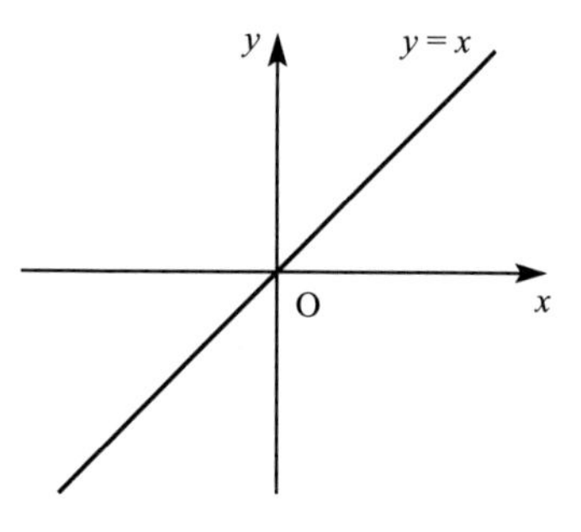

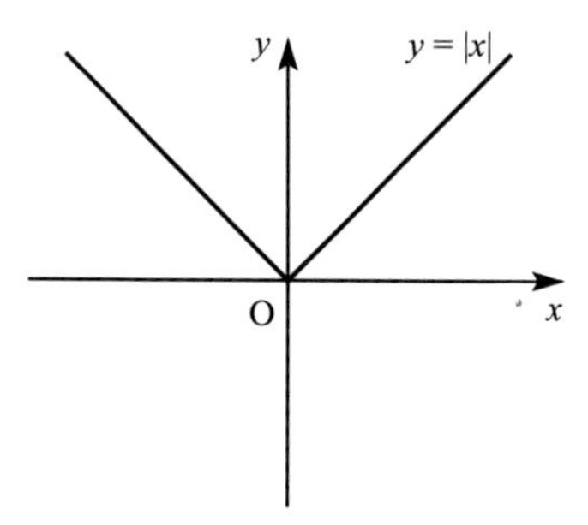

FIGURE 2.22

You see that the geometric effect is to reflect the part of the line below the x axis to above the x axis.

In general, the graph of $y = |f(x)|$ is found by drawing $y = f(x)$ and reflecting any part of it that lies below the x axis to above the x axis.

EXAMPLE 2.14

Sketch the graph of $y = |2x + 3|$.

Solution

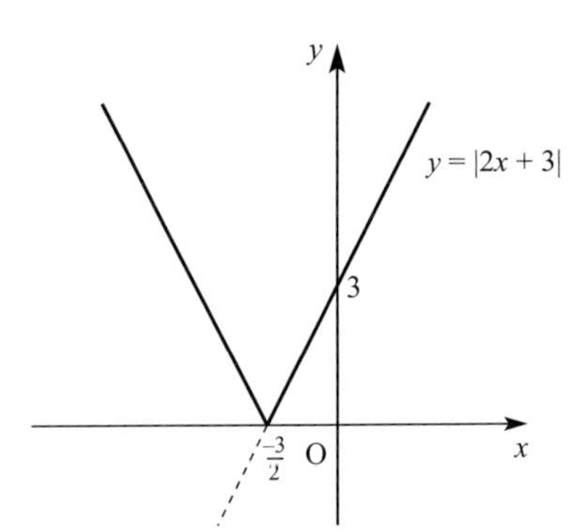

FIGURE 2.23

EXAMPLE 2.15

Sketch the graph of $y = |4 - x^2|$

Solution

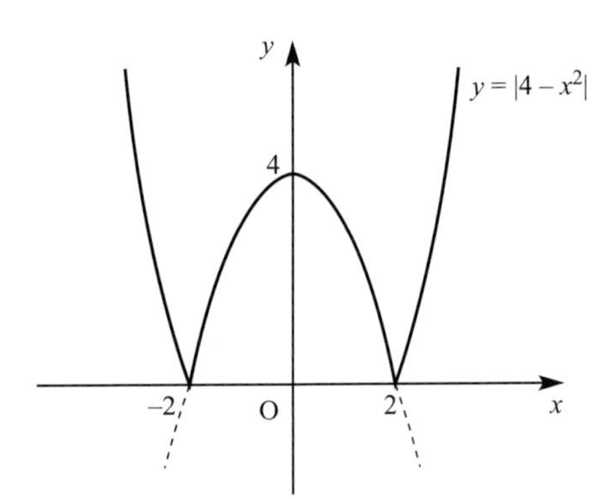

FIGURE 2.24

THE GRAPH OF $y = f(|x|)$

The function $f(|x|)$ has the same values whether the value of x is positive or negative. Since $|-x| = |x|$ the function $f(|x|)$ is an even one. Consequently, the graph of $y = f(|x|)$ will be symmetrical about the y axis.

To sketch the graph of $y = f(|x|)$ reflect the section of the graph of $y = f(x)$ for $x \geqslant 0$ in the y axis.

EXAMPLE 2.16

Sketch the graph of $y = 2|x| - 3$.

Solution

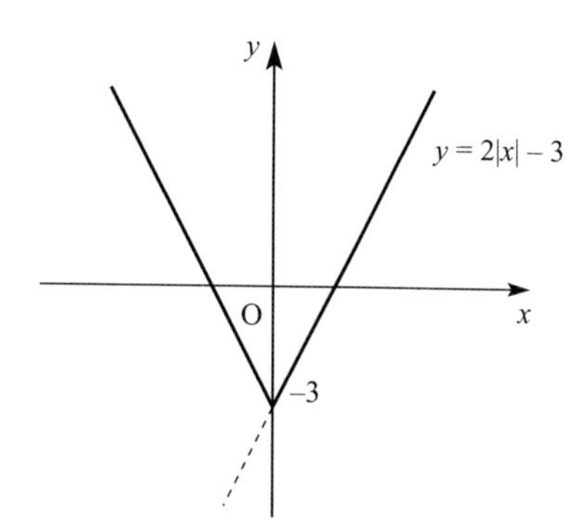

FIGURE 2.25

EXAMPLE 2.17

Figure 2.26 shows a sketch of the graph of $y = \mathrm{f}(x)$.
Sketch the graph of $y = \mathrm{f}(|x|)$.

Solution

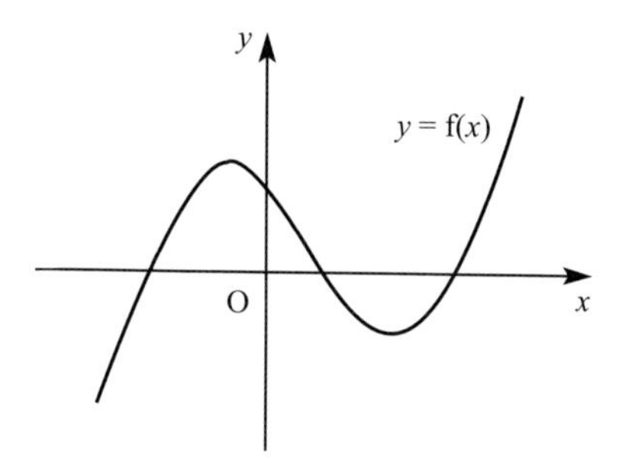

FIGURE 2.26

Reflect the section of $y = \mathrm{f}(x)$ for which $x \geqslant 0$ in the y axis to give the graph of $y = \mathrm{f}(|x|)$.

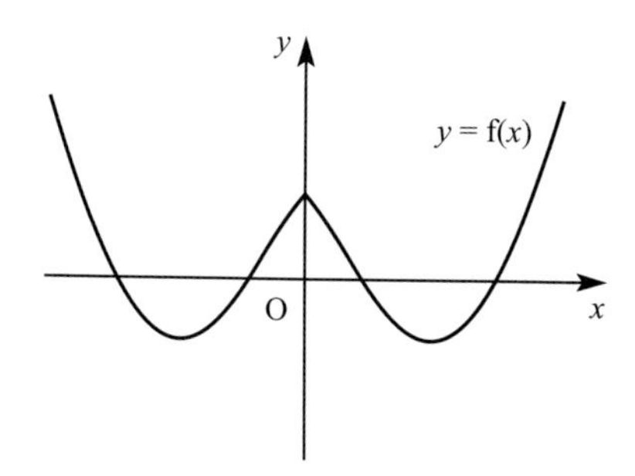

FIGURE 2.27

EXERCISE 2F

1 For each of the following sketches of $y = f(x)$ sketch the graph of $y = |f(x)|$.

(a)

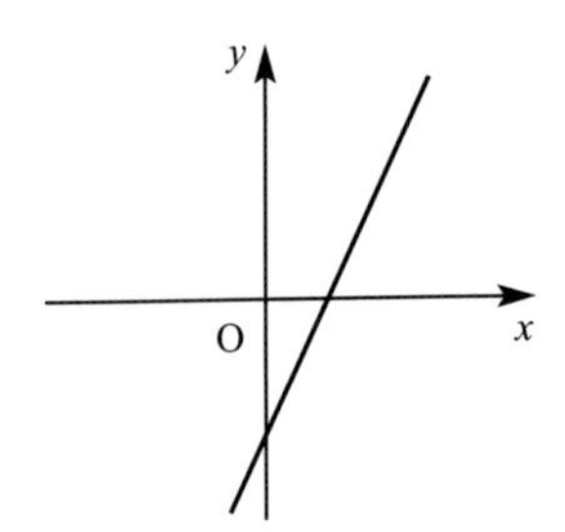

(b)

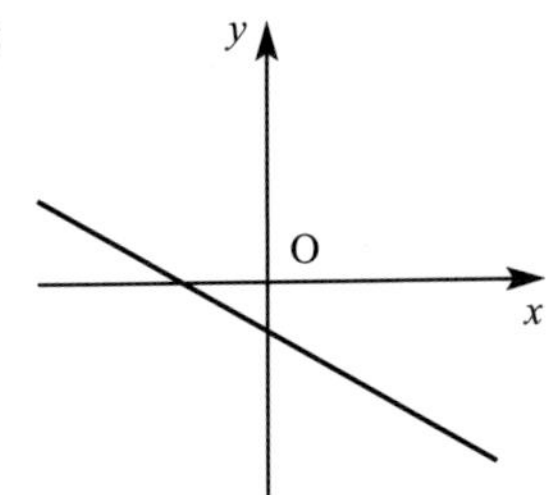

(c)

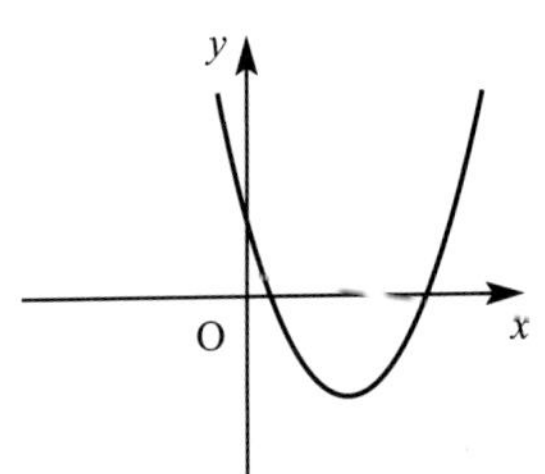

(d)

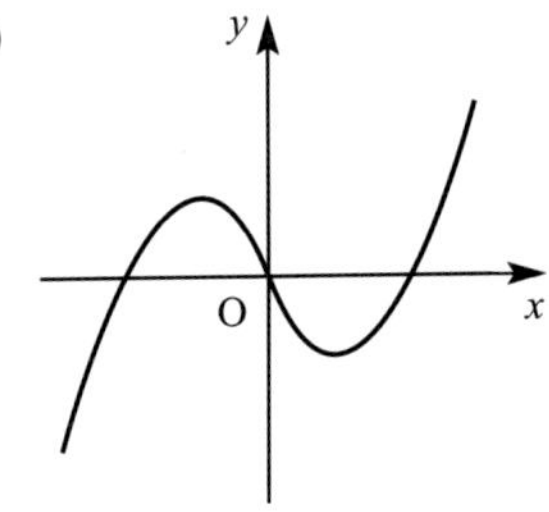

(e)

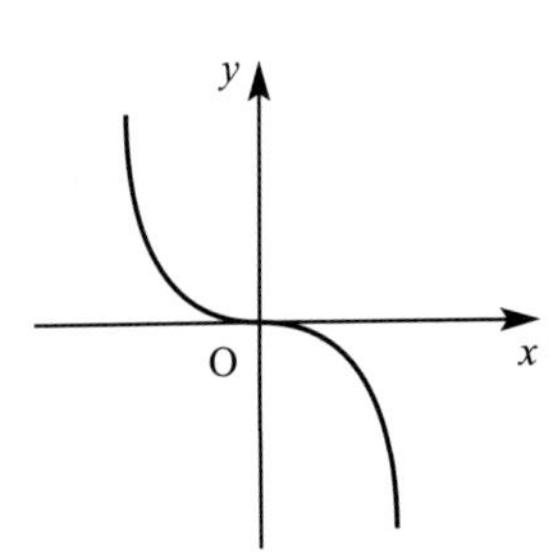

(f)

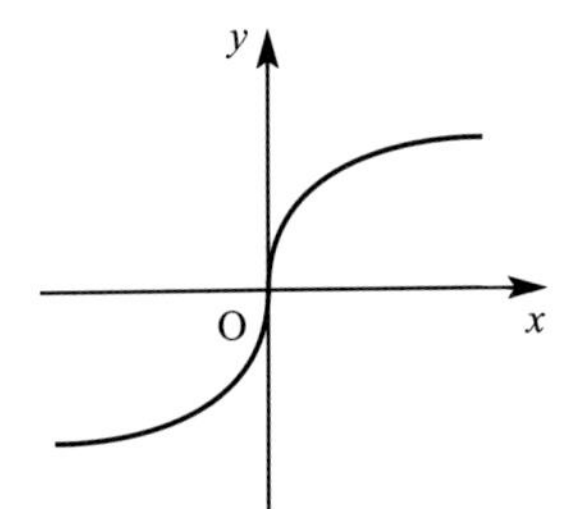

2 For each of the sketches of $y = f(x)$ in question 1 sketch the graph of $y = f(|x|)$.

3 Sketch the following graphs.

(a) $y = |x + 2|$

(b) $y = |3x - 2|$

(c) $y = |x| + 2$

(d) $y = |x^2 - 2|$

(e) $y = |x^3|$

(f) $y = |x^3| - 1$

(g) $y = \left|\frac{1}{x}\right|$

(h) $y = |3 - x|$

(i) $y = |9 - x^2|$

(j) $y = |\sin x|$.

4 The diagrams show sketches of $y = |f(x)|$. For each one draw possible sketches for $y = f(x)$.

(a)

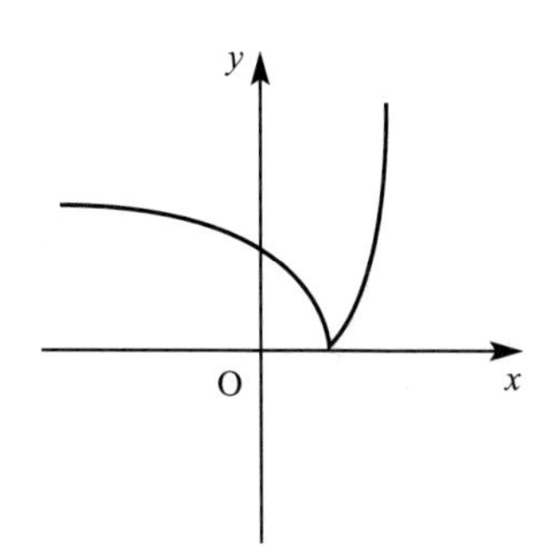

(b)

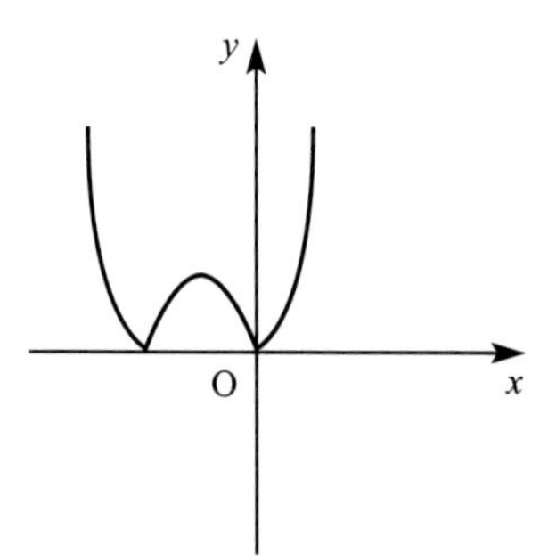

5 The diagrams show sketches of $y = f(|x|)$. For each one draw possible sketches for $y = f(x)$.

(a)

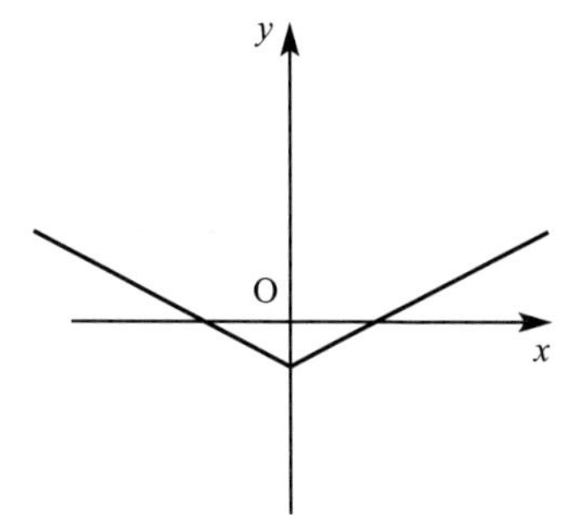

(b)

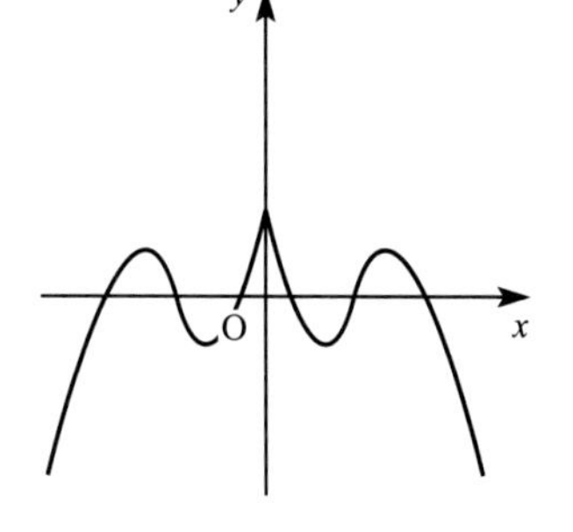

GEOMETRIC SOLUTIONS OF EQUATIONS

In *Pure Mathematics 1* you saw how to solve simultaneous equations by drawing graphs and findings their points of intersection.

EXAMPLE 2.18

By drawing the graphs of the lines solve the simultaneous equations

$$x + 2y = 7$$
$$4x - y = 1.$$

Solution $x + 2y = 7$ passes through $(0, 3\frac{1}{2})$ and $(7, 0)$.
$4x - y = 1$ passes through $(0, -1)$ and $(\frac{1}{4}, 0)$.

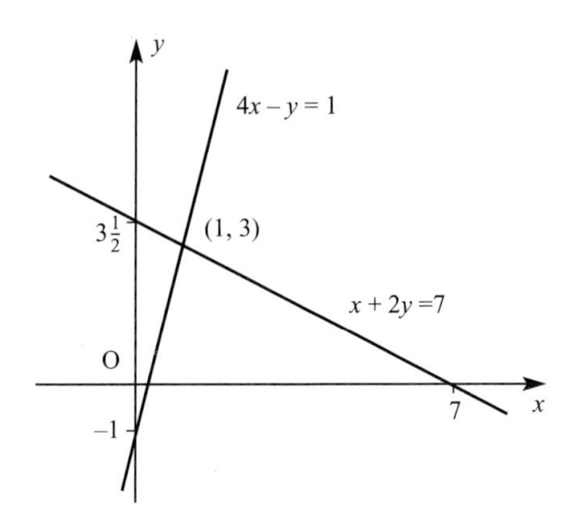

FIGURE 2.28

The lines cross at (1, 3) giving the solution to the simultaneous equations as $x = 1$ and $y = 3$.

Graphs can be used to solve other equations. You also saw in *Pure Mathematics 1* how to solve equations such as $4x^3 - 8x^2 - x + 2 = 0$ by drawing the graph of $y = 4x^3 - 8x^2 - x + 2$ and finding where it crossed the x axis.

EXAMPLE 2.19

Draw the graph of $y = x^3 - x^2 - 2x$.
Use your graph to solve

(a) $x^3 - x^2 - 2x = 0$

(b) $x^3 - x^2 - 2x = 2x$.

Solution Plot points on $y = x^3 - x^2 - 2x$.

x	−2	−1	0	1	2	3
y	−8	0	0	−2	0	12

Now sketch the graph of $y = x^3 - x^2 - 2x$.

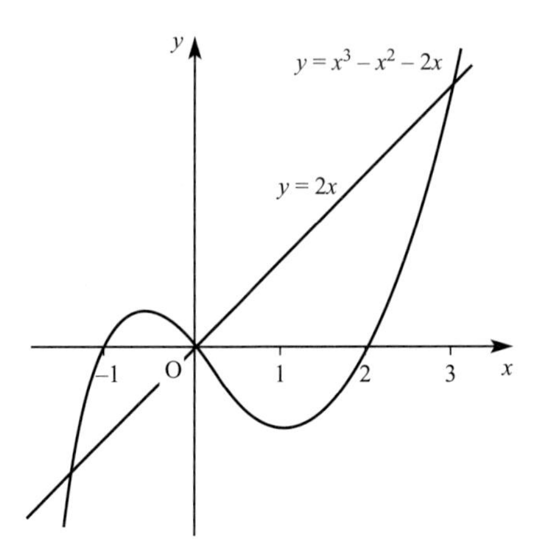

FIGURE 2.29

(a) To solve $x^3 - x^2 - 2x = 0$ you need to find the x values of the points of intersection between $y = x^3 - x^2 - 2x$ and $y = 0$, i.e. where the curve cuts the x axis.

So $\quad x = -1, 0$ or 2.

(b) To solve $x^3 - x^2 - 2x = 2x$ you need to find the x values of the points of intersection between $y = x^3 - x^2 - 2x$ and $y = 2x$. To do this draw the line $y = 2x$.

From the diagram the solutions are approximately:

$x = -1.6, \quad 0 \quad$ and $\quad 2.5.$

You should check these answers with a graphical calculator by drawing the graphs and using the trace function.

EXAMPLE 2.20

By drawing the graphs of $y = \dfrac{4}{x}$ and $y = \sqrt[3]{x}$ solve $x^{\frac{4}{3}} = 4$.

Solution $\quad y = \dfrac{4}{x}$ has points:

x	–4	–2	–1	1	2	4
y	–1	–2	–4	4	2	1

$y = \sqrt[3]{x}$ has points:

x	–8	–2	–1	1	2	8
y	–2	–1.26	–1	1	1.26	2

Plotting the points and drawing the graphs gives the result shown in figure 2.30.

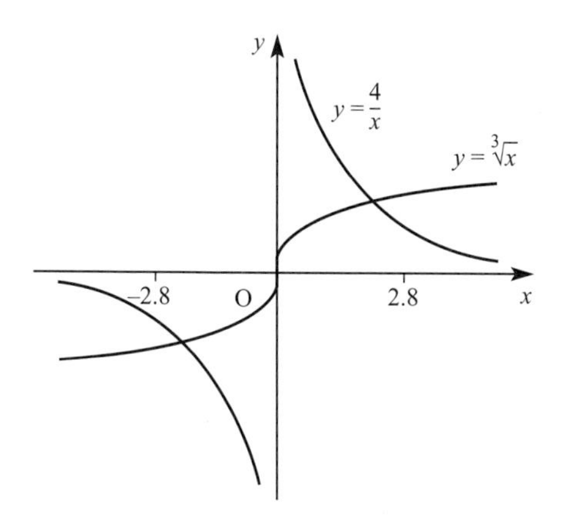

FIGURE 2.30

Where these graphs intersect gives $\sqrt[3]{x} = \frac{4}{x}$,

so $\quad x\sqrt[3]{x} = 4$

or $\quad x^{\frac{4}{3}} = 4$,

and the solution of this equation is found from the intersection of the graphs. That is

$$x = \pm 2.8.$$

EXERCISE 2G

1 Draw the graph of $y = x^2 - 1$ from $x = -3$ to $x = 3$.
From your graph solve
(a) $x^2 - 1 = 0$
(b) $x^2 - 1 = 3$.

2 Draw the graph of $y = x^2 - x - 1$.
(a) Use your graph to solve $x^2 - x - 1 = 0$.
(b) On the same diagram draw $y = x + 2$.
(c) Use your graph to solve $x^2 - x - 1 = x + 2$.
(d) For what values of x is $x^2 - x - 1 < x + 2$?

3 **(a)** On the same axes draw the graphs of $y = x(x + 3)$ and $y = x + 3$.
(b) Show that $x(x + 3) = x + 3$ can be re-written as $x^2 + 2x - 3 = 0$.
(c) Use your graph to find the solutions to $x^2 + 2x - 3 = 0$.

4 On the same diagram draw the graphs of $y = x^2 - 2x + 6$ and $y = 6x - x^2$.
Use your diagram to solve $6x - x^2 > x^2 - 2x + 6$.

5 Draw the graph of $y = \frac{1}{x^2}$ for $-3 < x < 3$.

On the same diagram draw the graph of $y = x + 3$.

Show that $\frac{1}{x^2} = x + 3$ can be re-written as $x^3 + 3x^2 - 1 = 0$.

State the approximate solutions to $x^3 + 3x^2 - 1 = 0$.

6 Show that $x^3 - 9x + 4 = 0$ can be written as $\frac{4}{x} = 9 - x^2$.

Draw the graphs of $y = \frac{4}{x}$ and $y = 9 - x^2$ and use them to find the solutions to $x^3 - 9x + 4 = 0$.

7 Find the x values of the points of intersection between the graphs of $y = 4\sqrt{x}$ and $y = 2x + 1$.

8 By drawing suitable graphs solve $|x| < 6 - x^2$.

9 On the same graph draw $y = \frac{1}{x^2}$ and $y = 2|x| - 3$.

(a) What can you say about both of these functions?

(b) Use your graph to solve $\frac{1}{x^2} = 2|x| - 3$.

10 The functions f and g are defined by

$$f: x \mapsto 2x - 3 \quad \text{and} \quad g: x \mapsto \sqrt{x} \text{ for } x > 0.$$

On the same diagram sketch $y = f^{-1}(x)$ and $y = g^{-1}(x)$.
Solve $f^{-1}(x) = g^{-1}(x)$.

TRANSFORMATIONS

You already know how to sketch the curves of many functions. Frequently, curve sketching can be made easier by relating the equation of the function to that of a standard function of the same form. This allows you to map the points on the standard curve to equivalent points on the curve you need to draw.

The mappings you will use for curve sketching are called *transformations*. There are several types of transformation, each with different effects, and you will find that by using them singly or in combination you can sketch a large variety of curves much more quickly.

TRANSLATIONS

Figure 2.31 shows the graphs of $y = x^2$ and $y = x^2 + 3$. You could draw this yourself, either on a graphics calculator or by hand. For any given value of x, the y coordinate for the second curve is 3 units more than the y coordinate for the first curve.

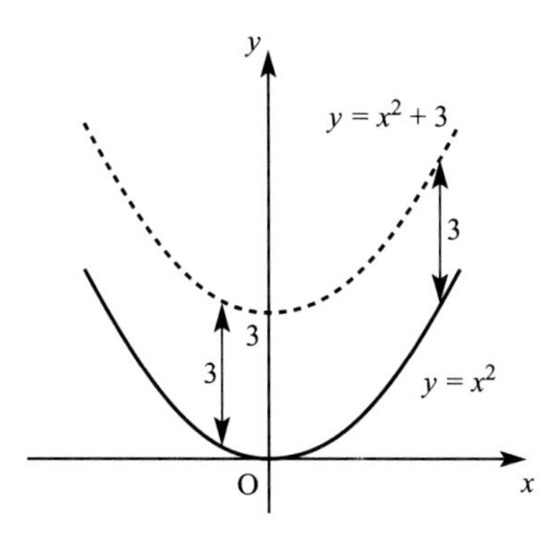

FIGURE 2.31

Although the curves appear to get closer together as you move away from the line of symmetry, their vertical separation is, in fact, constant. Each of the vertical arrows is 3 units long.

You can see that the graphs of $y = x^2 + 3$ and $y = x^2$ are exactly the same shape, but $y = x^2$ has been translated through 3 units in the positive y direction to obtain $y = x^2 + 3$.

Similarly, $y = x^2 - 2$ could be obtained by translating $y = x^2$ through 2 units in the negative y direction (i.e. –2 units in the positive y direction).

In general, for any function $f(x)$, the curve $y = f(x) + b$ can be obtained from that of $y = f(x)$ by translating it through b units in the positive y direction.

What about the relationship between the graphs of $y = x^2$ and $y = (x - 2)^2$? Figure 2.32 shows the graphs of these two functions. Again, these curves have exactly the same shape, but as you can see, this time they are separated by a constant 2 units in the x direction.

You may find it surprising that $y = x^2$ moves in the positive x direction when 2 is subtracted from x. It happens because x must be correspondingly larger if $(x - 2)$ is to give the same output that x did in the first mapping.

Notice that the axis of symmetry of the curve $y = (x - 2)^2$ is the line $x = 2$.

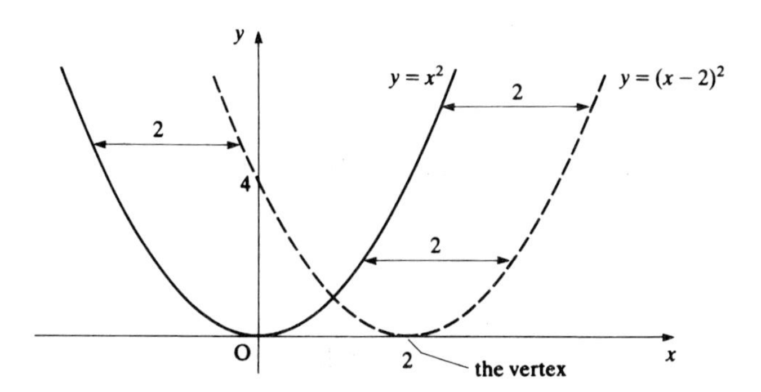

FIGURE 2.32

In general, the curve with equation $y = f(x - a)$ can be obtained from the curve with equation $y = f(x)$ by a translation of a units in the positive x direction.

Combining these results, $y = f(x - a) + b$ is obtained from $y = f(x)$ by a translation of b units in the positive y direction and a units in the positive x direction. This translation is represented by the vector $\begin{pmatrix} a \\ b \end{pmatrix}$.

$y = f(x)$ mapped on to $y = f(x - a) + b$
$\Rightarrow$ translation $\begin{pmatrix} a \\ b \end{pmatrix}$.

Note You would usually write $y = f(x - a) + b$ with y as the subject, but this is equivalent to $y - b = f(x - a)$. This form emphasises that subtracting a number from x or y moves the graph in the positive x or y direction.

EXAMPLE 2.21

Sketch the curve $y = \sin x$ for $0° \leqslant x \leqslant 180°$ and show how it can be used to obtain the graph of $y = \sin(x + 90°)$.

Solution Re-writing $\sin(x + 90°)$ as $\sin(x - (-90°))$, you can see that the graph of $y = \sin(x + 90°)$ is obtained from the graph of $y = \sin x$ by a translation of $-90°$ in the positive x direction, i.e. 90° in the negative x direction (see figure 2.33).

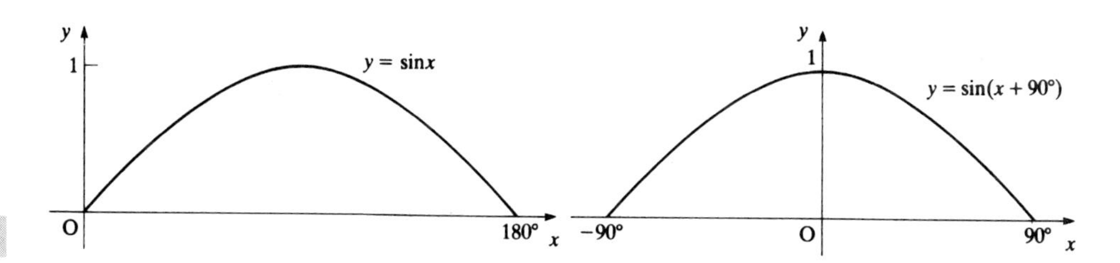

FIGURE 2.33

Notice that the resulting graph is the same as that of $y = \cos x$.

In *Pure Mathematics 1* you met the technique of completing the square of a quadratic function. Using this technique, any quadratic expression of the form $y = x^2 + bx + c$ can be written as $y = (x - p)^2 + q$, so its graph can be sketched by relating it to the graph of $y = x^2$.

EXAMPLE 2.22

(a) Find values of a and b such that $x^2 - 2x + 5 \equiv (x - a)^2 + b$.

(b) Sketch the graph of $y = x^2 - 2x + 5$ and state the position of its vertex and the equation of its axis of symmetry.

Solution Since the form of the right-hand side is given, it is easiest to expand this and then compare coefficients:

$$x^2 - 2x + 5 \equiv (x - a)^2 + b$$

$$\Rightarrow \quad x^2 - 2x + 5 \equiv x^2 - 2xa + a^2 + b.$$

Comparing coefficients gives $a = 1$ and $b = 4$.

Re-writing the equation as $y = (x - 1)^2 + 4$ or $(y - 4) = (x - 1)^2$ shows that the curve can be obtained from the graph of $y = x^2$ by a translation of 1 unit in the positive x direction, and 4 units in the positive y direction, i.e. a translation $\begin{pmatrix}1\\4\end{pmatrix}$.

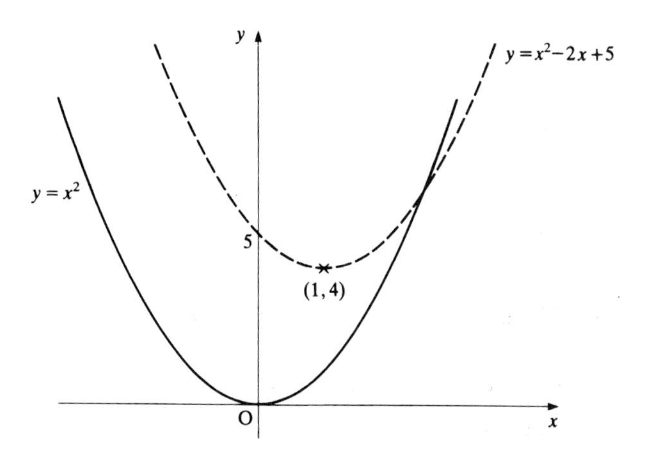

FIGURE 2.34

The vertex is (1, 4) and the axis of symmetry is the line $x = 1$ (see figure 2.34).

EXAMPLE 2.23

Figure 2.35 shows part of the scalloped bottom of a rollerblind.

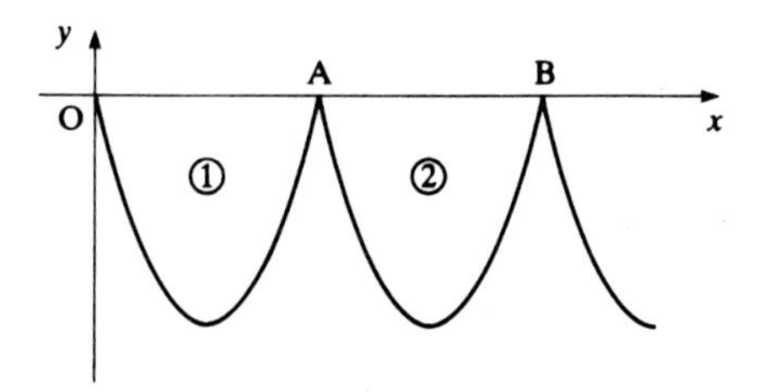

FIGURE 2.35

The equation of curve ① with respect to the x and y axes shown is $y = x^2 - 4x$ for $0 \leqslant x \leqslant 4$.

Find the equation of curve ②.

Solution The equation of curve ① can be factorised to give $y = x(x - 4)$, so when $y = 0$, $x = 0$ or 4.

Clearly, O is (0, 0) and A is (4, 0).

Curve ② is therefore obtained from curve ① by a translation of 4 units in the positive x direction.

To find the equation of curve ②, you replace x by $(x - 4)$ in the equation of curve ①.

Using the factorised form, the equation of curve ② is

$$y = (x - 4)[(x - 4) - 4] \text{ for } 0 \leqslant x - 4 \leqslant 4$$

$$\Rightarrow \quad y = (x - 4)(x - 8) \qquad \text{for } 4 \leqslant x \leqslant 8.$$

EXAMPLE 2.24

Sketch the graph of $y = \frac{1}{x}$.

On a separate diagram, marking any asymptotes, sketch the graph of $y = \frac{1}{x + 1} + 2$.

Solution The graph of $y = \frac{1}{x}$ looks like figure 2.36.

Its asymptotes are the x axis and the y axis (these are the lines that the curve gets closer and closer to but never quite reaches).

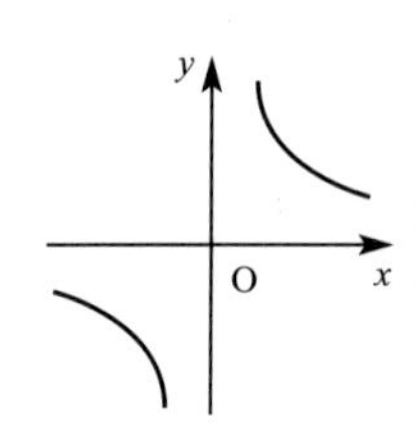

FIGURE 2.36

To obtain $y = \dfrac{1}{x+1} + 2$ the graph of $y = \dfrac{1}{x}$ is translated by $\begin{pmatrix} -1 \\ 2 \end{pmatrix}$. This gives:

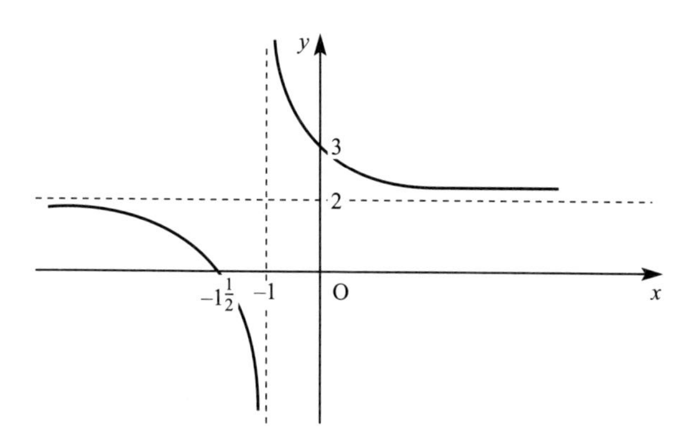

FIGURE 2.37

The asymptotes are the lines $x = -1$ and $y = 2$.

ONE-WAY STRETCHES

Figure 2.38 shows the graphs of $y = \sin x$ and $y = 2\sin x$ on the same axes, for $0° < x < 180°$. If you are using a graphics calculator or computer package you could easily add $y = \frac{1}{2}\sin x$ and $y = 3\sin x$ to the display.

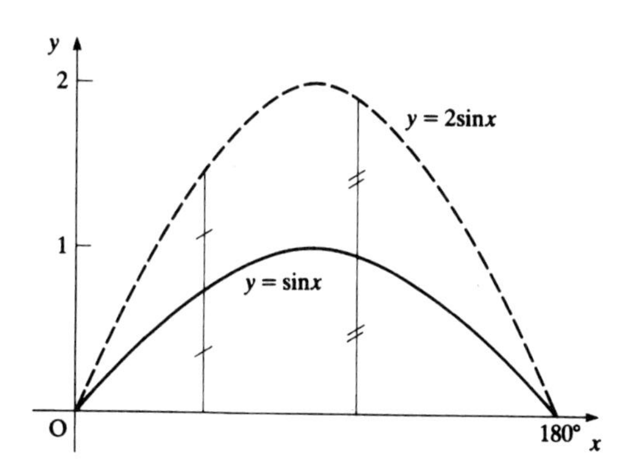

FIGURE 2.38

You will notice that for any value of x, the y coordinate of the point on the curve $y = 2\sin x$ is exactly double that on the curve $y = \sin x$.

This is the equivalent of the curve being stretched parallel to the y axis. Since each y coordinate is doubled, this is called a *stretch of scale factor 2 parallel to the y axis.*

Note

The equation $y = 2\sin x$ could also be written as $\frac{y}{2} = \sin x$, so dividing y by 2 gives a stretch of scale factor 2 in the y direction.

In general, for any curve $y = \mathrm{f}(x)$, and any value of a greater than 0, $y = a\mathrm{f}(x)$ is obtained from $y = \mathrm{f}(x)$ by a stretch of scale factor a parallel to the y axis.

Drawing the graphs of $y = \sin x$ and $y = \sin 2x$ for $0° \leqslant x \leqslant 180°$ on the same axes gives figure 2.39.

You will see that the graph of $y = \sin 2x$ is compressed parallel to the x axis, so that for any value of y, the x coordinate of the point on the curve $y = \sin 2x$ will be exactly half of that on the curve $y = \sin x$ (figure 2.39).

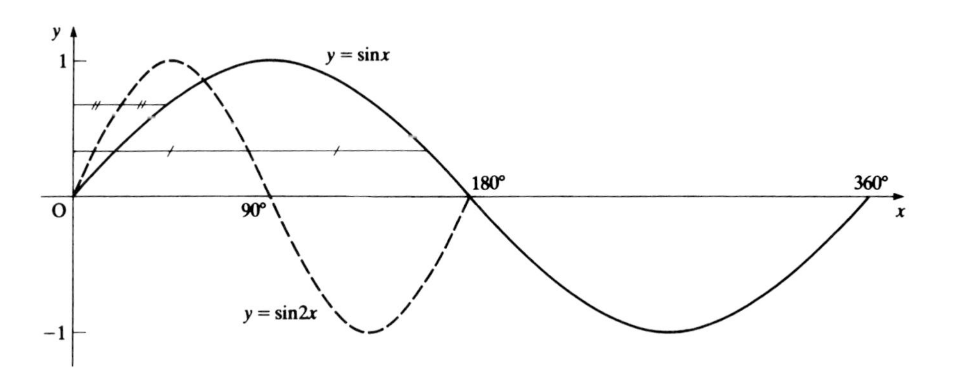

FIGURE 2.39

Since the curve is compressed to half its original size, this is referred to as a stretch of scale factor $\frac{1}{2}$ parallel to the x axis.

In general, for any curve $y = \mathrm{f}(x)$ and any value of a greater than 0, $y = \mathrm{f}(ax)$ is obtained from $y = \mathrm{f}(x)$ by a stretch of scale factor $\frac{1}{a}$ parallel to the x axis. Similarly $y = \mathrm{f}(\frac{x}{a})$ corresponds to a stretch of scale factor a parallel to the x axis.

Note This is as you would expect: dividing x by a gives a stretch of scale factor a in the x direction, just as dividing y by a gives a stretch of scale factor a in the y direction.

$y = \mathrm{f}(x)$ mapped onto $y = \mathrm{f}(ax) \Rightarrow$ stretch scale factor $\frac{1}{a}$, parallel to the x axis
$y = \mathrm{f}(x)$ mapped onto $y = a\mathrm{f}(x) \Rightarrow$ stretch scale factor a, parallel to the y axis.

EXAMPLE 2.25

Starting with the curve $y = \cos x$, show how transformations can be used to sketch the curves

(a) $y = 2\cos 3x$ **(b)** $y = 3 + \cos\frac{x}{2}$ **(c)** $y = \cos(2x - 60°)$.

Solution **(a)** The curve with equation $y = \cos 3x$ is obtained from the curve with equation $y = \cos x$ by a stretch of scale factor $\frac{1}{3}$ parallel to the x axis. There will therefore be one complete oscillation of the curve in 120° (instead of 360°).

The curve of $y = 2\cos 3x$ is obtained from that of $y = \cos 3x$ by a stretch of scale factor 2 parallel to the y axis. The curve therefore oscillates between $y = 2$ and $y = -2$ (instead of $y = 1$ and $y = -1$). This is shown in figure 2.40.

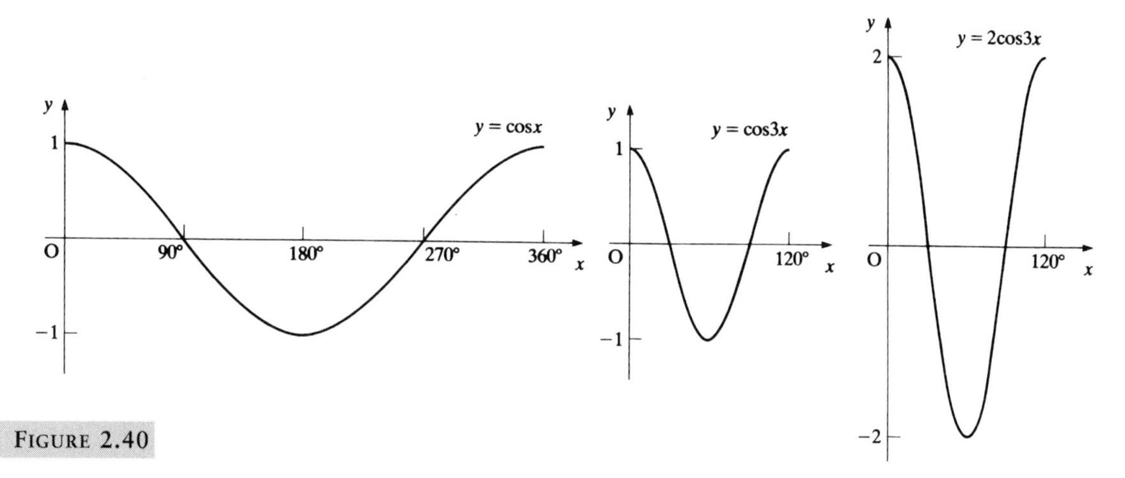

FIGURE 2.40

(b) The curve of $y = \cos\frac{x}{2}$ is obtained from that of $y = \cos x$ by a stretch of scale factor 2 in the x direction. There will therefore be one complete oscillation of the curve in 720° (instead of 360°).

The curve of $y = 3 + \cos\frac{x}{2}$ is obtained from that of $y = \cos\frac{x}{2}$ by a translation $\begin{pmatrix} 0 \\ 3 \end{pmatrix}$.

The curve therefore oscillates between $y = 4$ and $y = 2$ (see figure 2.41).

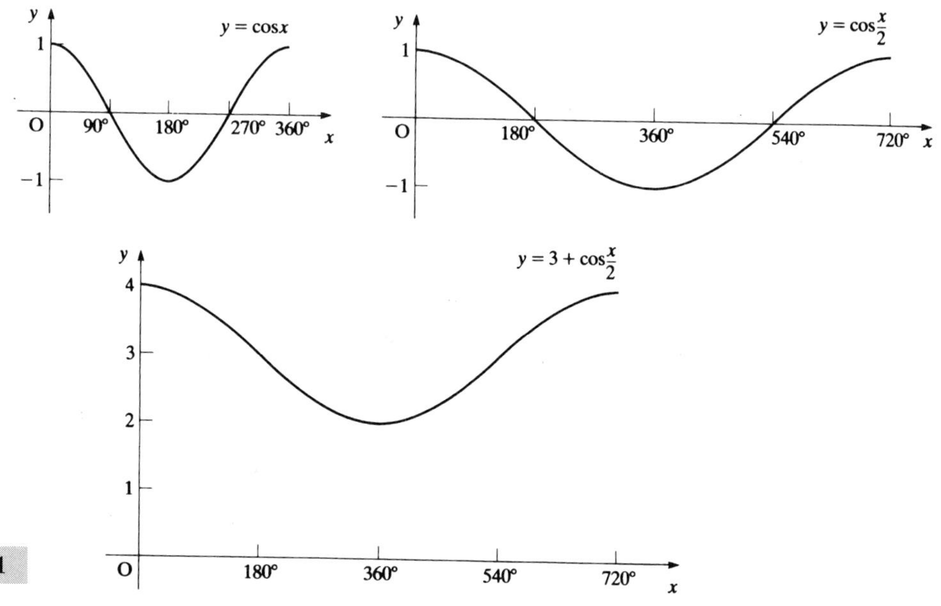

FIGURE 2.41

(c) The curve of $y = \cos(x - 60°)$ is obtained from that of $y = \cos x$ by a translation of $\begin{pmatrix} 60° \\ 0 \end{pmatrix}$.

The curve of $y = \cos(2x - 60°)$ is obtained from that of $y = \cos(x - 60°)$ by a stretch of scale factor $\frac{1}{2}$ parallel to the x axis (see figure 2.42).

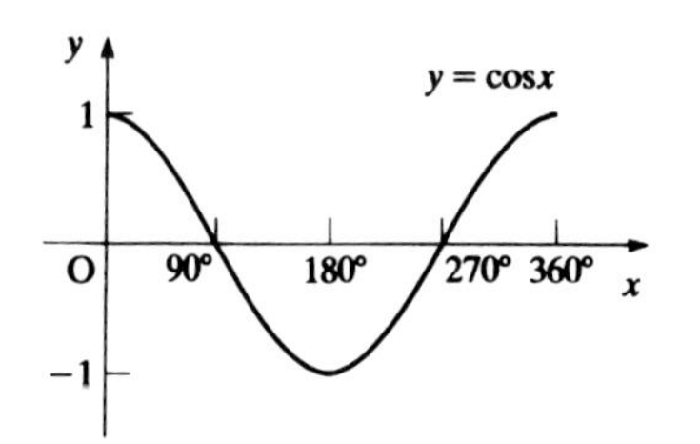

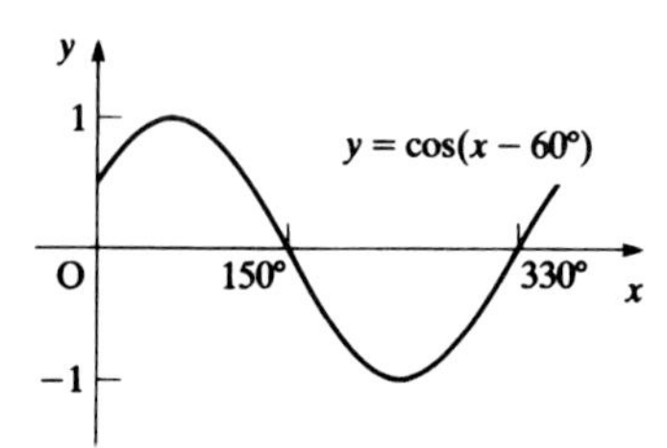

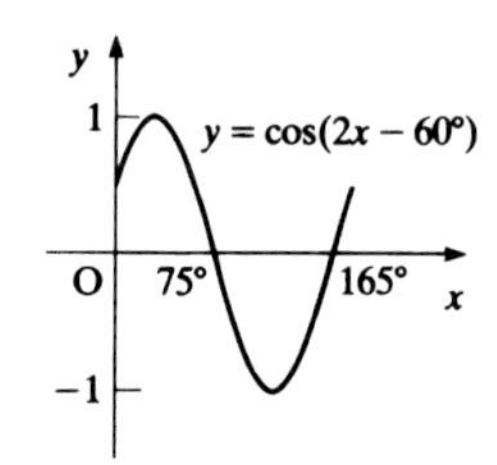

FIGURE 2.42

Note

You must be careful to perform the transformations in the correct order. It is always a good idea to check your results using a graphics calculator whenever possible.

EXAMPLE 2.26

Find the values of a, p and q when $y = 2x^2 + 4x - 1$ is written in the form $y = a[(x + p)^2 + q]$.

Show how the graph can be obtained from the graph of $y = x^2$ by successive transformations, and list the transformations in the order in which they are applied.

Solution Expanding the equivalent expression

$$\begin{aligned} a[(x + p)^2 + q] &= a[x^2 + 2px + p^2 + q] \\ &= ax^2 + 2apx + a(p^2 + q). \end{aligned}$$

Comparing the coefficients in $y = 2x^2 + 4x - 1$ with those above gives:

- coefficient of x^2: $a = 2$
- coefficient of x: $2ap = 4$, which gives $p = 1$
- constant term: $a(p^2 + q) = -1$, which gives $q = -1\frac{1}{2}$.

The equation of the curve can be written as $y = 2[(x + 1)^2 - 1\frac{1}{2}]$.

To sketch the graph, start with the curve $y = x^2$.
The curve $y = x^2$ becomes $y = (x + 1)^2 - 1\frac{1}{2}$ by applying the translation $\begin{pmatrix} -1 \\ -1\frac{1}{2} \end{pmatrix}$.

The curve $y = (x + 1)^2 - 1\frac{1}{2}$ becomes $y = 2[(x + 1)^2 - 1\frac{1}{2}]$ by applying a stretch of scale factor 2 parallel to the y axis.

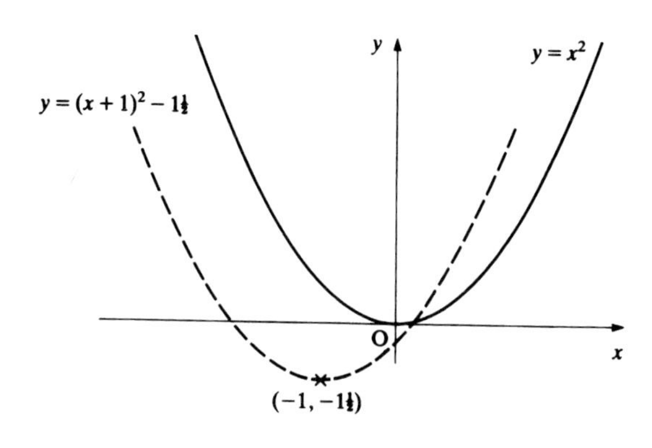

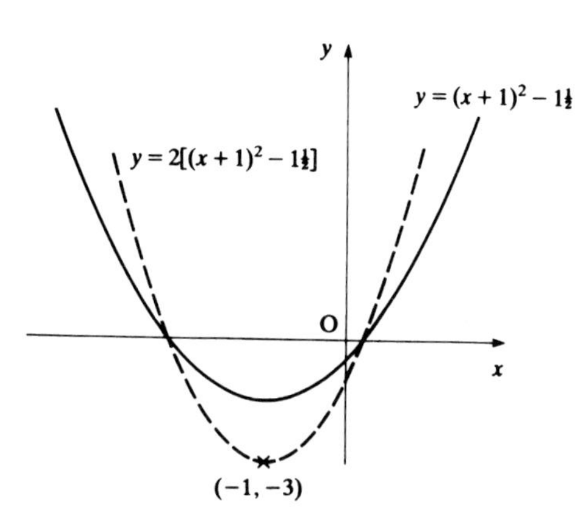

FIGURE 2.43

The translation $\begin{pmatrix} -1 \\ -1\frac{1}{2} \end{pmatrix}$

The stretch of scale factor 2 parallel to the y axis

Note

Notice on figure 2.43 how the stretch doubles the y coordinate of every point on the curve, including the turning point.

Points on the x axis have a y coordinate of 0, so are unchanged.

EXERCISE 2H

1 Starting with the graph of $y = x^2$, state transformations which can be used to sketch the following curves. Specify the transformations in the order in which they are used, and where there is more than one stage in the sketching of the curve, state each stage. State the equation of the line of symmetry.

(a) $y = x^2 - 2$ **(b)** $y = (x + 4)^2$

(c) $y = 4x^2$ **(d)** $3y = x^2$

(e) $y = (x - 3)^2 - 5$ **(f)** $y = x^2 - 2x$

(g) $y = x^2 - 4x + 3$ **(h)** $y = 2x^2 + 4x - 1$

(i) $y = 3x^2 - 6x - 2$

2 Starting with $y = \sin x$, state transformations which can be used to sketch the following curves. Specify the transformations in the order in which they are used, and where there is more than one stage in the sketching of the curve, state each stage.

(a) $y = \sin(x - 90°)$ **(b)** $y = \sin 3x$

(c) $2y = \sin x$ **(d)** $y = \sin\frac{x}{2}$

(e) $y = 2 + \sin 3x$

3 Starting with $y = \cos x$ state transformations which can be used to sketch the following curves. Specify the transformations in the order in which they are used, and where there is more than one stage in the sketching of the curve, state each stage.

(a) $y = \cos(x + 60°)$ **(b)** $3y = \cos x$

(c) $y = \cos x + 1$ **(d)** $y = \cos 2(x + 90°)$

4 For each of the following curves

(a) sketch the curve;

(b) identify the curve as being the same as one of the following:

$y = \pm\sin x$, $y = \pm\cos x$, or $y = \pm\tan x$.

(i) $y = \sin(x + 360°)$ **(ii)** $y = \sin(x + 90°)$

(iii) $y = \tan(x - 180°)$ **(iv)** $y = \cos(x - 90°)$

(v) $y = \cos(x + 180°)$

5 **(a)** Show that $x^2 + 6x + 5$ can be written in the form $(x + 3)^2 + a$ where a is a constant to be determined.

(b) Sketch the graph of $y = x^2 + 6x + 5$, giving the equation of the axis of symmetry and the coordinates of the vertex.

6 Given that $f(x) = x^2 - 6x + 11$, find values of p and q such that $f(x) \equiv (x - p)^2 + q$. On the same set of axes, sketch the curves

(a) $y = f(x)$ and **(b)** $y = f(x + 4)$, labelling clearly which is which.

7 The diagram shows the graph of $y = f(x)$ which has a maximum point at (–2, 2), a minimum point at (2, –2), and passes through the origin.

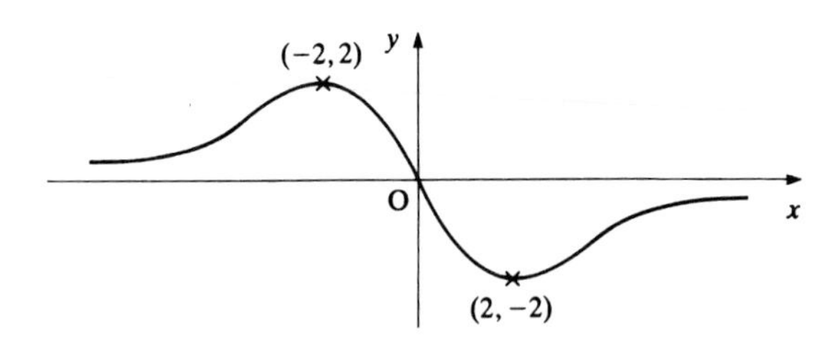

Sketch the following graphs, using a separate set of axes for each graph, and indicating the coordinates of the turning points.

(a) $y = 2f(x)$ **(b)** $y = f(x - 2)$

(c) $y = f(2x)$ **(d)** $y = 2 + f(x)$

(e) $y = f(x + 2) - 2$ **(f)** $y = 2f\left(\frac{x}{2}\right)$

8 A firm can produce a maximum of 50 machines per week, and its income and expenditure are given in pounds by the following equations:

total income = $40x$

total expenditure = $600 + 0.5x^2$

where x is the number of machines produced per week.

(a) Sketch both of these graphs on the same axes.

(b) Find the smallest number of machines which must be produced each week to ensure that the firm makes a profit.

9 The diagram shows the graph of $y = f(x)$.

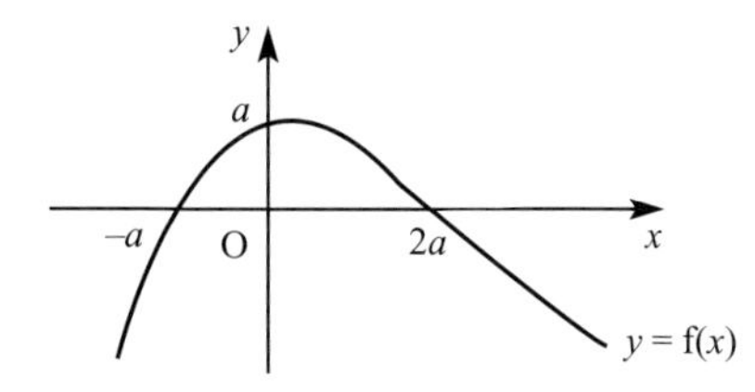

Sketch, on separate diagrams, the graphs of

(a) $y = f(x + a)$

(b) $y = f(2x)$

(c) $y = |f(x)|$.

10 Sketch the graph of $y = \frac{1}{x}$.

On separate diagrams, showing any asymptotes with dotted lines, sketch

(a) $y = \frac{1}{x - 1}$

(b) $y = \frac{1}{x} - 1$

(c) $y = \frac{1}{x - 2} + 3$.

REFLECTIONS

You have seen that the graph of $y = \sqrt{x}$ looks like that in figure 2.44.

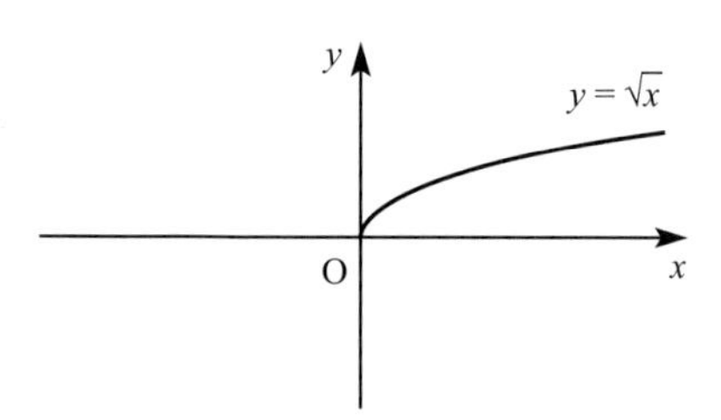

FIGURE 2.44

If you draw the graph of $y = \sqrt{-x}$ on a graphics calculator or by hand you will obtain that in figure 2.45.

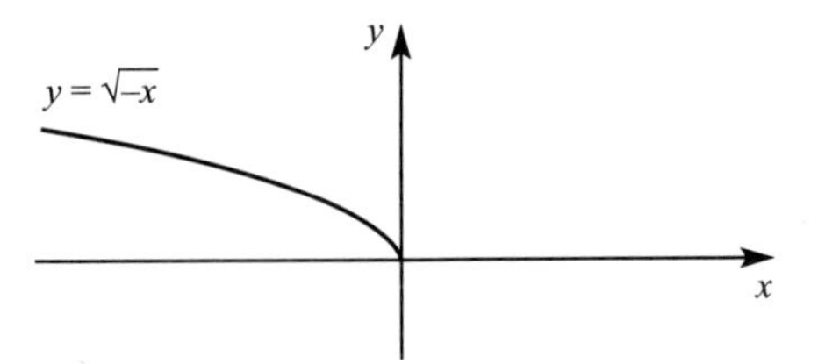

FIGURE 2.45

You can see that graphs of $y = \sqrt{x}$ and $y = \sqrt{-x}$ are reflections of each other in the y axis.

If you draw the graph of $y = -\sqrt{x}$ it looks like that in figure 2.46.

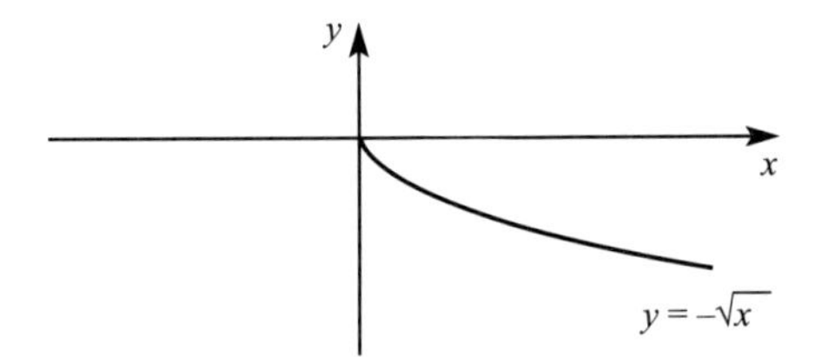

FIGURE 2.46

This time you will notice that $y = \sqrt{x}$ and $y = -\sqrt{x}$ are reflections of each other in the x axis.

In general:

- If $y = \text{f}(x)$ is mapped onto $y = \text{f}(-x)$ the transformation is a reflection in the y axis.
- If $y = \text{f}(x)$ is mapped onto $y = -\text{f}(x)$ the transformation is a reflection in the x axis.

EXAMPLE 2.27

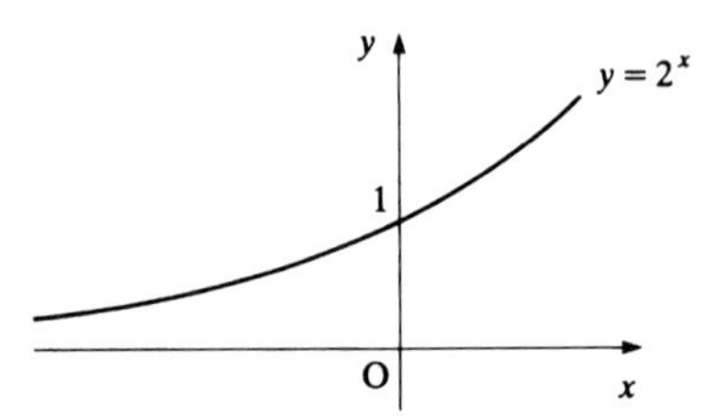

FIGURE 2.47

Figure 2.47 shows the graph of $y = 2^x$. The curve passes through the point (0, 1).

Sketch, on separate diagrams, the graphs of:

(a) $y = 2^{-x}$ (b) $y = -(2^x)$.

Solution (a) Replacing x by $-x$ reflects the curve in the y axis (see figure 2.48).

(b) Replacing 2^x by $-(2^x)$ reflects the curve in the x axis (see figure 2.49).

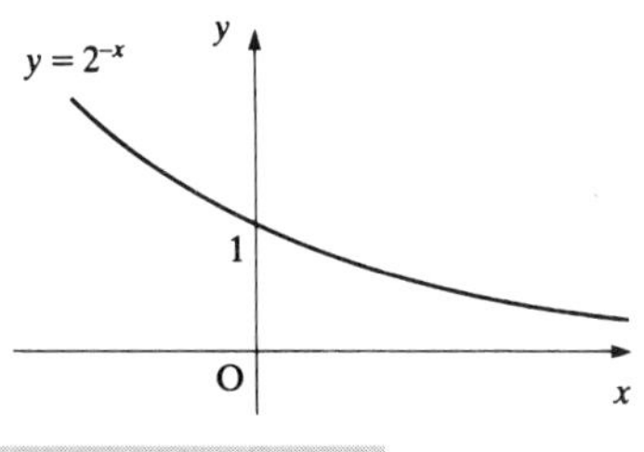

FIGURE 2.48

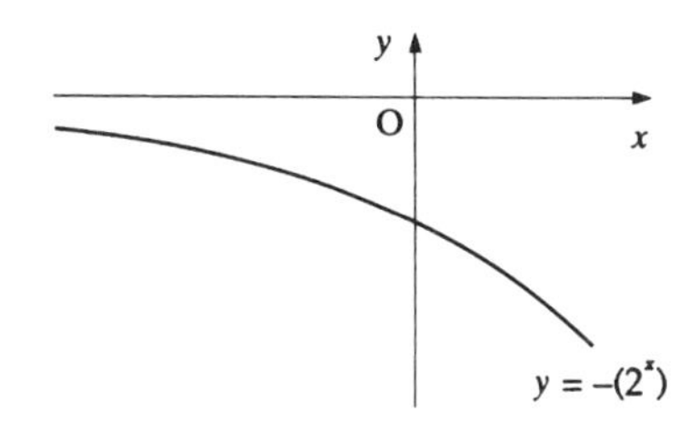

FIGURE 2.49

THE GENERAL QUADRATIC CURVE

You are now able to relate any quadratic curve to that of $y = x^2$.

EXAMPLE 2.28

(a) Write the equation $y = 1 + 4x - x^2$ in the form $y = a[(x + p)^2 + q]$.

(b) Show how the graph of $y = 1 + 4x - x^2$ can be obtained from the graph of $y = x^2$ by a succession of transformations, and list the transformations in the order in which they are applied.

(c) Sketch the graph.

Solution (a) If $1 + 4x - x^2 \equiv a[(x + p)^2 + q]$

then $-x^2 + 4x + 1 \equiv ax^2 + 2apx + a(p^2 + q)$

Comparing coefficients of x^2: $a = -1$.

Comparing coefficients of x: $2ap = 4$, giving $p = -2$.

Comparing constant terms: $a(p^2 + q) = 1$, giving $q = -5$.

The equation is $y = -[(x - 2)^2 - 5]$.

(b) The curve $y = x^2$ becomes the curve $y = (x - 2)^2 - 5$ by applying the translation $\begin{pmatrix} 2 \\ -5 \end{pmatrix}$ as shown in figure 2.50.

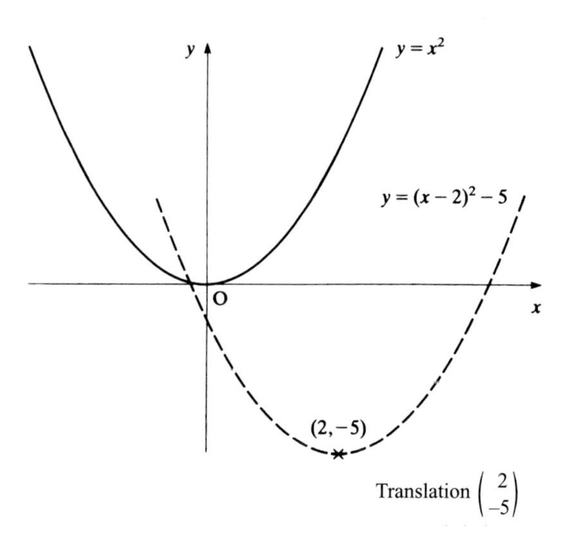

FIGURE 2.50

The curve $y = (x - 2)^2 - 5$ becomes the curve $y = -[(x - 2)^2 - 5]$ by applying a reflection in the x axis (see figure 2.51).

(c)

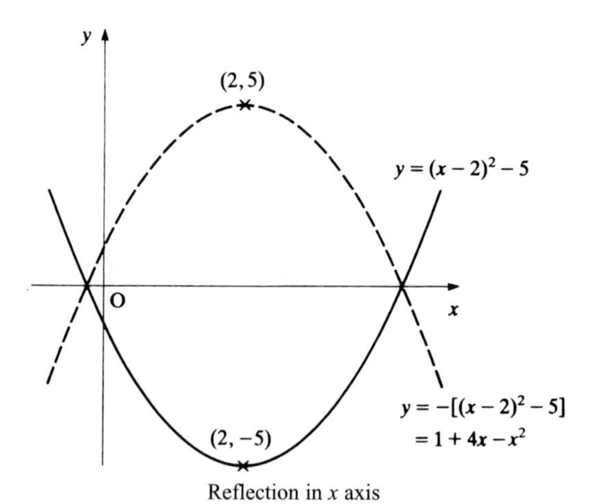

FIGURE 2.51

EXERCISE 2I

1 Starting with the graph of $y = x^2$, state transformations which can be used to sketch the following curves. Specify the transformations in the order in which they are used, and where there is more than one stage in the sketching of the curve, state each stage. State the equation of the line of symmetry.

(a) $y = -2x^2$ **(b)** $y = 4 - x^2$ **(c)** $y = 2x - 1 - x^2$.

2 For each of the following curves:

(a) sketch the curve

(b) identify the curve as being the same as one of the following:

$y = \pm\sin x$, $y = \pm\cos x$, or $y = \pm\tan x$.

(i) $y = \cos(-x)$ **(ii)** $y = \tan(-x)$

(iii) $y = \sin(180° - x)$ **(iv)** $y = \tan(180° - x)$

(v) $y = \sin(-x)$

3 (a) Write the expression $x^2 - 6x + 14$ in the form $(x - a)^2 + b$ where a and b are numbers which you are to find.

(b) Sketch the curves $y = x^2$ and $y = x^2 - 6x + 14$ and state the transformation which maps $y = x^2$ onto $y = x^2 - 6x + 14$.

(c) The curve $y = x^2 - 6x + 14$ is reflected in the x axis. Write down the equation of the image.

4 (a) Sketch the curve with equation $y = x^2$.

(b) Given that $f(x) = (x - 2)^2 + 1$ sketch the curves with the following equations on separate diagrams. Label each curve and give the coordinates of its vertex and the equation of its axis of symmetry.

(i) $y = f(x)$ **(ii)** $y = -f(x)$ **(iii)** $y = f(x + 1) + 2$.

[MEI]

5 Write the expression $2x^2 + 4x + 5$ in the form $a(x + b)^2 + c$ where a, b and c are numbers to be found.

Use your answer to *write down* the coordinates of the minimum point on the graph of $y = 2x^2 + 4x + 5$.

[OCR]

6 The diagram shows the graph of $y = f(x)$. The curve passes through the origin and has a maximum point at (1, 1).

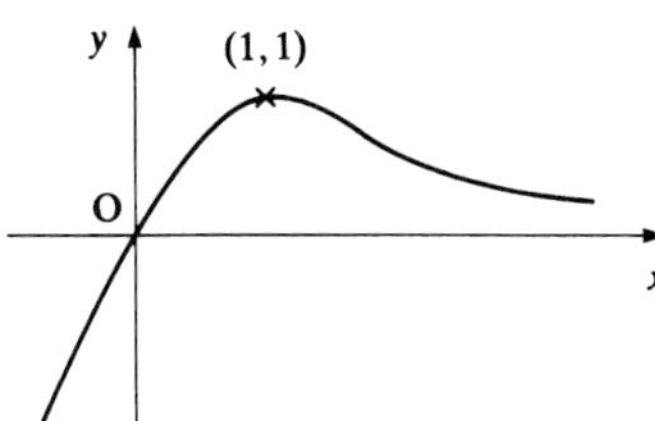

Sketch, on separate diagrams, the graphs of:

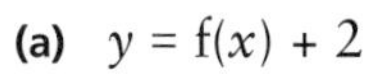

(a) $y = f(x) + 2$

(b) $y = f(x + 2)$

(c) $y = f(2x)$,

giving the coordinates of the maximum point in each case.

[OCR]

7 The circle with equation $x^2 + y^2 = 1$ is shown here. It is stretched with scale factor 3 parallel to the x axis and with scale factor 2 parallel to the y axis. Sketch both curves on the same graph, and write down the equation of the new curve. (It is an ellipse.)

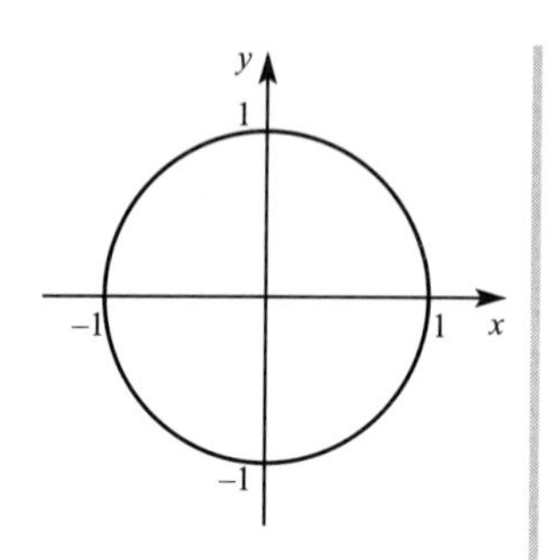

8 In each of the diagrams below, the curve drawn with a dashed line is obtained as a mapping of the curve $y = f(x)$ using a single transformation. It could be a translation, a one-way stretch or a reflection. In each case, write down the equation of the image (dashed) in terms of $f(x)$.

(a)

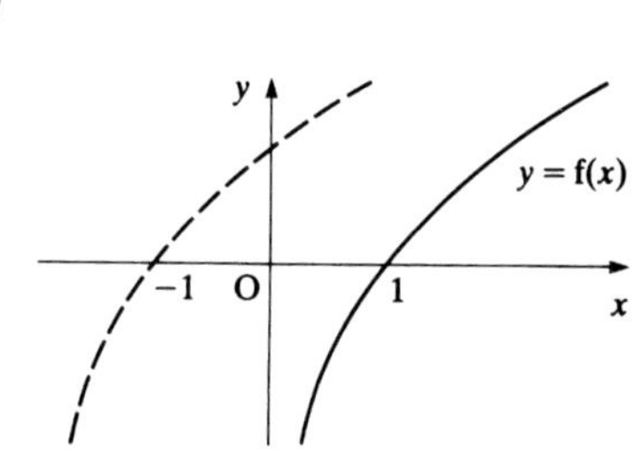

(b)

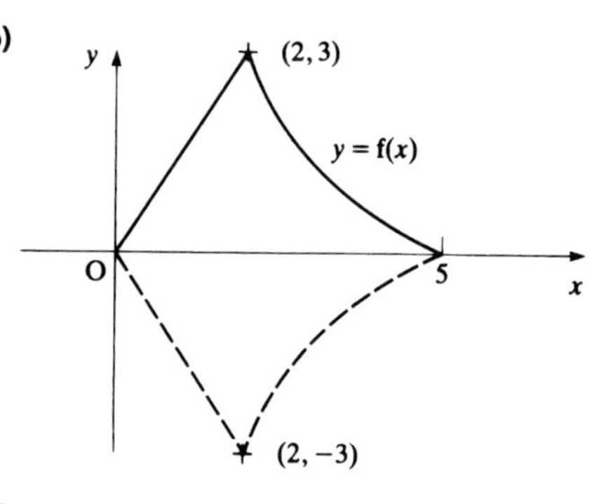

(c)

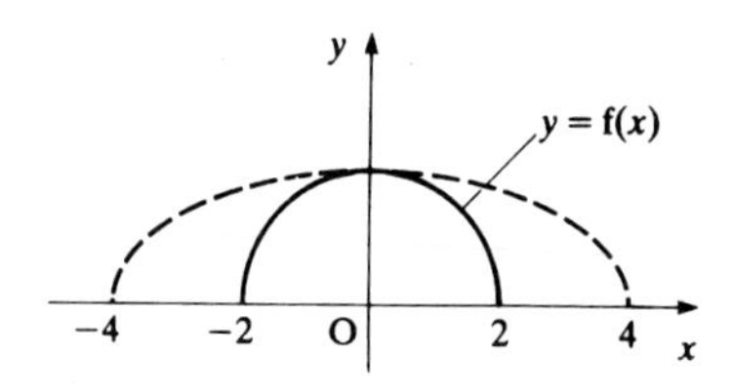

(d)

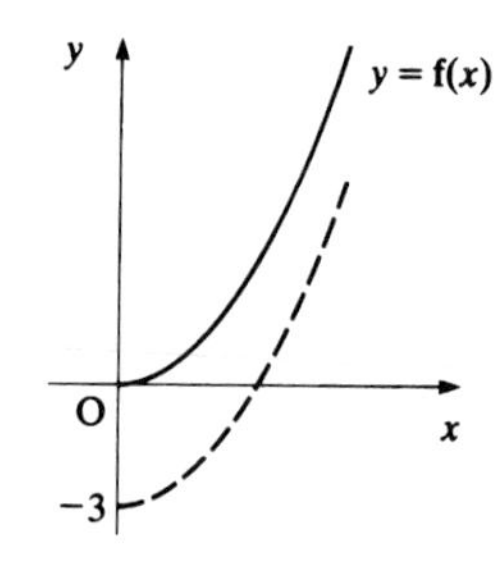

(e)

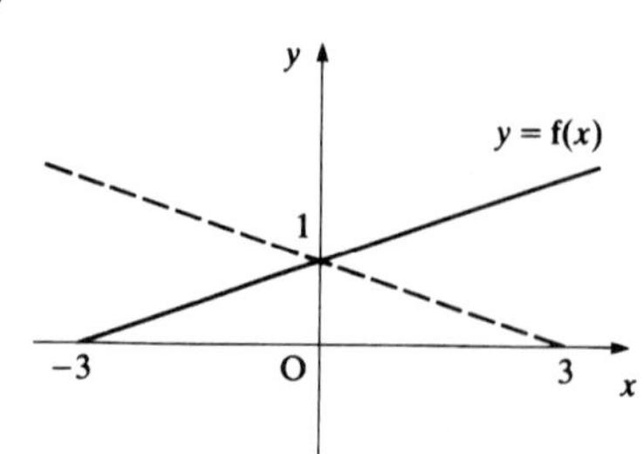

(f)

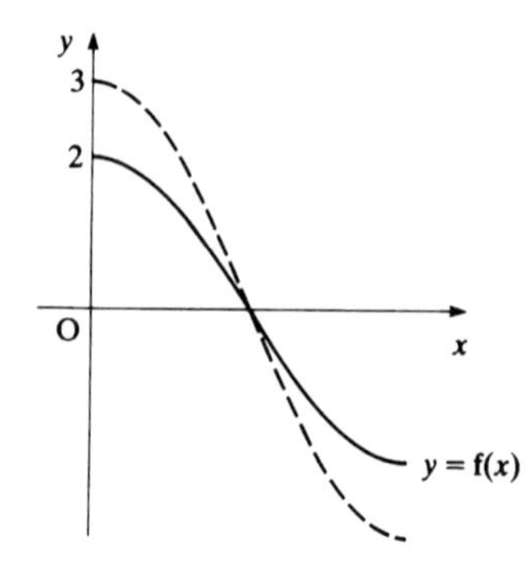

9 The sketch below shows the curve with equation $y = 2 - 6x - 3x^2$ and its axis of symmetry $x = -1$.

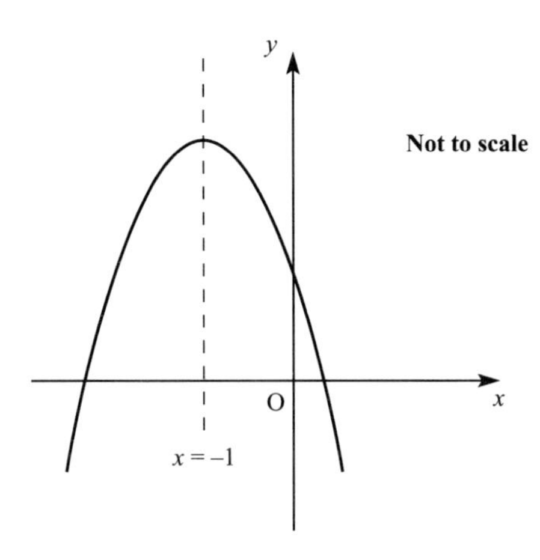

(a) Give the coordinates of the vertex and the value of y when $x = 0$.

(b) Find the values of the constants a, b such that $2 - 6x - 3x^2 = a(x + 1)^2 + b$.

(c) Copy the given sketch and draw in the reflection of the curve with equation $y = 2 - 6x - 3x^2$ in the line $y = 2$.

(d) Write down the equation of the new curve and give the coordinates of its vertex.

[MEI]

EXERCISE 2J Examination-style questions

1 The functions f and g are defined by

$f: x \mapsto x - 3, \quad x \in \mathbb{R}$

$g: x \mapsto 2x^2, \quad x \in \mathbb{R}$

(a) Find the range of g.

(b) Solve $gf(x) = 50$.

(c) Sketch the graph of $y = gf(x)$.

(d) Sketch the graph of $y = |f(x)|$ and hence solve $|f(x)| < 4$.

2 The functions f and g are defined by

$f: x \mapsto 2x + 1, \quad x \in \mathbb{R}$

$g: x \mapsto \sqrt{x}, \quad x \geqslant 0.$

(a) State the range of g.

(b) Find the inverses f^{-1} and g^{-1}.

(c) On the same diagram sketch $y = f^{-1}(x)$ and $y = g^{-1}(x)$.

(d) State with reasons the number of solutions to $f^{-1}(x) = g^{-1}(x)$.

(e) Evaluate $gf(5)$ to 3 decimal places.

3 **(a)** On the same axes draw the graphs of $y = |3x + 1|$ and $y = |x - 1|$.

(b) Use your diagram to solve $|3x + 1| = |x - 1|$.

4 **(a)** Express $x^2 - 4x + 5$ in the form $(x - a)^2 + b$, where a and b are integers to be defined.

(b) Sketch the graph of $y = x^2 - 4x + 5$.

(c) Define a domain for which $f(x) = x^2 - 4x + 5$ is a one-to-one function.

(d) Find the inverse function f^{-1}.

(e) For your domain, sketch f and f^{-1} on the same diagram showing the geometrical connection between them.

(f) From your graph find the value of x for which $f(x) = f^{-1}(x)$.

5 The functions f and g are defined by

$$f: x \mapsto 2x - 1, \quad x \in \mathbb{R}$$

$$g: x \mapsto \frac{1}{x + 1}, \quad x \neq -1.$$

(a) Find the inverse function g^{-1}.

(b) Find and simplify $fg(x)$.

(c) Sketch, on the same diagram, the graphs of $y = f(x)$ and $y = g(x)$.

(d) From your sketch find the approximate solutions to $f(x) = g(x)$.

(e) Solve $f(x) = g(x)$ algebraically, giving your answer to 3 significant figures.

6 The graph shows one wavelength of the curve of $y = A + B\sin(2x)$.

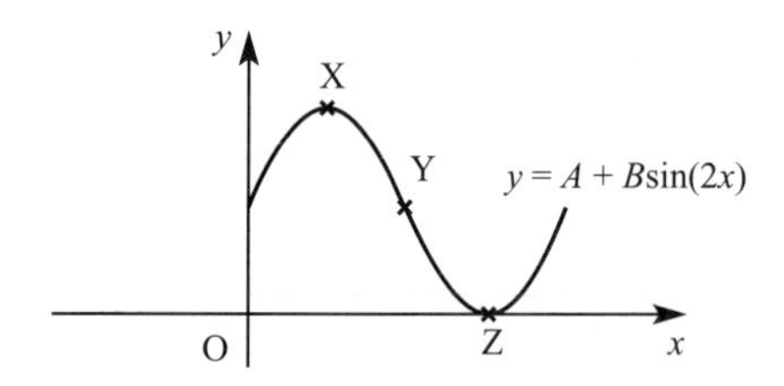

(a) Given that the point X has coordinates $(\frac{\pi}{4}, 4)$ state the coordinates of the central point Y and the minimum point Z and find the values of A and B.

(b) If $f(x) = A + B\sin(2x)$ sketch, on separate diagrams, the graphs of

(i) $y = f(x + \frac{\pi}{2})$,

(ii) $y = |f(x) - 2|$.

7 The diagram shows the graphs of $y = \mathrm{f}(x)$.

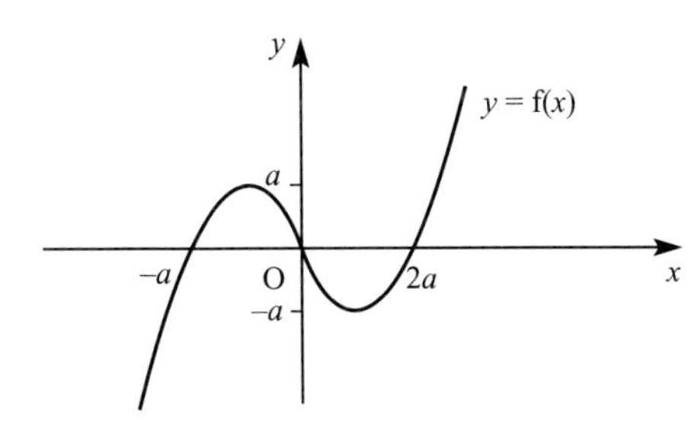

On separate diagrams sketch,

(a) $y = \mathrm{f}(-x)$,

(b) $y = -\mathrm{f}(x)$,

(c) $y = \mathrm{f}(x - a)$,

(d) $y = \mathrm{f}(x) - a$,

(e) $y = |\mathrm{f}(x)|$.

For what values of x is $\mathrm{f}(x) > 0$?

8 The functions f and g are defined by

$$\mathrm{f}: x \mapsto 4x - 1, \quad x \in \mathbb{R}$$

$$\mathrm{g}: x \mapsto \frac{3}{2x - 1}, \quad x \neq \tfrac{1}{2}$$

Find, in its simplest form,

(a) the inverse function f^{-1},

(b) the composite function gf, stating its domain.

(c) the values of x for which

$$2\mathrm{f}(x) = \mathrm{g}(x),$$

giving your answers to 3 decimal places.

9 The function f is an odd function defined on the interval [–2, 2]. Given that

$$\mathrm{f}(x) = -x, \quad 0 \leqslant x \leqslant 1,$$
$$\mathrm{f}(x) = x - 2, \quad 1 \leqslant x \leqslant 2,$$

(a) sketch the graph of f for $-2 \leqslant x \leqslant 2$,

(b) find the values of x for which $\mathrm{f}(x) = \frac{1}{2}$.

[Edexcel]

10 **(a)** Using the same scales and axes, sketch the graph of $y = |2x|$ and $y = |x - a|$, where $a > 0$.

(b) Write down the coordinates of the points where the graph of $y = |x - a|$ meets the axes.

(c) Show that the point with coordinates $(-a, 2a)$ lies on both graphs.

(d) Find the coordinates, in terms of a, of a second point which lies on both graphs.

[Edexcel]

KEY POINTS

1 The **domain** is the set of x values.

2 The **range** is the set of y values.

3 A **one-to-one** mapping means each value of x maps onto a unique value of y, e.g. $y = x^3$.

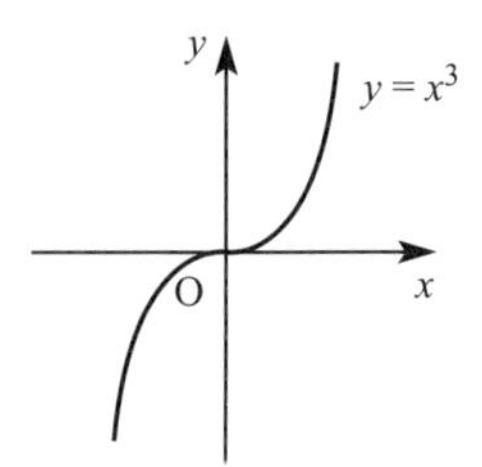

4 **Composite functions**
fg means fg(x) or f(g(x)),
i.e. replace x in f(x) by the function g(x).

5 **Inverse functions**
- For a function to have an inverse it must be one-to-one.
- Geometrically the inverse function f^{-1} is the reflection of the function f in the line $y = x$.
- The equation f^{-1} is found by swapping x and y around in the equation $y = f(x)$ and making y the subject.

6 **Even functions** are symmetrical about the y axis, i.e. $f(-x) = f(x)$.

7 **Odd functions** are symmetrical about the x axis, i.e. $f(-x) = -f(x)$.

8 **Modulus function**
$y = |f(x)|$ is found by reflecting any parts below the x axis to above the x axis.
$y = f(|x|)$ is found by reflecting any parts to the right of the y axis to the left of the y axis.

9 **Transformations of f(x)**
- $f(x - a) + b$ is a translation with vector $\begin{pmatrix} a \\ b \end{pmatrix}$.
- $f(ax)$ is a stretch, scale factor $\frac{1}{a}$ parallel to the x axis.
- $af(x)$ is a stretch, scale factor a parallel to the y axis.
- $f(-x)$ is reflection in the y axis.
- $-f(x)$ is a reflection in the x axis.

Chapter three

SEQUENCES AND SERIES

Again! Again! Again!
And the havoc did not slack.

Thomas Campbell

POSITION TO TERM AND TERM TO TERM RULES

In *Pure Mathematics 1* you met sequences and series that were generated in a variety of ways. For example, you saw position to term rules such as

$$u_n = 5n - 3.$$

In this equation you put $n = 1$ to find the first term, u_1, you put $n = 2$ to find the second term, u_2, and so on. The first five terms are

$$u_1 = 5 \times 1 - 3 = 2$$
$$u_2 = 5 \times 2 - 3 = 7$$
$$u_3 = 5 \times 3 - 3 = 12$$
$$u_4 = 5 \times 4 - 3 = 17$$
$$u_5 = 5 \times 5 - 3 = 22.$$

It is very easy to calculate any term, e.g. the hundredth term is

$$u_{100} = 5 \times 100 - 3 = 497.$$

The general term is the nth term, u_n, and the term following the nth is the $(n + 1)$th term, $u_{n+1.}$

You then studied the arithmetic progressions in which the nth term is

$$u_n = a + (n - 1)d$$

where a is the first term, d is the common difference and n is the number of the term. The nth term of a geometric progression, you will remember, is

$$u_n = a \times r^{n-1}.$$

In the same chapter you came across term to term rules. An example of such is

$$u_{n+1} = 2u_n + 1, \qquad u_1 = 1.$$

This means that if you know the nth term, u_n, you can find the next term u_{n+1} by multiplying the nth term by 2 and adding 1. So the first five terms are

$$u_1 = 1$$
$$u_2 = 2u_1 + 1 = 2 \times 1 + 1 = 3$$
$$u_3 = 2u_2 + 1 = 2 \times 3 + 1 = 7$$
$$u_4 = 2u_3 + 1 = 2 \times 7 + 1 = 15$$
$$u_5 = 2u_4 + 1 = 2 \times 15 + 1 = 31.$$

u_n is often used for the nth term of sequences but other letters are used, e.g. t_n or x_n are quite common.

EXERCISE 3A

For each of the following write out the first five terms.

1 $u_n = \frac{1}{2}n(n + 1)$.

2 $t_{n+1} = 2t_n - 1$, $t_1 = 2$.

3 $x_n = 3n + 2$.

4 $u_{n+1} = 10 - u_n$, $u_1 = 4$.

5 $x_{n+1} = \frac{1}{2}x_n$, $x_1 = 16$.

6 $t_n = 1 + \dfrac{1}{n}$.

7 $u_{n+1} = \dfrac{1}{1 + u_n}$, $u_1 = 1$.

8 $x_n = \sqrt{1 + n}$.

9 $u_n = \dfrac{5 + n}{2}$.

10 $t_{n+1} = \dfrac{5 + t_n}{2}$, $t_2 = 3$.

RECURRENCE RELATIONS $x_{n+1} = f(x_n)$

Recurrence relations are a form of term to term sequences. You start with a given term, usually the first term u_1, and use this to calculate the next term. The second term is then used to find the third term, and so on. The process in which consecutive terms are calculated in this way is often called *iteration* and you will learn how to use iterations to solve equations in Chapter 8.

EXAMPLE 3.1

Find the first six terms of the recurrence relation

$$x_{n+1} = \frac{x_n^2 + 10}{10}, \; x_1 = 1.$$

Solution

$x_1 = 1$

$$x_2 = \frac{x_1^2 + 10}{10} = 1.1$$

$$x_3 = \frac{x_2^2 + 10}{10} = 1.21$$

$$x_4 = \frac{x_3^2 + 10}{10} = 1.125\,664\,1$$

$$x_5 = \frac{x_4^2 + 10}{10} = 1.126\,711\,967$$

$$x_6 = \frac{x_5^2 + 10}{10} = 1.126\,947\,986.$$

Hint: on your calculator start by keying in 1 = (or EXE) then (Ans^2 + 10) ÷ 10 followed by repeated pressing of the = (or EXE) key.

Note In Example 3.1 the sequence of the terms appears to converge to some limit.

EXAMPLE 3.2

A recurrence relation is defined by $u_{n+1} = 2u_n - 1$.

(a) What happens if $u_1 = 1$?

(b) What happens if $u_1 = -1$?

(c) A second recurrence relation is defined by $x_n = u_n + 1$. Show that $x_{n+1} = 2(x_n - 1)$.

Solution

(a) $u_1 = 1$ gives the sequence 1, 1, 1, 1, 1, …
So 1 is a fixed point of the iteration.

(b) $u_1 = -1$ gives the sequence −1, −3, −7, −15, −31, …
In this case the sequence diverges.

(c) If $x_n = u_n + 1$ then

$$\begin{aligned} x_{n+1} &= u_{n+1} + 1 \\ &= 2u_n - 1 + 1 \\ &= 2u_n. \end{aligned}$$

So $x_{n+1} = 2(x_n - 1)$.

Exercise 3B

1 A recurrence relation is given by $x_{n+1} = \frac{1}{4}x_n$ where $x_1 = 2$. Write the values of x_2, x_3, x_4 and x_5 leaving your answers as fractions. State the value of x_n as $n \to \infty$.

2 Given that $x_{n+1} = \sqrt{x_n + 4}$ and $x_1 = 2.6$ write down the values of x_2, x_3, x_4 and x_5 giving each to 4 decimal places.

3 A recurrence relation has equation

$$x_{n+1} = \frac{2}{x_n + 5}.$$

Given that $x_1 = 1$ find x_5 to 5 significant figures showing all intermediate iterations.

4 A sequence of terms $\{u_n\}$ is defined for $n \geqslant 1$, by the recurrence relation

$$u_{n+1} = u_n{}^2.$$

(a) Determine what happens if $u_1 = 0$, $u_1 = 1$ and $u_1 = 2$.
(b) Express u_2 and u_3 in terms of u_1.
(c) Express u_n in terms of u_1.

5 A sequence of terms $u_1, u_2, \ldots u_n, \ldots$ is given by the formula

$$u_{n+1} = (u_n - 1)^2$$

If $u_1 = 3$ find u_5.

6 **(a)** Write down the first five terms of $t_n = 2(n + 1)$.
(b) Write down the first five terms of $u_{n+1} = u_n + 4$, $u_1 = 8$.
(c) Using your answer to parts (a) and (b) express u_n in terms of n. Hence state the value of u_{100}.

7 A recurrence relation is $u_{n+1} = \frac{1}{2}u_n$.
(a) Write down u_1, u_2, u_3, u_4 and u_5 given that $u_0 = 1$.
(b) If an infinite number of terms of this sequence is taken what is their sum?

8 A sequence of terms $x_1, x_2, \ldots x_n, \ldots$ is given by the formula

$$x_{n+1} = x_n{}^2 - 6.$$

(a) Find the iterations x_1 to x_4 in the case when $x_0 = 1$, $x_0 = 2$ and $x_0 = 3$.
(b) When $x_{n+1} = x_n = x$ obtain a quadratic in x and solve it.
(c) Comment on your answers to parts (a) and (b).

9 A recurrence relation has equation $u_{n+1} = \dfrac{5}{u_n + 3}$.

(a) With $u_1 = 2$ find u_{10} to 5 decimal places.
(b) By putting $u_n = u_{n+1} = x$ show that $x^2 + 3x - 5 = 0$.
(c) Solve the quadratic equation in part (b) and compare your solutions with your answer to part (a).

10 A sequence of terms $\{u_n\}$ is defined for $n \geqslant 1$, by the recurrence relation

$$u_{n+1} = \frac{u_n}{3}.$$

(a) Find u_2, u_3, u_4 and u_5 in terms of u_1.

(b) Express u_n in terms of u_1.

(c) Given that $u_1 = 1$ find u_∞ and find $\sum_{n=1}^{\infty} u_n$.

BINOMIAL EXPANSIONS

A special type of polynomial is produced when a binomial (i.e. two part) expression like $(1 + x)$ is raised to a power. The resulting polynomial is often called a *binomial expansion.*

The simplest binomial expansion is $(1 + x)$ itself. This, and other expansions, are given below:

$$(1 + x)^1 = 1 + x$$
$$(1 + x)^2 = (1 + x)(1 + x) = 1 + 2x + x^2$$
$$(1 + x)^3 = (1 + x)(1 + 2x + x^2) = 1 + 3x + 3x^2 + x^3$$
$$(1 + x)^4 = (1 + x)(1 + 3x + 3x^2 + x^3) = 1 + 4x + 6x^2 + 4x^3 + x^4$$
$$(1 + x)^5 = (1 + x)(1 + 4x + 6x^2 + 4x^3 + x^4) = 1 + 5x + 10x^2 + 10x^3 + 5x^4 + x^5.$$

You will notice that the numbers, which are called coefficients, form the following pattern:

```
                    1
                 1     1
              1     2     1
           1     3     3     1
        1     4     6     4     1
     1     5    10    10     5     1
```

This is called *Pascal's triangle* (or the *Chinese triangle*). Each number is obtained by adding the two above it, for example

$$\begin{matrix} 4 & + & 6 \\ & = 10. & \end{matrix}$$

Following the pattern the next few rows would be:

```
        1     6    15    20    15     6     1
     1     7    21    35    35    21     7     1
  1     8    28    56    70    56    28     8     1
```

Note Pascal's triangle can give you hours of fun. For example, if you look at diagonal lines from the top you first have a series of 1's, then the counting numbers 1, 2, 3, 4, 5 ..., the next diagonal has triangular numbers, the next has pyramid numbers and so on. If you add all the numbers on each row you will have 1, 2, 4, 8, 16 ... that is the powers of 2. There are many other fascinating patterns to be found. Can you spot the powers of 11 or can you find a way to generate the Fibonacci series?

You will use Pascal's triangle to give the coefficients in your expansions. For example:

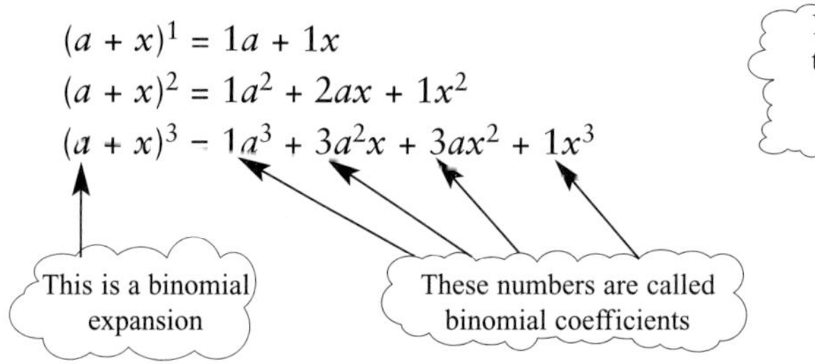

EXAMPLE 3.3

Write out the binomial expansion of $(2 + x)^4$.

Solution The binomial coefficients from Pascal's triangle are 1 4 6 4 1. In each term the sum of the powers of 2 and x must be equal to 4.
So the expansion is:

$$(2 + x)^4 = 1 \times 2^4 + 4 \times 2^3 \times x + 6 \times 2^2 \times x^2 + 4 \times 2 \times x^3 + 1 \times x^4$$
$$\therefore\ (2 + x)^4 = 16 + 32x + 24x^2 + 8x^3 + x^4.$$

EXAMPLE 3.4

Write out the binomial expansion of $(2a - 3b)^5$.

Solution The binomial coefficients for power 5 are 1 5 10 10 5 1.
The expression $(2a - 3b)$ is treated as $(2a + (-3b))$.
So the expansion is

$$1 \times (2a)^5 + 5 \times (2a)^4 \times (-3b) + 10 \times (2a)^3 \times (-3b)^2 + 10 \times (2a)^2 \times (-3b)^3 + 5 \times (2a) \times (-3b)^4 + 1 \times (-3b)^5.$$

Note that $(2a)^5 = 2^5 \times a^5 = 32a^5$ and $(-3b)^3 = (-3)^3 \times b^3 = -27b^3$, etc.

$$\therefore\ (2a - 3b)^5 = 32a^5 - 240a^4b + 720a^3b^2 - 1080a^2b^3 + 810ab^4 - 243b^5.$$

EXAMPLE 3.5

Find the first four terms in the expansion of $(1 - 2x)^8$.

Solution The binomial coefficients are 1 8 28 56 ... Hence

$$\begin{aligned}(1-2x)^8 &= (1+(-2x))^8 \\ &= 1^8 + 8 \times 1^7 \times (-2x)^1 + 28 \times 1^6 \times (-2x)^2 + 56 \times 1^5 \times (-2x)^3 \\ &= 1 - 16x + 112x^2 - 448x^3 \ldots\end{aligned}$$

EXAMPLE 3.6

Find the term independent of x in the expansion of $\left(x + \frac{1}{x}\right)^6$.

Solution Expanding gives:

$$\left(x + \frac{1}{x}\right)^6 = x^6 + 6x^5 \times \frac{1}{x} + 15x^4 \times \frac{1}{x^2} + 20x^3 \times \frac{1}{x^3} + \ldots$$

You do not need to look at any more terms because you see that in the last term written the x cancels out, i.e. the term independent of x is 20.

EXAMPLE 3.7

Expand $(1 + x + x^2)^2$.

Solution

$$\begin{aligned}(1 + x + x^2)^2 &= (1 + (x + x^2))^2 \\ &= 1 + 2\,(x + x^2) + (x + x^2)^2 \\ &= 1 + 2\,(x + x^2) + x^2 + 2x^3 + x^4 \\ &= 1 + 2x + 3x^2 + 2x^3 + x^4.\end{aligned}$$

Historical note

Blaise Pascal has been described as the greatest might-have-been in the history of mathematics. Born in France in 1623, he was making discoveries in geometry by the age of 16 and had developed the first computing machine before he was 20.

Pascal suffered from poor health and religious anxiety, so that for periods of his life he gave up mathematics in favour of religious contemplation. The second of these periods was brought on when he was riding in his carriage: his runaway horses dashed over the parapet of a bridge, and he was only saved by the miraculous breaking of the traces. He took this to be a sign of God's disapproval of his mathematical work. A few years later a toothache subsided when he was thinking about geometry and this, he decided, was God's way of telling him to return to mathematics.

Pascal's triangle (and the binomial theorem) had actually been discovered by Chinese mathematicians several centuries earlier, and can be found in the works of Yang Hui (around 1270 A.D.) and Chu Shi-kie (in 1303 A.D.). Pascal is remembered for his application of the triangle to elementary probability, and for his study of the relationships between binomial coefficients.

Pascal died at the early age of 39.

EXERCISE 3C

1 Expand and simplify

(a) $(x + 3)^4$ (b) $(x - 2)^5$ (c) $(1 + 2x)^4$

(d) $(2 - 3x)^3$ (e) $(2 + x)^6$ (f) $(2x - 3)^4$

2 Find the first four terms in the expansion of

(a) $(1 + x)^7$ (b) $\left(x - \frac{1}{2}\right)^8$ (c) $\left(2 + \frac{x}{2}\right)^6$

3 Find the term independent of x in the following expansions

(a) $\left(x + \frac{1}{x}\right)^4$ (b) $\left(x^2 - \frac{1}{x}\right)^3$ (c) $\left(\frac{1}{2x} + 2x\right)^8$

4 (a) Expand $(1 + x)^5$ and $(1 - x)^5$ as far as the term in x^3.

(b) Expand $(1 - x^2)^5$ as far as the term in x^3.

(c) Multiply your two expansions in part (a) together to give a series expansion as far as the term x^3. Explain why the answer is the same as that from part (b).

5 Expand $(1 + x + x^2)^3$.

THE FORMULA FOR A BINOMIAL COEFFICIENT

There will be times when you need to find binomial coefficients that are outside the range of your Pascal's triangle. What happens if you need to find the power of x^{17} in the expansion of $(x + 2)^{25}$? Clearly you need a formula that gives binomial coefficients.

The first thing you need is a notation for identifying binomial coefficients. It is usual to denote the power of the binomial expression by n, and the position in the row of binomial coefficients by r, where r can take any value from 0 to n. So for row 5 of Pascal's triangle

$n = 5$:	1	5	10	10	5	1
	$r = 0$	$r = 1$	$r = 2$	$r = 3$	$r = 4$	$r = 5$

The general binomial coefficient corresponding to values of n and r is written as ${}^nC_r$ and said as 'N C R'.

An alternative notation is $\binom{n}{r}$. Thus ${}^5C_3 = \binom{5}{3} = 10$.

The next step is to find a formula for the general binomial coefficient ^nC_r or $\binom{n}{r}$. However, to do this you must be familiar with the term *factorial*.

The quantity '8 factorial', written 8!, is

$$8! = 8 \times 7 \times 6 \times 5 \times 4 \times 3 \times 2 \times 1 = 40\,320.$$

Similarly, $12! = 12 \times 11 \times 10 \times 9 \times 8 \times 7 \times 6 \times 5 \times 4 \times 3 \times 2 \times 1 = 479\,001\,600$,

and $\quad n! = n \times (n-1) \times (n-2) \times \ldots \times 1$, where n is a positive integer.

Note 0! is defined to be 1. You will see the need for this when you use the formula for $\binom{n}{r}$.

The table shows an alternative way of laying out Pascal's triangle.

		Column (r)									
		0	1	2	3	4	5	6	7	8	r
	1	1	1								
	2	1	2	1							
Row (n)	3	1	3	3	1						
	4	1	4	6	4	1					
	5	1	5	10	10	5	1				
	6	1	6	15	20	15	6	1			
	7	1	7	21	35	35	21	7	1		
	8	1	8	28	56	70	56	28	8	1	
	...	...	...	...	...	...	...	...	...	...	...
	n	1	n	?	?	?	?	?	?	?	

Note The numbers in each row of Pascal's triangle are symmetrical about the middle number or middle pair of numbers.

From the table it is possible to see that:

when	the coefficient is
$r = 0$	1
$r = 1$	n
$r = 2$	$\dfrac{n(n-1)}{2}$
$r = 3$	$\dfrac{n(n-1)(n-2)}{3 \times 2}$
$r = 4$	$\dfrac{n(n-1)(n-2)(n-3)}{4 \times 3 \times 2}$

and, in general, the rth term is

$$\frac{n(n-1)(n-2)(n-3)\ldots(n-r+1)}{r} = \frac{n!}{r!(n-r)!}$$

i.e. the rth term of the nth row of Pascal's triangle is:

$${}^nC_r = \binom{n}{r} = \frac{n!}{r!(n-r)!}.$$

THE FORMULA FOR THE BINOMIAL EXPANSION OF $(a + b)^n$

This allows us to write the binomial expansion when n is a positive integer as:

$$(a+b)^n = a^n + \binom{n}{1}a^{n-1}b + \binom{n}{2}a^{n-2}b^2 + \ldots + \binom{n}{r}a^{n-r}b^r + \ldots + b^n.$$

Note The values of the coefficients can be found from writing out Pascal's triangle, from using the formula for $\binom{n}{r}$ or by using the ${}^nC_r$ button on your calculator.

It follows that the expansion of $(a + bx)^n$ is

$$(a+bx)^n = a^n + \binom{n}{1}a^{n-1}bx + \binom{n}{2}a^{n-2}b^2x^2 + \ldots + \binom{n}{r}a^{n-r}b^rx^r + \ldots + b^nx^n.$$

EXAMPLE 3.8

Use the formula $\binom{n}{r} = \frac{n!}{r!(n-r)!}$ to calculate:

(a) $\binom{5}{0}$ **(b)** $\binom{5}{1}$ **(c)** $\binom{5}{2}$ **(d)** $\binom{5}{3}$ **(e)** $\binom{5}{4}$ **(f)** $\binom{5}{5}$.

Solution

(a) $\binom{5}{0} = \frac{5!}{0!(5-0)!} = \frac{120}{1 \times 120} = 1.$

(b) $\binom{5}{1} = \frac{5!}{1!4!} = \frac{120}{1 \times 24} = 5.$

(c) $\binom{5}{2} = \frac{5!}{2!3!} = \frac{120}{2 \times 6} = 10.$

(d) $\binom{5}{3} = \frac{5!}{3!2!} = \frac{120}{6 \times 2} = 10.$

(e) $\binom{5}{4} = \frac{5!}{4!1!} = \frac{120}{24 \times 1} = 5.$

(f) $\binom{5}{5} = \frac{5!}{5!0!} = \frac{120}{120 \times 1} = 1.$

Note You can see that these numbers, 1, 5, 10, 10, 5, 1, are row 5 of Pascal's triangle.

EXAMPLE 3.9

Find the coefficient of x^{17} in the expansion of $(x + 2)^{25}$.

Solution

$$(x+2)^{25} = \binom{25}{0}x^{25} + \binom{25}{1}x^{24}2^1 + \binom{25}{2}x^{23}2^2 + \ldots + \binom{25}{8}x^{17}2^8 + \ldots \binom{25}{25}2^{25}$$

So the required term is $\binom{25}{8} \times 2^8 \times x^{17}$

$$\binom{25}{8} = \frac{25!}{8!17!} = \frac{25 \times 24 \times 23 \times 22 \times 21 \times 20 \times 19 \times 18 \times \cancel{17!}}{8! \times \cancel{17!}}$$

$$= 1\,081\,575.$$

So the coefficient of x^{17} is $1\,081\,575 \times 2^8 = 276\,883\,200$.

Note Notice how 17! was cancelled in working out ${}^{25}C_8$. Factorials become large numbers very quickly and you should keep a look-out for such opportunities to simplify calculations.

EXAMPLE 3.10

Expand $(2 - x)^{10}$ as far as the term in x^3.

Solution

$$(2-x)^{10} = 2^{10} + 10 \times 2^9 \times (-x) + \binom{10}{2} \times 2^8 \times (-x)^2 + \binom{10}{3} \times 2^7 \times (-x)^3 + \ldots$$

but $\binom{10}{2} = \dfrac{10 \times 9 \times 8!}{2! \times 8!} = 45$ and $\binom{10}{3} = \dfrac{10 \times 9 \times 8 \times 7!}{3! \times 7!} = 120$

so

$$(2-x)^{10} = 1024 - 5120x + 11\,520x^2 - 15\,360x^3 + \ldots$$

EXAMPLE 3.11

Find the coefficient of x^6 in the expansion of $(1 + 2x)^{15}$.

Solution You require the term when $r = 6$, that is:

$$\binom{15}{6} \times 1^9 \times (2x)^6 = \frac{15!}{6! \times 9!} \times 2^6 \times x^6 = 320\,320x^6$$

$\therefore$ the coefficient of x^6 is 320 320.

EXAMPLE 3.12

Given that the coefficient of x^4 in the expansion of $(3 - 2x)^n$ is 2160 find the value of n.

Solution You require the term when $r = 4$ so

$$\binom{n}{4} \times 3^{n-4} \times (-2)^4 = 2160$$

$$\text{so } \binom{n}{4} \times 3^{n-4} = 135.$$

Try different values of n

$n = 4$ $\binom{4}{4} \times 3^0 = 1$ too small

$n = 5$ $\binom{5}{4} \times 3^1 = 15$ too small

$n = 6$ $\binom{6}{4} \times 3^2 = 135$ correct.

Hence the value of n is 6.

EXAMPLE 3.13

Find the binomial expansion of $\left(2 + \frac{x}{10}\right)^4$. Use your expansions with a suitable value of x to work out 2.1^4.

Solution

$$\left(2 + \frac{x}{10}\right)^4 = 2^4 + 4 \times 2^3 \times \frac{x}{10} + 6 \times 2^2 \times \left(\frac{x}{10}\right)^2 + 4 \times 2^1 \times \left(\frac{x}{10}\right)^3 + \left(\frac{x}{10}\right)^4$$

$$= 16 + 3.2x + 0.24x^2 + 0.008x^3 + 0.0001x^4.$$

To find 2.1^4 put $\frac{x}{10} = 0.1$ so $x = 1$ giving

$$2.1^4 = 16 + 3.2 + 0.24 + 0.008 + 0.0001 = 19.4481.$$

THE EXPANSION OF $(1 + x)^n$

When deriving the result for $\binom{n}{r}$ we found the binomial coefficients in the form

$$1 \quad n \quad \frac{n(n-1)}{2!} \quad \frac{n(n-1)(n-2)}{3!} \quad \frac{n(n-1)(n-2)(n-3)}{4!} \ldots$$

This form is commonly used in the expansion of expressions of the type $(1 + x)^n$.

$$(1 + x)^n = 1 + nx + \frac{n(n-1)x^2}{1 \times 2} + \frac{n(n-1)(n-2)x^3}{1 \times 2 \times 3} + \frac{n(n-1)(n-2)(n-3)x^4}{1 \times 2 \times 3 \times 4} + \ldots$$

$$+ \frac{n(n-1)x^{n-2}}{1 \times 2} + nx^{n-1} + 1x^n$$

EXAMPLE 3.14

Use the binomial expansion to write down the first four terms of $(1 + x)^9$.

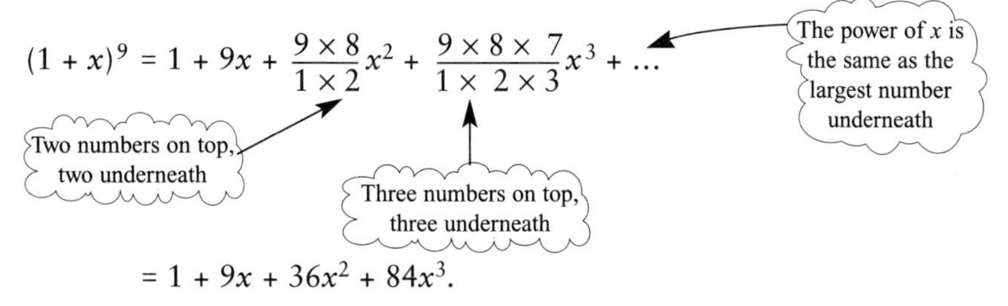

$$(1 + x)^9 = 1 + 9x + \frac{9 \times 8}{1 \times 2}x^2 + \frac{9 \times 8 \times 7}{1 \times 2 \times 3}x^3 + \ldots$$

$$= 1 + 9x + 36x^2 + 84x^3.$$

EXAMPLE 3.15

Use the binomial expansion to write down the first four terms of $(1 - 3x)^7$. Simplify the terms.

Solution Think of $(1 - 3x)^7$ as $(1 + (-3x))^7$. Keep the brackets while you write out the terms.

$$(1 + (-3x))^7 = 1 + 7(-3x) + \frac{7 \times 6}{1 \times 2}(-3x)^2 + \frac{7 \times 6 \times 5}{1 \times 2 \times 3}(-3x)^3 + \ldots$$

$$= 1 - 21x + 189x^2 - 945x^3 + \ldots$$

Note how the signs alternate

EXAMPLE 3.16

(a) Write down the binomial expansion of $(1 - x)^4$.

(b) Using $x = 0.03$ and the first three terms of the expansion find an approximate value for $(0.97)^4$.

(c) Use your calculator to find the percentage error in your answer.

Solution

(a)
$$\begin{aligned}(1 - x)^4 &= (1 + (-x))^4 \\ &= 1 + 4(-x) + 6(-x)^2 + 4(-x)^3 + (-x)^4 \\ &= 1 - 4x + 6x^2 - 4x^3 + x^4\end{aligned}$$

(b)
$$\begin{aligned}(0.97)^4 &\approx 1 - 4(0.03) + 6(0.03)^2 \\ &= 0.8854\end{aligned}$$

(c)
$$\begin{aligned}(0.97)^4 &= 0.885\,292\,81 \\ \text{Error} &= 0.8854 - 0.885\,292\,81 \\ &= 0.000\,107\,19.\end{aligned}$$

$$\begin{aligned}\text{Percentage error} &= \frac{\text{error}}{\text{true value}} \times 100 \\ &= \frac{0.000\,107\,19}{0.885\,292\,81} \times 100 \\ &= 0.0121\%.\end{aligned}$$

EXERCISE 3D

1 Write out the following binomial expansions:

(a) $(x + 1)^4$ **(b)** $(1 + x)^7$ **(c)** $(x + 2)^5$

(d) $(2x + 1)^6$ **(e)** $(2x - 3)^5$ **(f)** $(2x + 3y)^3$

2 Calculate the following binomial coefficients

(a) $\binom{4}{2}$ **(b)** $\binom{6}{2}$ **(c)** $\binom{6}{3}$ **(d)** $\binom{6}{4}$ **(e)** $\binom{6}{0}$

(f) $\binom{12}{9}$ **(g)** $\binom{12}{3}$ **(h)** $\binom{15}{11}$ **(i)** $\binom{8}{8}$

3 Find:

(a) the coefficient of x^5 in the expansion of $(1 + x)^8$;

(b) the coefficient of x^4 in the expansion of $(1 - x)^{10}$;

(c) the coefficient of x^6 in the expansion of $(1 + 3x)^{12}$;

(d) the coefficient of x^7 in the expansion of $(1 - 2x)^{15}$;

(e) the value of the term in the expansion of $\left(x - \frac{1}{x}\right)^8$ which is independent of x.

4 (a) Write down the binomial expansion of $(1 + x)^4$.

(b) Use the first two terms of the expansion to find an approximate value for $(1.002)^4$, substituting $x = 0.002$.

(c) Find, using your calculator, the percentage error in making this approximation.

5 (a) Simplify $(1 + x)^3 - (1 - x)^3$.

(b) Show that $a^3 - b^3 = (a - b)(a^2 + ab + b^2)$.

(c) Substitute $a = 1 + x$ and $b = 1 - x$ in the result in part (b) and show that your answer is the same as that for part (a).

6 (a) Write down the binomial expansion of $(2 - x)^5$.

(b) By substituting $x = 0.01$ in the first three terms of your expansion, obtain an approximate value for 1.99^5.

(c) Use your calculator to find the percentage error in your answer.

7 Expand and simplify the following:

(a) $\left(x + \frac{1}{x}\right)^6$

(b) $\left(2x - \frac{1}{2x}\right)^4$

(c) $\left(1 + \frac{2}{x}\right)^5$

8 A sum of money, £P, is invested such that compound interest is earned at a rate of r% per year. The amount, £A, in the account n years later is given by

$$A = P\left(1 + \frac{r}{100}\right)^n$$

(a) Write down the first four terms in this expansion when $P = 1000$, $r = 10$, $n = 10$ and add them to get an approximate value for A.

(b) Compare your result with what you get using your calculator for 1000×1.1^{10}.

(c) Calculate the percentage error in using the sum of these four terms instead of the true value.

9 **(a)** Show that $(2 + x)^4 = 16 + 32x + 24x^2 + 8x^3 + x^4$ for all x.

(b) Find the values of x for which $(2 + x)^4 = 16 + 16x + x^4$.

[MEI]

10 Given that $f(x) = (1 + 2x)^{12}$,

(a) expand $f(x)$ as far as the term in x^3

(b) use your expansion to evaluate 1.02^{12}.

(c) Find the percentage error between your answer to part (b) and the answer from your calculator.

EXERCISE 3E

Examination-style questions

1 Use the iteration formula $x_{n+1} = 5^{\frac{1}{x_n}}$ to find x_5 to 4 decimal places given that $x_1 = 2.1$. Show the intermediate iterations in your solution.

2 A sequence of terms $\{u_n\}$ is defined for $n \geqslant 1$, by the recurrence relation

$$u_{n+1} = u_n^2 - 2.$$

(a) Describe the sequence when $u_1 = 1$.

(b) Describe the sequence when $u_1 = 2$.

(c) If $u_1 = u_2 = x$ find a quadratic equation in x and solve it.

(d) Find an expression for u_3 in terms of u_1.

3 A sequence of terms is defined by $t_n = \frac{1}{2}n(n + 1)$,

(a) find the sum of the first five terms of the sequence

(b) find $t_1 + t_2$, $t_2 + t_3$, $t_3 + t_4$ and $t_4 + t_5$.

(c) Express $t_n + t_{n+1}$ in terms of n and hence prove that the sum of two consecutive terms of the sequence is always a square number.

4 A sequence of terms $u_1, u_2, \ldots, u_n, \ldots$ is given by the formula

$$u_n = 3\left(\frac{2}{3}\right)^n - 1$$

where n is a positive integer.

(a) Find the values of u_1, u_2 and u_3.

(b) Find $\sum_{n=1}^{15} 3\left(\frac{2}{3}\right)^n$ and hence show that $\sum_{n=1}^{15} u_n = -9.014$ to 4 significant figures.

(c) Prove that $3u_{n+1} = 2u_n - 1$.

[Edexcel]

5 A sequence of terms $\{u_n\}$ is defined for $n \geqslant 1$, by the recurrence relation

$$u_{n+2} = 2ku_{n+1} + 15u_n,$$

where k is a constant. Given that $u_1 = 1$ and $u_2 = -2$,

(a) find an expression, in terms of k, for u_3.

(b) Hence find an expression, in terms of k, for u_4.

Given also that $u_4 = -38$,

(c) find the possible values of k.

[Edexcel]

6 Expand and simplify $\left(3x - \frac{1}{3x}\right)^4$

7 Given that $(2 + x)^{12} \equiv A + Bx + Cx^2 + Dx^3 + \ldots$ find the values of the integers A, B, C and D.

8 **(a)** Expand $(1 - 4x)^{11}$ as far as the term in x^3.

(b) Use your expansion to find an approximate value for 0.96^{11}.

9 **(a)** Expand $(3 + 2x)^4$ in ascending powers of x, giving each coefficient as an integer.

(b) Hence, or otherwise, write down the expansion of $(3 - 2x)^4$ in ascending powers of x.

(c) Hence, by choosing a suitable value for x show that $(3 + 2\sqrt{2})^4 + (3 - 2\sqrt{2})^4$ is an integer and state its value.

[Edexcel]

10 The coefficient of x^2 in the expansion of $\left(1 + \frac{x}{2}\right)^n$, where n is a positive integer, is 7.

(a) Find the value of n.

(b) Using the value of n found in (a), find the coefficient of x^4.

[Edexcel]

KEY POINTS

1 Recurrence relations

$x_{n+1} = \mathrm{f}(x_n) \Rightarrow$ start with the value of x_1 and substitute it to find x_2 then use x_2 to find x_3 and so on.

2 Binomial expansions

$$(a + b)^n = a^n + \binom{n}{1}a^{n-1}b + \binom{n}{2}a^{n-2}b^2 + \dots + \binom{n}{r}a^{n-r}b^r + \dots + b^n$$

The coefficients are found using

$$\binom{n}{r} = {}^nC_r = \frac{n!}{r!(n-r)!}$$

or using Pascal's triangle

1 1

1 2 1

1 3 3 1

1 4 6 4 1

1 5 10 10 5 1

1 6 15 20 15 6 1

1 7 21 35 35 21 7 1

1 8 28 56 70 56 28 8 1

Chapter four

TRIGONOMETRY

The young light-hearted Masters of the Waves.

Matthew Arnold

TRIGONOMETRIC FUNCTIONS

In *Pure Mathematics 1* you studied the trigonometric functions sine ($\sin\theta$), cosine ($\cos\theta$) and tangent ($\tan\theta$). You will also remember their graphs which are drawn again for you in figure 4.1.

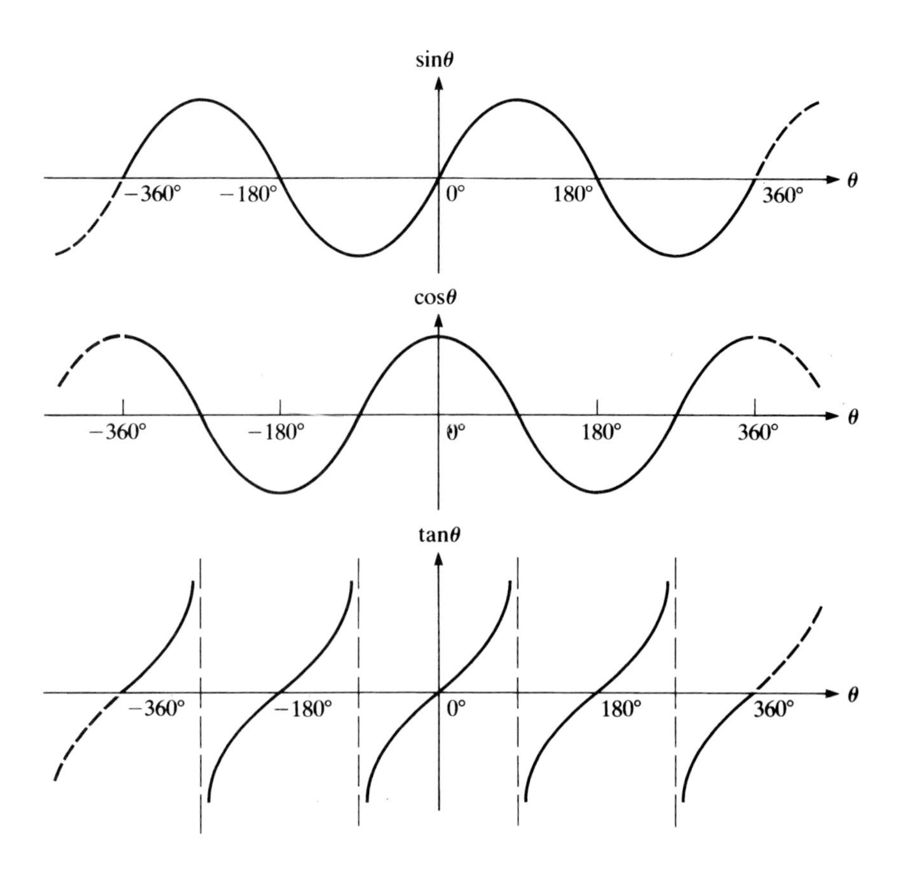

FIGURE 4.1

You will also recall the identites

$$\tan\theta = \frac{\sin\theta}{\cos\theta}$$

and

$$\sin^2\theta + \cos^2\theta = 1.$$

RECIPROCAL TRIGONOMETRIC FUNCTIONS

As well as the three main trigonometric functions, there are three more functions which are commonly used. These are the reciprocals of $\sin\theta$, $\cos\theta$ and $\tan\theta$ and they are called $\text{cosec}\theta$ (short for cosecant), $\sec\theta$ (secant) and $\cot\theta$ (cotangent). These functions are defined by

$$\text{cosec}\theta = \frac{1}{\sin\theta}, \quad \sec\theta = \frac{1}{\cos\theta}, \quad \cot\theta = \frac{1}{\tan\theta}.$$

The graphs of the functions can be found by taking the corresponding sine, cosine and tangent graph and calculating the reciprocals of each point on the graph. The results are shown in figure 4.2.

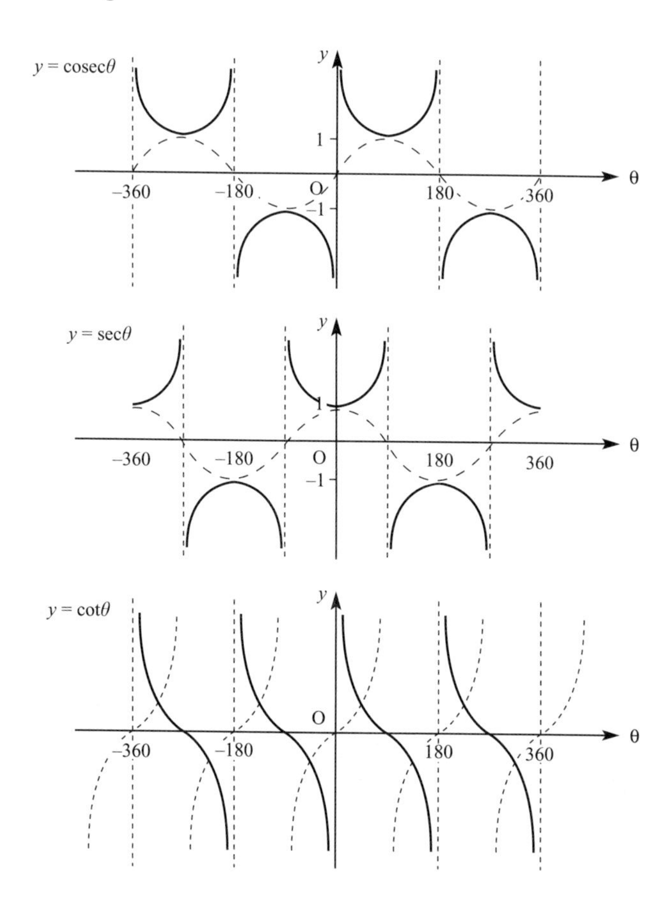

FIGURE 4.2

(To help you remember which reciprocal function relates to which of sine, cosine and tangent take the third letter, so, for example, cosecθ has third letter *s* and it relates to sinθ.)

EXAMPLE 4.1

Solve $\sec\theta = 2$ for $-2\pi \leqslant \theta \leqslant 2\pi$.

Solution You could solve this by using the graph of $y = \sec\theta$ but it is easier to write it in terms of cosθ.

So $\quad \sec\theta = 2.$

Taking the reciprocal of both sides gives

$$\cos\theta = \tfrac{1}{2}$$

which has principal solution

$$\theta = \frac{\pi}{3}.$$

Using the graph of $y = \cos\theta$ (see figure 4.3) gives the set of solutions from $-2\pi \leqslant \theta \leqslant 2\pi$ as

$$\theta = \pm\frac{\pi}{3}, \pm\frac{5\pi}{3}.$$

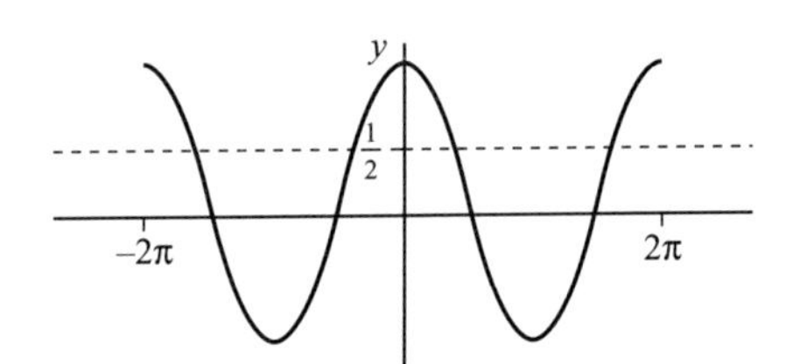

FIGURE 4.3

EXAMPLE 4.2

Solve $3\cot^2\theta - 4\cot\theta + 1 = 0$ for $0° \leqslant \theta \leqslant 360°$.

Solution Factorise:

$$3\cot^2\theta - 4\cot\theta + 1 = 0$$
$$(3\cot\theta - 1)(\cot\theta - 1) = 0$$
$$\therefore \quad 3\cot\theta - 1 = 0 \quad \text{or} \quad \cot\theta - 1 = 0$$
$$\text{so} \quad \cot\theta = \tfrac{1}{3} \quad \text{or} \quad \cot\theta = 1$$
$$\therefore \quad \tan\theta = 3 \quad \text{or} \quad \tan\theta = 1$$

and using the graph of $y = \tan\theta$ gives $\theta = 71.6°, 251.6°, 45°$ or $225°$.

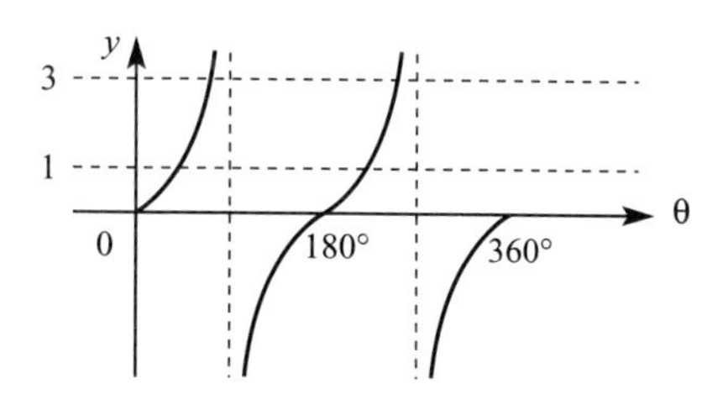

FIGURE 4.4

EXAMPLE 4.3

Solve $\cot 3x = \sqrt{3}$ for $0° \leqslant x \leqslant 360°$.

Solution $\cot 3x = \sqrt{3}$

Take the reciprocal of both sides

$$\tan 3x = \frac{1}{\sqrt{3}}$$

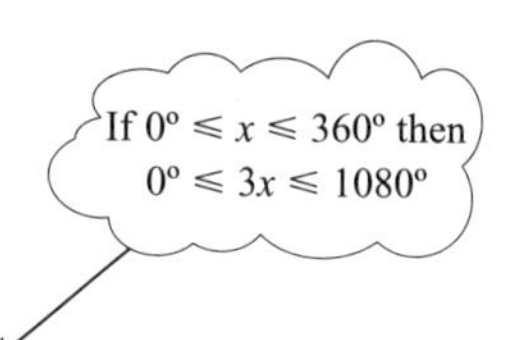

$\therefore\ 3x = 30°, 210°, 390°, 570°, 750°, 930°$

$\therefore\ x = 10°, 70°, 130°, 190°, 250°, 310°.$

COTθ IN MORE DETAIL

Since $\tan\theta = \dfrac{\sin\theta}{\cos\theta}$, if you take the reciprocal of both sides you have

$$\cot\theta = \frac{\cos\theta}{\sin\theta}.$$

EXAMPLE 4.4

Solve $2\cos\theta = \cot\theta$ for $0 \leqslant \theta \leqslant 2\pi$

Solution

$$2\cos\theta = \cot\theta$$

$$2\cos\theta = \frac{\cos\theta}{\sin\theta}$$

$$2\cos\theta\sin\theta - \cos\theta = 0$$

$$\cos\theta(2\sin\theta - 1) = 0$$

$$\cos\theta = 0 \text{ or } \sin\theta = \tfrac{1}{2}$$

$$\theta = \frac{\pi}{2}, \frac{3\pi}{2}, \frac{\pi}{6} \text{ or } \frac{5\pi}{6}.$$

EXERCISE 4A

1 Solve the following.

(a) $\operatorname{cosec} x = 2$ for $0° \leqslant x \leqslant 360°$

(b) $\sec\theta = 5$ for $0° \leqslant \theta \leqslant 360°$

(c) $\cot x = 1$ for $0 \leqslant x \leqslant 2\pi$

(d) $\operatorname{cosec} x = -\dfrac{2}{\sqrt{3}}$ for $-2\pi \leqslant x \leqslant 2\pi$

(e) $\cot\theta = 0.5$ for $0° \leqslant \theta \leqslant 360°$

2 Solve the following.

(a) $\sec 2x = 2$ for $0° \leqslant x \leqslant 360°$

(b) $\operatorname{cosec} 3\theta = -1$ for $-2\pi \leqslant \theta \leqslant 2\pi$

(c) $\cot 2x = -0.5$ for $-180° \leqslant x \leqslant 180°$

(d) $\operatorname{cosec} 4\theta = \dfrac{2}{\sqrt{3}}$ for $\pi \leqslant \theta \leqslant \pi$

(e) $\cot 2x = \dfrac{1}{\sqrt{3}}$ for $-\pi \leqslant \theta \leqslant 3\pi$

3 Solve the following quadratics.

(a) $\sec^2 x - 3\sec x + 2 = 0$ for $0° \leqslant x \leqslant 360°$

(b) $\operatorname{cosec}^2 x - 2\operatorname{cosec} x - 3 = 0$ for $0° \leqslant x \leqslant 360°$

(c) $2\cot^2\theta - 3\cot\theta + 1 = 0$ for $0° \leqslant \theta \leqslant 360°$

(d) $2\sec^2 2\theta + \sec 2\theta - 1 = 0$ for $0 \leqslant \theta \leqslant 2\pi$

(e) $3\cot^2 3\theta - 10\cot 3\theta + 3 = 0$ for $0° \leqslant \theta \leqslant 180°$

4 Sketch the graphs of the following functions over the given values.

(a) $y = \sec x$ for $0° \leqslant x \leqslant 360°$

(b) $y = \operatorname{cosec} x$ for $-360° \leqslant x \leqslant 360°$

(c) $y = \cot x$ for $0° \leqslant x \leqslant 720°$

(d) $y = \sec 2x$ for $0° \leqslant x \leqslant 360°$

(e) $y = \operatorname{cosec} 3x$ for $0° \leqslant x \leqslant 360°$

5 Without using a calculator find the following.

(a) $\sec 120°$

(b) $\operatorname{cosec} 210°$

(c) $\cot 210°$

(d) $\sec 300°$

(e) $\operatorname{cosec}(-150°)$

INVERSE TRIGONOMETRIC FUNCTIONS

When asked to solve $\sin\theta = 0.5$ you have used the inverse sine function to obtain the angle, that is

$$\sin\theta = 0.5 \Rightarrow \quad \theta = \sin^{-1}0.5 = 30°.$$

$\sin^{-1}\theta$ is the inverse sine of θ which is sometimes written as $\arcsin\theta$.

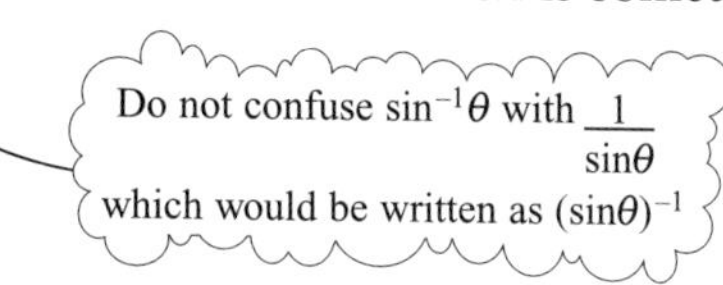

Similarly:

- the inverse cosine of θ is written as $\cos^{-1}\theta$ or $\arccos\theta$
- the inverse tangent of θ is written as $\tan^{-1}\theta$ or $\arctan\theta$.

Using functional notation if

$$\text{f}: x \mapsto \sin x$$

then its inverse function f^{-1} is found by

$$\text{let} \quad y = \sin x$$

and swapping x with y gives

$$x = \sin y$$
$$\therefore\ y = \sin^{-1}x$$

so the inverse function is

$$\text{f}^{-1}: x \mapsto \sin^{-1}x.$$

However for a function to have an inverse it must be a one-to-one mapping.

The functions sine, cosine and tangent are all many-to-one mappings, so their inverse mappings are one-to-one. Thus the problem 'find sin30°' has only one solution, 0.5, whilst 'find θ such that $\sin\theta = 0.5$' has infinitely many solutions. You can see this from the graph of $y = \sin\theta$ (figure 4.5).

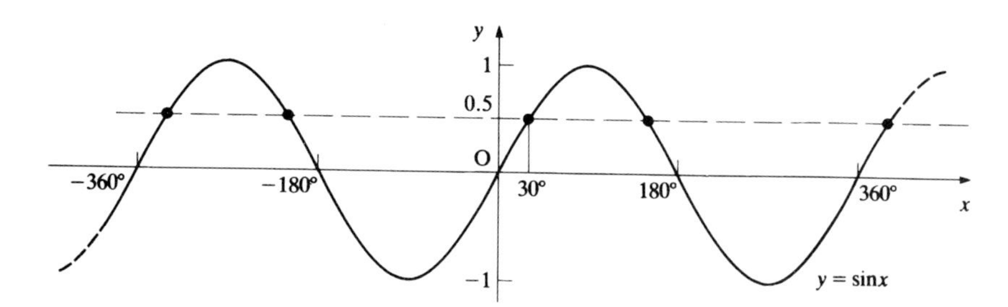

FIGURE 4.5

In order to define inverse functions for sine, cosine and tangent, a restriction has to be placed on the domain of each so that it becomes a one-to-one mapping.

The restriction of the domain determines the principal values for that trigonometrical function. The restricted domains are not all the same. They are listed below.

Function	Domain (degrees)	Domain (radians)
$y = \sin\theta$	$-90° \leqslant \theta \leqslant 90°$	$-\frac{\pi}{2} \leqslant \theta \leqslant \frac{\pi}{2}$
$y = \cos\theta$	$0° \leqslant \theta \leqslant 180°$	$0 \leqslant \theta \leqslant \pi$
$y = \tan\theta$	$-90° < \theta < 90°$	$-\frac{\pi}{2} < \theta < \frac{\pi}{2}$

Not only do these domains define parts of each curve to be one-to-one functions they are also the sets of values within which your calculator always gives an answer.

The graphs of the trigonometric functions over the defined domains are shown, with their inverse functions, in figure 4.6. The inverses are found by reflection in the line $y = x$.

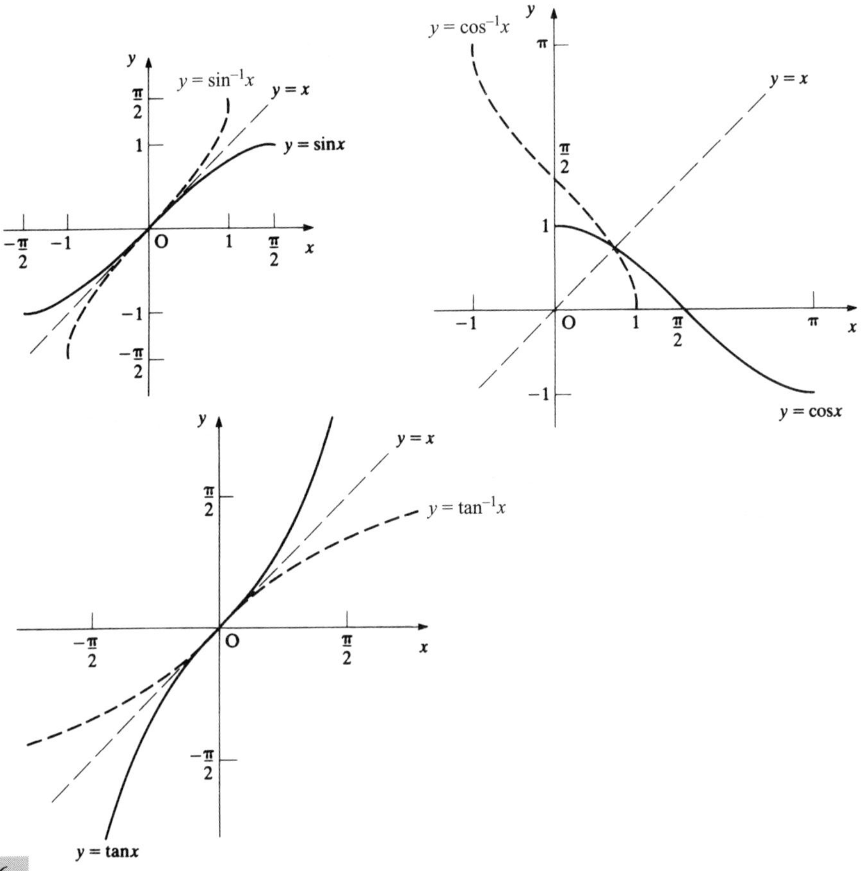

FIGURE 4.6

EXAMPLE 4.5

In a triangle OAB, OA = 5 cm, OB = 12 cm and angle AOB = 90°. Find $\cos^{-1}A$ and $\tan^{-1}B$.

Solution Using Pythagoras' theorem

$$AB^2 = 5^2 + 12^2$$

so $\quad AB = 13.$

Hence,

$$\cos^{-1}A = \frac{5}{13}$$

and

$$\tan^{-1}B = \frac{5}{12}.$$

FIGURE 4.7

STANDARD TRIANGLES

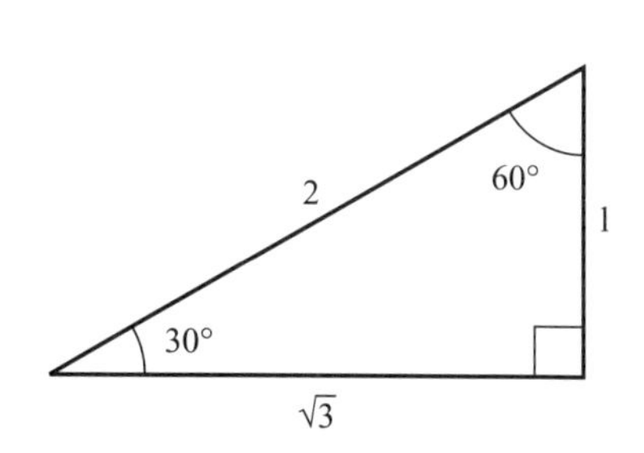

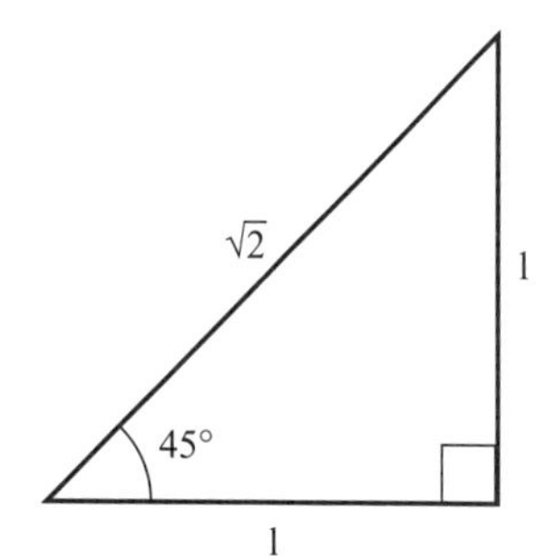

FIGURE 4.8

The two standard triangles are shown in figure 4.8.

Tabulating the information given in these tables gives:

	30°	60°	45°	$\frac{\pi}{6}$	$\frac{\pi}{3}$	$\frac{\pi}{4}$
sin$x$	$\frac{1}{2}$	$\frac{\sqrt{3}}{2}$	$\frac{1}{\sqrt{2}}$	$\frac{1}{2}$	$\frac{\sqrt{3}}{2}$	$\frac{1}{\sqrt{2}}$
cos$x$	$\frac{\sqrt{3}}{2}$	$\frac{1}{2}$	$\frac{1}{\sqrt{2}}$	$\frac{\sqrt{3}}{2}$	$\frac{1}{2}$	$\frac{1}{\sqrt{2}}$
tan$x$	$\frac{1}{\sqrt{3}}$	$\sqrt{3}$	1	$\frac{1}{\sqrt{3}}$	$\sqrt{3}$	1

The inverse functions are:

$\sin^{-1}\frac{1}{2} = 30°$ or $\frac{\pi}{6}$ $\qquad$ $\sin^{-1}\frac{\sqrt{3}}{2} = 60°$ or $\frac{\pi}{3}$ $\qquad$ $\sin^{-1}\frac{1}{\sqrt{2}} = 45°$ or $\frac{\pi}{4}$

$\cos^{-1}\frac{1}{2} = 60°$ or $\frac{\pi}{3}$ $\qquad$ $\cos^{-1}\frac{\sqrt{3}}{2} = 30°$ or $\frac{\pi}{6}$ $\qquad$ $\cos^{-1}\frac{1}{\sqrt{2}} = 45°$ or $\frac{\pi}{4}$

$\tan^{-1}\frac{1}{\sqrt{3}} = 30°$ or $\frac{\pi}{6}$ $\qquad$ $\tan^{-1}\sqrt{3} = 60°$ or $\frac{\pi}{3}$ $\qquad$ $\tan^{-1}1 = 45°$ or $\frac{\pi}{4}$.

FUNDAMENTAL TRIGONOMETRIC IDENTITY

From *Pure Mathematics 1* we had the fundamental trigonometric identity

$$\sin^2\theta + \cos^2\theta = 1.$$

If you take this equation and divide throughout by $\cos^2\theta$ then

$$\frac{\sin^2\theta}{\cos^2\theta} + \frac{\cos^2\theta}{\cos^2\theta} = \frac{1}{\cos^2\theta}$$

giving $\quad \tan^2\theta + 1 = \sec^2\theta.$

Dividing both sides of the original equation by $\sin^2\theta$ gives

$$\frac{\sin^2\theta}{\sin^2\theta} + \frac{\cos^2\theta}{\sin^2\theta} = \frac{1}{\sin^2\theta}.$$

giving $\quad 1 + \cot^2\theta = \mathrm{cosec}^2\theta.$

EXAMPLE 4.6

Find values of θ in the interval $0° \leqslant \theta \leqslant 360°$ for which $\sec^2\theta = 3 + \tan\theta$.

Solution First it is necessary to obtain an equation containing only one trigonometrical function.

$$\sec^2\theta = 3 + \tan\theta$$
$$\Rightarrow \quad \tan^2\theta + 1 = 3 + \tan\theta$$
$$\Rightarrow \quad \tan^2\theta - \tan\theta - 2 = 0$$

This is a quadratic equation like $x^2 - x - 2 = 0$

$$\Rightarrow (\tan\theta - 2)(\tan\theta + 1) = 0$$
$$\Rightarrow \quad \tan\theta = 2 \text{ or } \tan\theta = -1$$

$\tan\theta = 2 \Rightarrow \quad \theta = 63.4°$ (calculated)

or $\theta = 63.4° + 180°$ (graph)

$= 243.4°.$

$\tan\theta = -1 \Rightarrow \quad \theta = -45°$ (calculated).

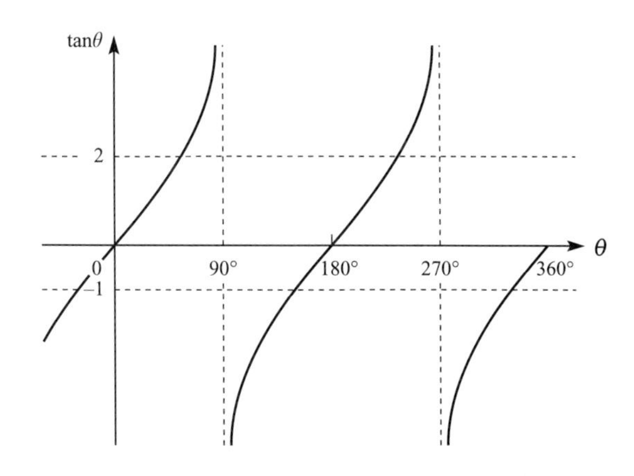

FIGURE 4.9

This is not in the range $0° \leqslant \theta \leqslant 360°$ so figure 4.9 is used to give

$$\theta = -45° + 180° = 135°$$

or $\theta = -45° + 360° = 315°$.

The values of θ are 63.4°, 135°, 243.4°, 315°.

EXAMPLE 4.7

Solve $3\sec^2 x - 5\tan x - 4 = 0$ for $0° \leqslant x \leqslant 360°$.

Solution Using $\sec^2 x = \tan^2 x + 1$ gives

$$3(\tan^2 x + 1) - 5\tan x - 4 = 0$$
$$3\tan^2 x - 5\tan x - 1 = 0.$$

This does not factorise so use the formula $\dfrac{-b \pm \sqrt{b^2 - 4ac}}{2a}$ to give

$$\tan x = -0.1805 \text{ or } \tan x = 1.847$$

so $x = 61.6°, 169.8°, 241.6°, 349.8°$.

EXAMPLE 4.8

Prove that $(1 - \cos\theta)(1 + \sec\theta) \equiv \sin\theta\tan\theta$.

Solution Left-hand side $= (1 - \cos\theta)(1 + \sec\theta)$

$$= 1 + \sec\theta - \cos\theta - \cos\theta\sec\theta$$

But $\cos\theta\sec\theta = \cos\theta \times \dfrac{1}{\cos\theta} = 1$

$$= \frac{1}{\cos\theta} - \cos\theta$$

$$= \frac{1 - \cos^2\theta}{\cos\theta}$$

$$= \frac{\sin^2\theta}{\cos\theta}$$

$$= \sin\theta \times \frac{\sin\theta}{\cos\theta}$$

$$= \sin\theta\tan\theta$$

= right-hand side.

EXERCISE 4B

1 Solve the following.

(a) $2\cos^2 x + \sin x - 1 = 0$ for $0° \leqslant x \leqslant 360°$

(b) $2\text{cosec}^2\theta - 3\cot\theta - 1 = 0$ for $0° \leqslant \theta \leqslant 180°$

(c) $2\tan^2 x - 7\sec x + 8 = 0$ for $0° \leqslant x \leqslant 360°$

(d) $\text{cosec}^2\theta = 3\cot\theta - 1$ for $0° \leqslant \theta \leqslant 360°$

(e) $2\sec^2 x + \tan x - 3 = 0$ for $0° \leqslant x \leqslant 360°$

(f) $\tan^2 x - 5\sec x + 7 = 0$ for $0° \leqslant x \leqslant 180°$

(g) $5\cot\theta = 1 + 2\text{cosec}^2\theta$ for $0° \leqslant \theta \leqslant 180°$

(h) $\text{cosec}x + 1 = \cot^2 x$ for $0 \leqslant x \leqslant 2\pi$

(i) $\tan^2 x + \sec x - 1 = 0$ for $0 \leqslant x \leqslant 2\pi$

(j) $6\text{cosec}^2 x + \cot x = 8$ for $0° \leqslant x \leqslant 360°$

2 Prove the following identities.

(a) $\sin\theta - \text{cosec}\theta \equiv -\cos\theta\cot\theta$

(b) $\cos^4\theta - \sin^4\theta \equiv 2\cos^2\theta - 1$

(c) $\text{cosec}^2\theta + \sec^2\theta \equiv \text{cosec}^2\theta\,\sec^2\theta$

(d) $\dfrac{\sec\theta}{\text{cosec}^2\theta} \equiv \cos\theta\tan^2\theta$

3 In triangle ABC, angle $A = 90°$ and $\sec B = 2$.

(a) Find the angles B and C.

(b) Find $\tan B$.

(c) Show that $1 + \tan^2 B = \sec^2 B$.

4 In triangle LMN, angle $M = 90°$ and $\cot N = 1$.

(a) Find the angles L and N.

(b) Find $\sec L$, $\text{cosec} L$, and $\tan L$.

(c) Show that $1 + \tan^2 L = \sec^2 L$.

5 Solve the quadratic equation

$$8x^2 + 2x - 3 = 0$$

Hence find the values of θ between 0° and 360° which satisfy the equation

$$8\sin^2\theta + 2\sin\theta - 3 = 0.$$

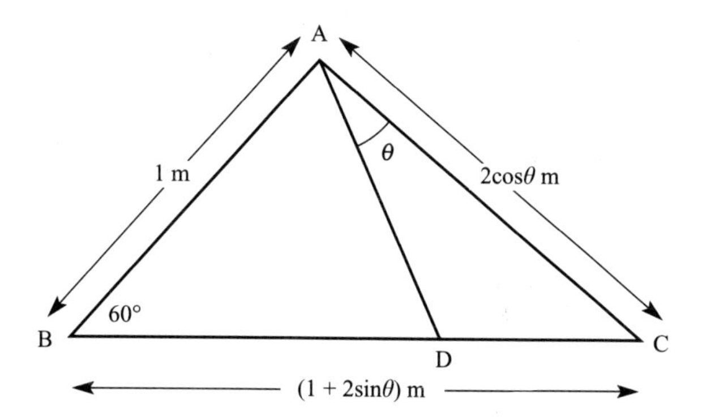

The diagram, which is not to scale, shows a metal framework in which rod AD is at an angle θ to rod AC. The length of AB is 1 m and the angle ABC is 60°. For structural reasons, AC = $2\cos\theta$ m and BC = $(1 + 2\sin\theta)$ m.

By applying the cosine rule to triangle ABC, show that

$$4\cos^2\theta = 4\sin^2\theta + 2\sin\theta + 1.$$

Write this equation as a quadratic equation in $\sin\theta$. Find the angle θ.

[MEI]

COMPOUND-ANGLE FORMULAE

You might think that $\sin(\theta + 60°)$ should equal $\sin\theta + \sin 60°$, but this is not so, as you can see by substituting a numerical value of θ. For example, putting $\theta = 30°$ gives $\sin(\theta + 60°) = 1$, but $\sin\theta + \sin 60° \approx 1.366$.

To find an expression for $\sin(\theta + 60°)$, you would use the *compound-angle formula*:

$$\sin(\theta + \phi) = \sin\theta\cos\phi + \cos\theta\sin\phi.$$

This is proved below in the case when θ and ϕ are acute angles. It is, however, true for all values of the angles. It is an *identity*.

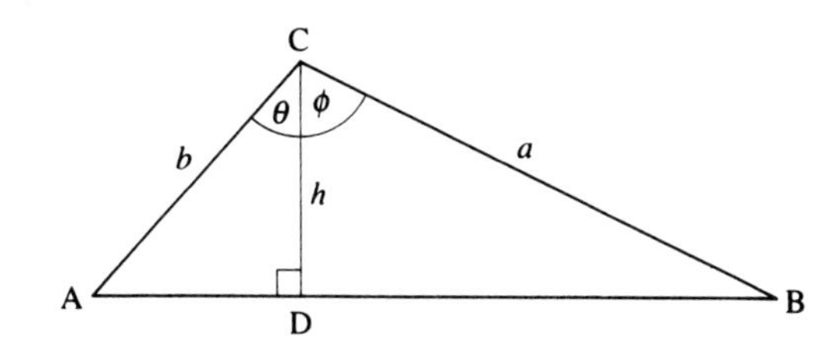

FIGURE 4.10

Using the trigonometric formula for the area of a triangle in figure 4.10

$$\text{area ABC} = \text{area ADC} + \text{area DBC}$$

$$\tfrac{1}{2}ab\sin(\theta + \phi) = \tfrac{1}{2}bh\sin\theta + \tfrac{1}{2}ah\sin\phi$$

$h = a\cos\phi$ from $\triangle$ DBC

$h = b\cos\theta$ from $\triangle$ ADC

$$ab\sin(\theta + \phi) = ab\sin\theta\cos\phi + ab\cos\theta\sin\phi$$

which gives

$$\sin(\theta + \phi) = \sin\theta\cos\phi + \cos\theta\sin\phi. \quad ①$$

This is the first of the compound-angle formulae (or expansions), and it can be used to prove several more. These are true for all values of θ and ϕ.

Replacing ϕ by $-\phi$ in ① gives

$$\sin(\theta - \phi) = \sin\theta\cos(-\phi) + \cos\theta\sin(-\phi)$$

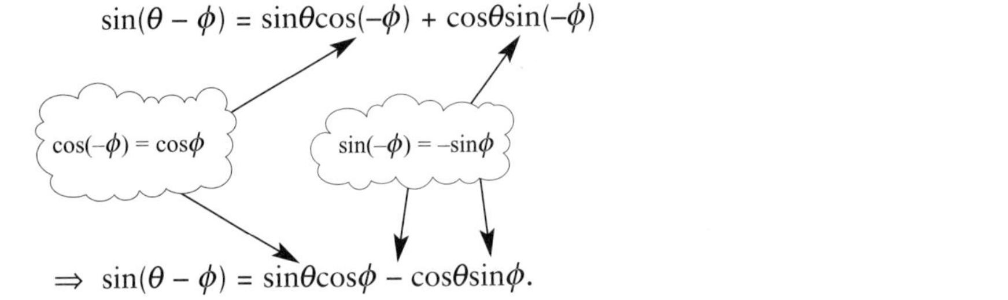

$$\Rightarrow \sin(\theta - \phi) = \sin\theta\cos\phi - \cos\theta\sin\phi. \quad ②$$

From the shape of the sine graph and the cosine graph you can show that

$$\sin(90° - \theta) = \cos\theta$$

and $\quad \cos(90° - \theta) = \sin\theta.$

Hence replacing θ by $90° - \theta$ in $\sin(\theta + \phi) = \sin\theta\cos\phi + \cos\theta\sin\phi$ (equation ① above) gives

$$\sin(90° - \theta + \phi) = \sin(90° - \theta)\cos\phi + \cos(90° - \theta)\sin\phi$$
$$\sin(90° - (\theta - \phi)) = \cos\theta\cos\phi + \sin\theta\sin\phi$$

so

$$\cos(\theta - \phi) = \cos\theta\cos\phi + \sin\theta\sin\phi.$$

Replacing ϕ by $-\phi$ gives

$$\cos(\theta + \phi) = \cos\theta\cos\phi - \sin\theta\sin\phi.$$

From these can be obtained the equivalent formulae for tangents. For example, put

$$\tan(\theta + \phi) = \frac{\sin(\theta + \phi)}{\cos(\theta + \phi)}$$

and expand and simplify.

The results are summarised as follows:

$$\sin(\theta + \phi) = \sin\theta\cos\phi + \cos\theta\sin\phi$$

$$\sin(\theta - \phi) = \sin\theta\cos\phi - \cos\theta\sin\phi$$

$$\cos(\theta + \phi) = \cos\theta\cos\phi - \sin\theta\sin\phi$$

$$\cos(\theta - \phi) = \cos\theta\cos\phi + \sin\theta\sin\phi$$

$$\tan(\theta + \phi) = \frac{\tan\theta + \tan\phi}{1 - \tan\theta\tan\phi}$$

$$\tan(\theta - \phi) = \frac{\tan\theta - \tan\phi}{1 + \tan\theta\tan\phi}.$$

EXAMPLE 4.9

By putting $75° = 45° + 30°$ find the exact value of $\cos 75°$.

Solution

$$\begin{aligned}\cos 75° &= \cos(45° + 30°)\\ &= \cos 45°\cos 30° - \sin 45°\sin 30°\\ &= \frac{1}{\sqrt{2}}\frac{\sqrt{3}}{2} - \frac{1}{\sqrt{2}}\frac{1}{2}\\ &= \frac{\sqrt{3} - 1}{2\sqrt{2}}.\end{aligned}$$

EXAMPLE 4.10

Find the acute angle for which $\sin(\theta + 60°) = \cos(\theta - 60°)$.

Solution

To find an acute angle θ such that $\sin(\theta + 60°) = \cos(\theta - 60°)$, you expand each side using the compound-angle formulae:

$$\begin{aligned}\sin(\theta + 60°) &= \sin\theta\cos 60° + \cos\theta\sin 60°\\ &= \frac{1}{2}\sin\theta + \frac{\sqrt{3}}{2}\cos\theta \qquad ①\end{aligned}$$

$$\begin{aligned}\cos(\theta - 60°) &= \cos\theta\cos 60° + \sin\theta\sin 60°\\ &= \frac{1}{2}\cos\theta + \frac{\sqrt{3}}{2}\sin\theta \qquad ②\end{aligned}$$

From ① and ②

$$\frac{1}{2}\sin\theta + \frac{\sqrt{3}}{2}\cos\theta = \frac{1}{2}\cos\theta + \frac{\sqrt{3}}{2}\sin\theta$$

$$\sin\theta + \sqrt{3}\cos\theta = \cos\theta + \sqrt{3}\sin\theta$$

Collect like terms:

$$\Rightarrow \quad (\sqrt{3} - 1)\cos\theta = (\sqrt{3} - 1)\sin\theta$$

$$\cos\theta = \sin\theta.$$

Divide by $\cos\theta$:

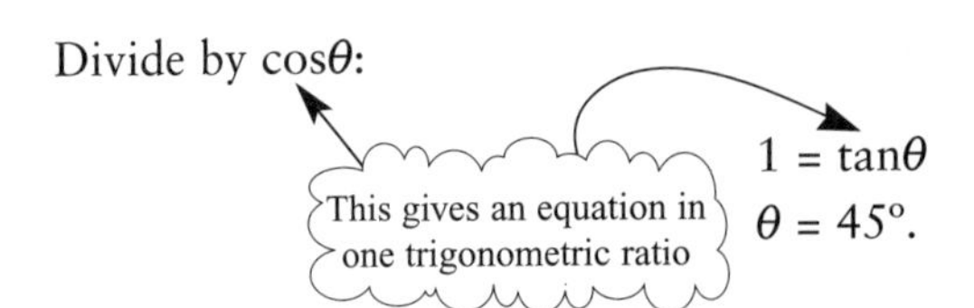

$$1 = \tan\theta$$
$$\theta = 45°.$$

Since an acute angle was required, this is the only root.

EXAMPLE 4.11

Simplify $\cos\theta\cos3\theta - \sin\theta\sin3\theta$.

Solution The formula which has the same pattern of cos cos – sin sin is

$$\cos(\theta + \phi) = \cos\theta\cos\phi - \sin\theta\sin\phi.$$

Using this, and replacing ϕ by 3θ, gives

$$\cos\theta\cos3\theta - \sin\theta\sin3\theta = \cos(\theta + 3\theta)$$
$$= \cos4\theta.$$

EXERCISE 4C

1 Use the compound-angle formulae to write the following as surds.

(a) $\sin75° = \sin(45° + 30°)$ **(b)** $\cos135° = \cos(90° + 45°)$

(c) $\tan15° = \tan(45° - 30°)$ **(d)** $\tan75° = \tan(45° + 30°)$

2 Expand each of the following expressions.

(a) $\sin(\theta + 45°)$ **(b)** $\cos(\theta - 30°)$

(c) $\sin(60° - \theta)$ **(d)** $\cos(2\theta + 45°)$

(e) $\tan(\theta + 45°)$ **(f)** $\tan(\theta - 45°)$

3 Simplify each of the following expressions.

(a) $\sin2\theta\cos\theta - \cos2\theta\sin\theta$

(b) $\cos\phi\cos3\phi - \sin\phi\sin3\phi$

(c) $\sin120°\cos60° + \cos120°\sin60°$

(d) $\cos\theta\cos\theta - \sin\theta\sin\theta$

4 Solve the following equations for values of θ in the range $0° \leqslant \theta \leqslant 180°$.

(a) $\cos(60° + \theta) = \sin\theta$ **(b)** $\sin(45° - \theta) = \cos\theta$

(c) $\tan(45° + \theta) = \tan(45° - \theta)$ **(d)** $2\sin\theta = 3\cos(\theta - 60°)$

(e) $\sin\theta = \cos(\theta + 120°)$

5 Solve the following equations for values of θ in the range $0 \leqslant \theta \leqslant \pi$. (When the range is given in radians, the solutions should be in radians, using multiples of π where appropriate.)

(a) $\sin\left(\theta + \frac{\pi}{4}\right) = \cos\theta$

(b) $2\cos\left(\theta - \frac{\pi}{3}\right) = \cos\left(\theta + \frac{\pi}{2}\right)$

6 Prove the following identities.

(a) $\dfrac{\sin(A + B)}{\cos A\cos B} \equiv \tan A + \tan B$

(b) $\sin(A + B) + \cos(A - B) \equiv (\sin A + \cos A)(\sin B + \cos B)$

(c) $\cos(45° + A) - \cos(45° - A) \equiv \sqrt{2}\sin A$

(d) $\tan(A + B) \equiv \dfrac{\tan A + \tan B}{1 - \tan A\tan B}$.

DOUBLE-ANGLE FORMULAE

Substituting $\phi = \theta$ in the relevant compound-angle formulae leads immediately to expressions for $\sin 2\theta$, $\cos 2\theta$ and $\tan 2\theta$, as follows.

(a) $\sin(\theta + \phi) = \sin\theta\cos\phi + \cos\theta\sin\phi$

When $\phi = \theta$, this becomes

$\sin(\theta + \theta) = \sin\theta\cos\theta + \cos\theta\sin\theta$

giving $\sin 2\theta = 2\sin\theta\cos\theta$.

(b) $\cos(\theta + \phi) = \cos\theta\cos\phi - \sin\theta\sin\phi$

When $\phi = \theta$, this becomes

$\cos(\theta + \theta) = \cos\theta\cos\theta - \sin\theta\sin\theta$

giving $\cos 2\theta = \cos^2\theta - \sin^2\theta$.

Using the fundamental trigonometric identity $\cos^2\theta + \sin^2\theta = 1$, two other forms for $\cos 2\theta$ can be obtained:

$$\cos 2\theta = (1 - \sin^2\theta) - \sin^2\theta \quad \Rightarrow \quad \cos 2\theta = 1 - 2\sin^2\theta$$

$$\cos 2\theta = \cos^2\theta - (1 - \cos^2\theta) \quad \Rightarrow \quad \cos 2\theta = 2\cos^2\theta - 1.$$

These alternative forms are often more useful since thay contain only one trigonometric function.

(c) $\tan(\theta + \phi) = \dfrac{\tan\theta + \tan\phi}{1 - \tan\theta\tan\phi}$

When $\phi = \theta$ this becomes

$$\tan(\theta + \theta) = \frac{\tan\theta + \tan\theta}{1 - \tan\theta\tan\theta}$$

giving $\tan 2\theta = \dfrac{2\tan\theta}{1 - \tan^2\theta}$.

In summary:

$$\sin 2\theta = 2\sin\theta\cos\theta$$

$$\cos 2\theta = \cos^2\theta - \sin^2\theta$$

$$= 2\cos^2\theta - 1$$

$$= 1 - 2\sin^2\theta$$

$$\tan 2\theta = \frac{2\tan\theta}{1 - \tan^2\theta}.$$

EXAMPLE 4.12

Solve the equation $\sin 2\theta = \sin\theta$ for $0° \leqslant \theta \leqslant 360°$

Solution

$$\sin 2\theta = \sin\theta$$

$$\Rightarrow \quad 2\sin\theta\cos\theta = \sin\theta$$

Be careful here: don't cancel by $\sin\theta$ or some roots will be lost

$$\Rightarrow \quad 2\sin\theta\cos\theta - \sin\theta = 0$$

$$\Rightarrow \quad \sin\theta(2\cos\theta - 1) = 0$$

$$\Rightarrow \quad \sin\theta = 0 \text{ or } \cos\theta = \tfrac{1}{2}$$

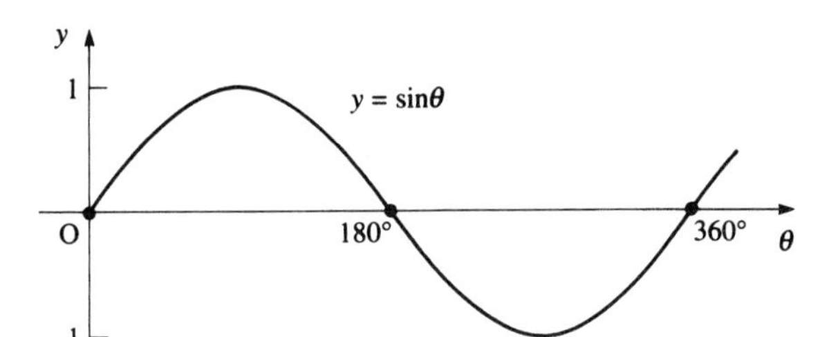

FIGURE 4.11

The principal value is the one which comes from your calculator

$\sin\theta = 0 \Rightarrow \theta = 0°$ (principal value) or 180° or 360° (see figure 4.11).

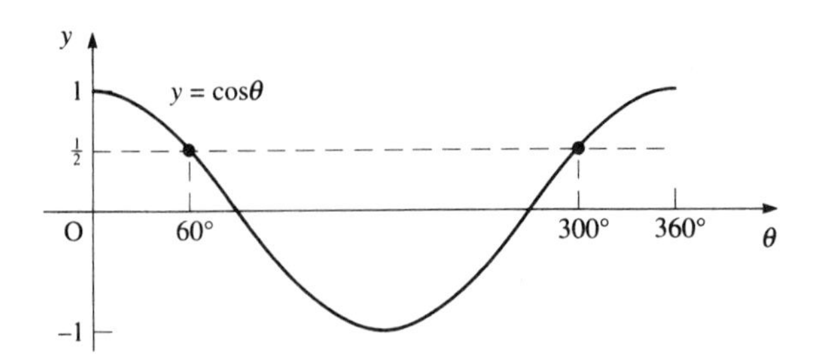

FIGURE 4.12

$\cos\theta = \frac{1}{2} \Rightarrow \theta = 60°$ (principal value) or 300° (see figure 4.12).

The full set of roots for $0 \leqslant \theta \leqslant 360°$ is $\theta = 0°, 60°, 180°, 300°, 360°$.

When an equation contains $\cos 2\theta$, you will save time if you take care to choose the most suitable expansion.

EXAMPLE 4.13

Solve $2 + \cos 2\theta = \sin\theta$ for $0 \leqslant \theta \leqslant 2\pi$. (Notice that the request for θ for $0 \leqslant \theta \leqslant 2\pi$, i.e. in radians, is an invitation to give the answer in radians.)

Solution

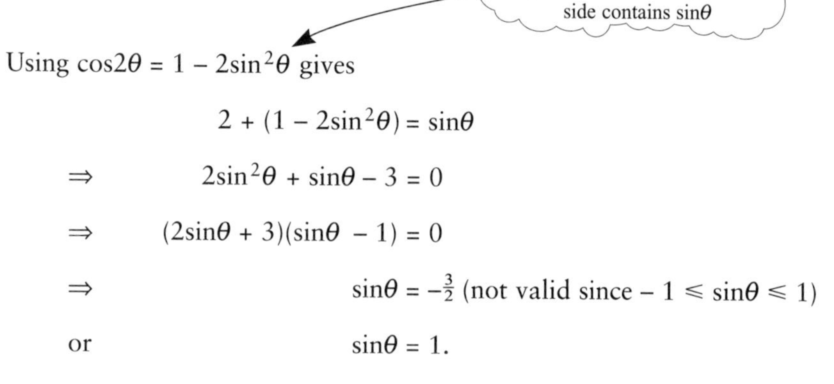

Using $\cos 2\theta = 1 - 2\sin^2\theta$ gives

$$2 + (1 - 2\sin^2\theta) = \sin\theta$$

$$\Rightarrow \quad 2\sin^2\theta + \sin\theta - 3 = 0$$

$$\Rightarrow \quad (2\sin\theta + 3)(\sin\theta - 1) = 0$$

$$\Rightarrow \quad \sin\theta = -\tfrac{3}{2} \text{ (not valid since } -1 \leqslant \sin\theta \leqslant 1)$$

or $\sin\theta = 1$.

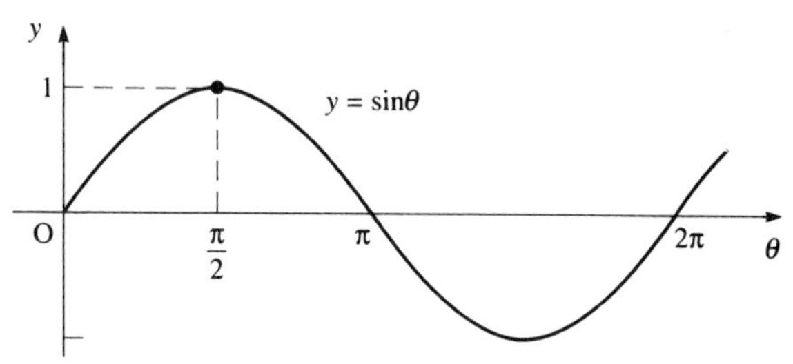

FIGURE 4.13

Figure 4.13 shows that the principal value $\theta = \frac{\pi}{2}$ is the only root for $0 \leqslant \theta \leqslant 2\pi$.

EXAMPLE 4.14

Prove that $\cos 3\theta \equiv 4\cos^3\theta - 3\cos\theta$.

Solution

$$\begin{aligned}\cos 3\theta &= \cos(\theta + 2\theta)\\ &= \cos\theta\cos 2\theta - \sin\theta\sin 2\theta\\ &= \cos\theta(2\cos^2\theta - 1) - \sin\theta \times 2\sin\theta\cos\theta\\ &= 2\cos^3\theta - \cos\theta - 2\cos\theta\sin^2\theta\\ &= 2\cos^3\theta - \cos\theta - 2\cos\theta(1 - \cos^2\theta)\\ &= 4\cos^3\theta - 3\cos\theta.\end{aligned}$$

HALF-ANGLE FORMULAE

The double-angle formulae can be written in terms of θ and its half-angle $\frac{1}{2}\theta$.

Since $\sin 2\theta = 2\sin\theta\cos\theta$

replacing θ by $\frac{1}{2}\theta$ gives

$$\sin\theta = 2\sin\tfrac{1}{2}\theta\cos\tfrac{1}{2}\theta.$$

Similarly
$$\begin{aligned}\cos\theta &= \cos^2\tfrac{1}{2}\theta - \sin^2\tfrac{1}{2}\theta \\ &= 2\cos^2\tfrac{1}{2}\theta - 1 \\ &= 1 - 2\sin^2\tfrac{1}{2}\theta\end{aligned}$$

and
$$\tan\theta = \frac{2\tan\frac{1}{2}\theta}{1 - \tan^2\frac{1}{2}\theta}.$$

EXAMPLE 4.15

Solve $\sin\theta = \cot\frac{1}{2}\theta$ for $0° \leqslant \theta \leqslant 360°$.

Solution

$$\begin{aligned}\sin\theta &= \cot\tfrac{1}{2}\theta \\ 2\sin\tfrac{1}{2}\theta\cos\tfrac{1}{2}\theta &= \frac{\cos\frac{1}{2}\theta}{\sin\frac{1}{2}\theta} \\ 2\sin^2\tfrac{1}{2}\theta\cos\tfrac{1}{2}\theta - \cos\tfrac{1}{2}\theta &= 0 \\ \cos\tfrac{1}{2}\theta(2\sin^2\tfrac{1}{2}\theta - 1) &= 0 \\ \cos\tfrac{1}{2}\theta = 0 \quad \text{or} \quad \sin\tfrac{1}{2}\theta &= \pm\frac{1}{\sqrt{2}} \\ \tfrac{1}{2}\theta = 90° \text{ or } \tfrac{1}{2}\theta &= 45°, 135° \\ \theta &= 90°, 180°, 270°.\end{aligned}$$

Note if $0° \leqslant \theta \leqslant 360°$, $0° \leqslant \frac{1}{2}\theta \leqslant 180°$

Note θ can be replaced by any other multiple of θ. For example, replacing θ by 2θ would give $\sin 4\theta = 2\sin 2\theta\cos 2\theta$.

t-FORMULAE

From the half-angle formulae you have seen that

$$\sin\theta = 2\sin\tfrac{1}{2}\theta\cos\tfrac{1}{2}\theta$$

and re-writing this gives

$$\begin{aligned}\sin\theta &= \frac{2\sin\frac{1}{2}\theta}{\cos\frac{1}{2}\theta} \times \cos^2\tfrac{1}{2}\theta \\ &= \frac{2\tan\frac{1}{2}\theta}{\sec^2\frac{1}{2}\theta}\end{aligned}$$

$$= \frac{2\tan\frac{1}{2}\theta}{1 + \tan^{2}\frac{1}{2}\theta}.$$

Putting $\tan^{2}\frac{1}{2}\theta = t$ gives the t-formula for $\sin\theta$, that is

$$\sin\theta = \frac{2t}{1 + t^{2}}. \qquad ①$$

Starting with $\cos\theta$ gives

$$\cos\theta = \cos^{2}\tfrac{1}{2}\theta - \sin^{2}\tfrac{1}{2}\theta$$

$$= \cos^{2}\tfrac{1}{2}\theta(1 - \tan^{2}\tfrac{1}{2}\theta)$$

$$= \frac{1 - \tan^{2}\frac{1}{2}\theta}{\sec^{2}\frac{1}{2}\theta}$$

$$= \frac{1 - \tan^{2}\frac{1}{2}\theta}{1 + \tan^{2}\frac{1}{2}\theta}$$

so $$\cos\theta = \frac{1 - t^{2}}{1 + t^{2}}. \qquad ②$$

Dividing equation ① by equation ② gives

$$\tan\theta = \frac{2t}{1 - t^{2}}.$$

One of the uses of these formulae is to solve trigonometric equations. The method produces polynomials in t. Having solved for t and hence for θ, substitute your values of θ back into the original equation to find the ones that give a valid solution.

EXAMPLE 4.16

Use the t-formulae to solve $\sin\theta + \cos\theta = 1$ for $0 \leqslant \theta \leqslant 2\pi$.

Solution

$$\sin\theta + \cos\theta = 1$$

putting this in terms of t gives

$$\frac{2t}{1 + t^{2}} + \frac{1 - t^{2}}{1 + t^{2}} = 1$$

$$2t + 1 - t^{2} = 1 + t^{2}$$

$$2t^{2} - 2t = 0$$

$$2t(t - 1) = 0$$

Remember $t = \tan\frac{1}{2}\theta$

$$\tan\tfrac{1}{2}\theta = 0 \text{ or } \tan\tfrac{1}{2}\theta = 1$$

$$\tfrac{1}{2}\theta = 0, \pi \text{ or } \tfrac{1}{2}\theta = \frac{\pi}{4}$$

Note that $0 \leqslant \frac{1}{2}\theta \leqslant \pi$

so $$\theta = 0, 2\pi, \frac{\pi}{2}.$$

Checking the answers in the original equation shows that they are all valid.

EXERCISE 4D

1 Solve the following equations for $0° \leqslant \theta \leqslant 360°$.

(a) $2\sin 2\theta = \cos\theta$ **(b)** $\tan 2\theta = 4\tan\theta$

(c) $\cos 2\theta + \sin\theta = 0$ **(d)** $\tan\theta\tan 2\theta = 1$

(e) $2\cos 2\theta = 1 + \cos\theta$

2 Solve the following equations for $-\pi \leqslant \theta \leqslant \pi$.

(a) $\sin 2\theta = 2\sin\theta$ **(b)** $\tan 2\theta = 2\tan\theta$

(c) $\cos 2\theta - \cos\theta = 0$ **(d)** $1 + \cos 2\theta = 2\sin^2\theta$

(e) $\sin 4\theta = \cos 2\theta$ (**Hint:** Express this as an equation in 2θ.)

3 By first writing $\sin 3\theta$ as $\sin(2\theta + \theta)$, express $\sin 3\theta$ in terms of $\sin\theta$. Hence solve the equation $\sin 3\theta = \sin\theta$ for $0 \leqslant \theta \leqslant 2\pi$.

4 Solve $\cos 3\theta = 1 - 3\cos\theta$ for $0° \leqslant \theta \leqslant 360°$.

5 Simplify $\dfrac{1 + \cos 2\theta}{\sin 2\theta}$.

6 Express $\tan 3\theta$ in terms of $\tan\theta$.

7 Show that $\dfrac{1 - \tan^2\theta}{1 + \tan^2\theta} = \cos 2\theta$.

8 **(a)** Show that $\tan\left(\frac{\pi}{4} + \theta\right)\tan\left(\frac{\pi}{4} - \theta\right) = 1$.

(b) Given that $\tan 26.6° = 0.5$, solve $\tan\theta = 2$ without using your calculator. Give θ to 1 decimal place, where $0° < \theta < 90°$.

9 Use the half-angle formulae to solve the following equations for $0° \leqslant \theta \leqslant 360°$.

(a) $\sin\theta + \cos\frac{1}{2}\theta = 0$ **(b)** $\sin\theta = 2(\cos\theta + 1)$

(c) $3\tan\theta = 8\tan\frac{1}{2}\theta$ **(d)** $4\cos\theta = 1 - 2\sin\frac{1}{2}\theta$

10 Prove the following identities.

(a) $\dfrac{\sin\theta}{1 + \cos\theta} \equiv \tan\frac{1}{2}\theta$ **(b)** $\cot\theta + \operatorname{cosec}\theta \equiv \cot\frac{1}{2}\theta$

(c) $\sin\theta \equiv \dfrac{2\tan\frac{1}{2}\theta}{1 + \tan^2\frac{1}{2}\theta}$ **(d)** $\sec\theta + \tan\theta \equiv \dfrac{\cos\frac{1}{2}\theta + \sin\frac{1}{2}\theta}{\cos\frac{1}{2}\theta - \sin\frac{1}{2}\theta}$

11 Use the t-formulae to solve the the following equations in the range 0° to 360°.

(a) $\sin\theta + 2\cos\theta = 1$ **(b)** $3\cos\theta + 2\sin\theta = 3$

(c) $5\tan\theta + \sec\theta + 5 = 0$ **(d)** $3\cot\theta + 2\operatorname{cosec}\theta = 2$

12 Express $1 - \sin\theta$ as a single fraction in terms of t, where $t = \tan\frac{1}{2}\theta$.
Prove that

$$\sqrt{\frac{1 - \sin\theta}{1 + \sin\theta}} \equiv \frac{1 - t}{1 + t}.$$

$a\cos\theta + b\sin\theta$ IN THE FORM $r\cos(\theta \pm \alpha)$ OR $r\sin(\theta \pm \alpha)$

$a\cos\theta + b\sin\theta$ can be combined to form a single function $r\cos(\theta \pm \alpha)$ or $r\sin(\theta \pm \alpha)$ using the compound-angle formulae. The following examples illustrate the process.

EXAMPLE 4.17

Express $4\cos\theta + 3\sin\theta$ in the form $r\sin(\theta + \alpha)$.

Solution Let $4\cos\theta + 3\sin\theta \equiv r\sin(\theta+\alpha)$

$$4\cos\theta + 3\sin\theta \equiv r\sin\theta\cos\alpha + r\cos\theta\sin\alpha$$

Comparing the coefficients of $\cos\theta$ and $\sin\theta$ gives

$\cos\theta$: $4 = r\sin\alpha$ ①

$\sin\theta$: $3 = r\cos\alpha$ ②

To find r

①² + ②²

$$r^2\sin^2\alpha + r^2\cos^2\alpha = 4^2 + 3^2$$
$$r^2(\sin^2\alpha + \cos^2\alpha) = 25$$
$$r^2 = 25$$
$$r = 5.$$

Since $\sin^2\alpha + \cos^2\alpha = 1$

To find α

① ÷ ②

$$\frac{r\sin\alpha}{r\cos\alpha} = \tan\alpha = \frac{4}{3}$$
$$\alpha = 53.1°.$$

So, the answer is

$$4\cos\theta + 3\sin\theta \equiv 5\sin(\theta + 53.1°).$$

Note

From equations ① and ② we could have calculated r and α by drawing a right-angled triangle (figure 4.14):

$4 = r\sin\alpha$

$3 = r\cos\alpha$

from which r and α can be found.

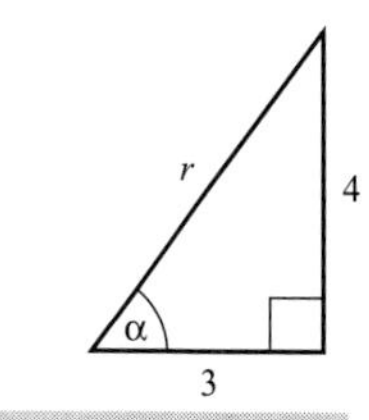

FIGURE 4.14

The value of r is always positive but $\cos\alpha$ and $\sin\alpha$ may be positive or negative, depending on the values of a and b. In all cases, it is possible to find an angle α for which $-180° \leqslant \alpha \leqslant 180°$. If the question does not specify which expression of $r\cos(\theta \pm \alpha)$ or $r\sin(\theta \pm \alpha)$ to use then choose either the sine or the cosine function with the signs of their expansions matching the signs of $a\cos\theta + b\sin\theta$.

For example $3\cos\theta - 4\sin\theta$ matches up with $r\cos(\theta + \alpha)$ or $-r\sin(\theta - \alpha)$.

The result of replacing $a\cos\theta + b\sin\theta$ with $r\cos(\theta \pm \alpha)$ or $r\sin(\theta \pm \alpha)$ is the combination of two waves to form a single sine wave. In Example 4.17 the amplitude of the single wave is 5 and the phase shift is 53.1°. This is illustrated in figure 4.15 over the range 0° to 360°.

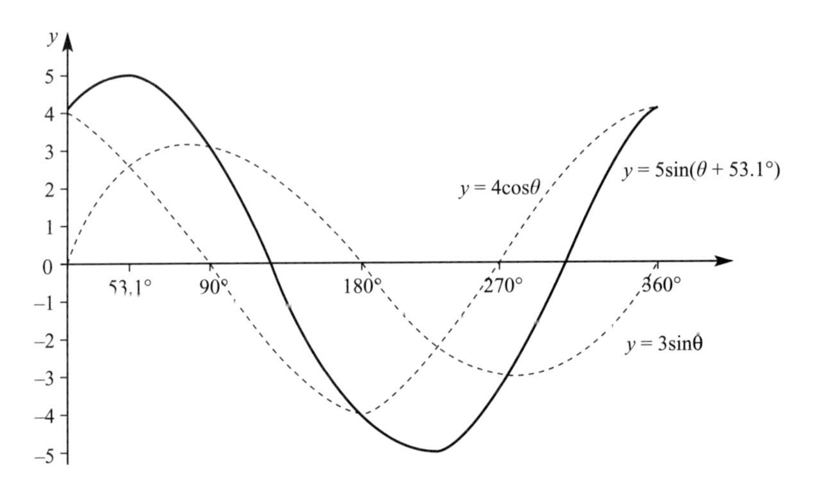

FIGURE 4.15

EXAMPLE 4.18

Express $12\cos\theta + 5\sin\theta$ in the form $r\cos(\theta - \alpha)$.
Use your answer to solve $12\cos\theta + 5\sin\theta = 4$ giving θ in the range –180° to 180°.

Solution Let $12\cos\theta + 5\sin\theta \equiv r\cos(\theta - \alpha)$

$$12\cos\theta + 5\sin\theta \equiv r\cos\theta\cos\alpha + r\sin\theta\sin\alpha$$

$\cos\theta$: $\quad 12 = r\cos\alpha$

$\sin\theta$: $\quad 5 = r\sin\alpha$

$$r = \sqrt{12^2 + 5^2} = 13$$

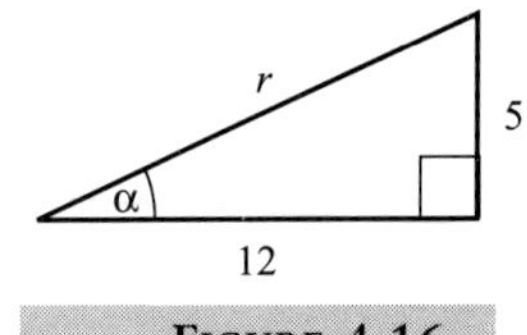

FIGURE 4.16

$$\tan\alpha = \frac{5}{12}$$

so $\quad \alpha = 22.6°$

hence

$$12\cos\theta + 5\sin\theta \equiv 13\cos(\theta - 22.6°).$$

To solve $\quad 12\cos\theta + 5\sin\theta = 4$

put $\quad 13\cos(\theta - 22.6°) = 4$

$$\therefore \cos(\theta - 22.6°) = \frac{4}{13}$$

$$\therefore \theta - 22.6° = \pm 72.1°$$

$$\therefore \theta = -49.5° \text{ or } 94.7°.$$

EXAMPLE 4.19

(a) Express $\sqrt{3}\sin\theta - \cos\theta$ in the form $r\sin(\theta - \alpha)$, where $r > 0$ and $0 < \alpha < \frac{\pi}{2}$.

(b) State the maximum and minimum values of $\sqrt{3}\sin\theta - \cos\theta$.

(c) Sketch the graph of $y = \sqrt{3}\sin\theta - \cos\theta$ for $0 \leqslant \theta \leqslant 2\pi$.

(d) Solve the equation $\sqrt{3}\sin\theta - \cos\theta = 1$ for $0 \leqslant \theta \leqslant 2\pi$.

Solution

(a) Let

$$\sqrt{3}\sin\theta - \cos\theta \equiv r\sin(\theta - \alpha)$$

$$\sqrt{3}\sin\theta - \cos\theta \equiv r\sin\theta\cos\alpha - r\cos\theta\sin\alpha$$

Comparing coefficients

$$\sin\theta: \quad \sqrt{3} = r\cos\alpha$$

$$\cos\theta: \quad 1 = r\cos\alpha$$

From the triangle (figure 4.17)

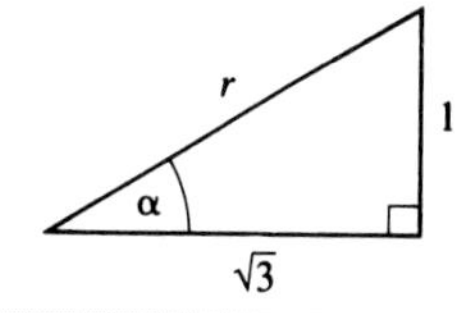

FIGURE 4.17

$$r = \sqrt{1 + 3} = 2 \quad \text{and} \quad \tan\alpha = \frac{1}{\sqrt{3}} \Rightarrow \quad \alpha = \frac{\pi}{6}$$

so $$\sqrt{3}\sin\theta - \cos\theta \equiv 2\sin\left(\theta - \frac{\pi}{6}\right).$$

(b) The sine function oscillates between 1 and –1, so $2\sin\left(\theta - \frac{\pi}{6}\right)$ oscillates between 2 and –2.

Maximum value = 2.
Minimum value = –2.

(c) The graph of $y = 2\sin\left(\theta - \frac{\pi}{6}\right)$ is obtained from the graph of $y = \sin\theta$ by a translation $\begin{pmatrix} \frac{\pi}{6} \\ 0 \end{pmatrix}$ and a stretch of factor 2 parallel to the y axis.

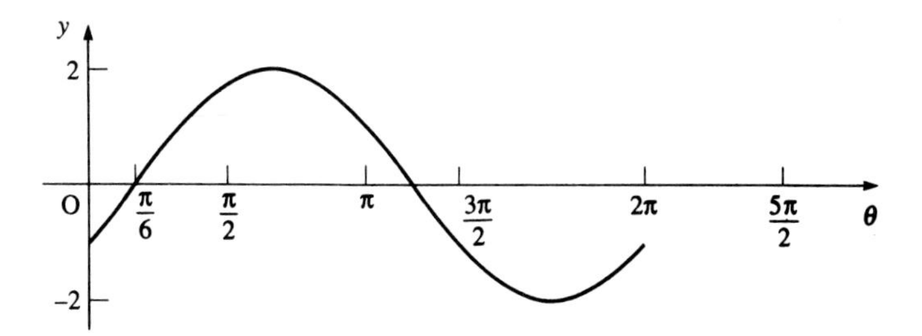

FIGURE 4.18

(d) The equation $\sqrt{3}\sin\theta - \cos\theta = 1$ is equivalent to

$$2\sin(\theta - \tfrac{\pi}{6}) = 1$$
$$\Rightarrow \quad \sin(\theta - \tfrac{\pi}{6}) = \tfrac{1}{2}$$

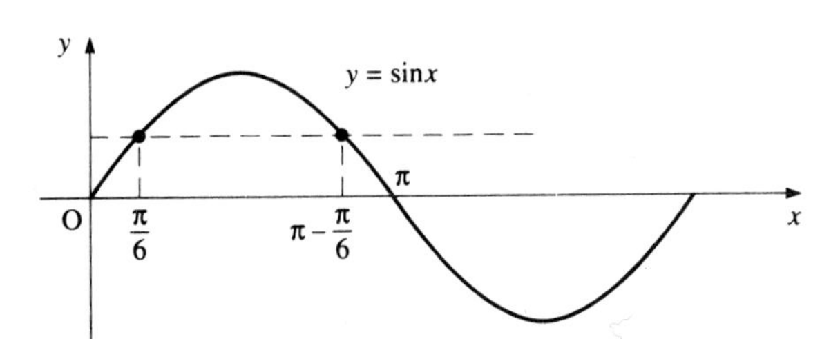

FIGURE 4.19

Solving $\sin x = \frac{1}{2}$ gives

$$x = \tfrac{\pi}{6} \text{ (principal value) or } x = \pi - \tfrac{\pi}{6} = \tfrac{5\pi}{6} \quad \text{(from the graph)}$$

giving $\quad \theta = \frac{\pi}{6} + \frac{\pi}{6} = \frac{\pi}{3} \qquad$ or $\qquad \theta = \frac{5\pi}{6} + \frac{\pi}{6} = \pi.$

The roots in $0 \leqslant \theta \leqslant 2\pi$ are $\theta = \frac{\pi}{3}$ and π.

Note

Always check (for example, by reference to a sketch graph) that the number of roots you have found is consistent with the number you are expecting. When solving equations of the form $\sin(\theta - \alpha) = c$ by considering $\sin x = c$, it is sometimes necessary to go outside the range specified for θ since, for example, $0 \leqslant \theta \leqslant 2\pi$ is the same as $-\alpha \leqslant x \leqslant 2\pi - \alpha$.

EXERCISE 4E

1 Express each of the following in the form $r\cos(\theta - \alpha)$, where $r > 0$ and $0 < \alpha < 90°$.

(a) $\cos\theta + \sin\theta$

(b) $3\cos\theta + 4\sin\theta$

(c) $\cos\theta + \sqrt{3}\sin\theta$

(d) $\sqrt{5}\cos\theta + 2\sin\theta$

2 Express each of the following in the form $r\cos(\theta + \alpha)$, where $r > 0$ and $0 < \alpha < \frac{\pi}{2}$.

(a) $\cos\theta - \sin\theta$

(b) $\sqrt{3}\cos\theta - \sin\theta$

3 Express each of the following in the form $r\sin(\theta + \alpha)$, where $r > 0$ and $0 < \alpha < 90°$.

(a) $\sin\theta + 2\cos\theta$

(b) $3\sin\theta + 4\cos\theta$

4 Express each of the following in the form $r\sin(\theta - \alpha)$, where $r > 0$ and $0 < \alpha < \frac{\pi}{2}$.

(a) $\sin\theta - \cos\theta$

(b) $\sqrt{3}\sin\theta - \cos\theta$

5 Express each of the following in the form $r\cos(\theta - \alpha)$, where $r > 0$ and $-180° < \alpha < 180°$.

(a) $\cos\theta - \sqrt{3}\sin\theta$

(b) $2\sqrt{2}\cos\theta - 2\sqrt{2}\sin\theta$

(c) $\sin\theta + \sqrt{3}\cos\theta$

(d) $5\sin\theta + 12\cos\theta$

(e) $\sin\theta - \sqrt{3}\cos\theta$

(f) $\sqrt{2}\sin\theta - \sqrt{2}\cos\theta$

6 **(a)** Express $5\cos\theta - 12\sin\theta$ in the form $r\cos(\theta + \alpha)$, where $r > 0$ and $0 < \alpha < 90°$.

(b) State the maximum and minimum values of $5\cos\theta - 12\sin\theta$.

(c) Sketch the graph of $y = 5\cos\theta - 12\sin\theta$ for $0 \leqslant \theta \leqslant 360°$

(d) Solve the equation $5\cos\theta - 12\sin\theta = 4$ for $0 \leqslant \theta \leqslant 360°$.

7 **(a)** Express $3\sin\theta - \sqrt{3}\cos\theta$ in the form $r\sin(\theta - \alpha)$, where $r > 0$ and $0 < \alpha < \frac{\pi}{2}$.

(b) State the maximum and minimum values of $3\sin\theta - \sqrt{3}\cos\theta$ and the smallest positive values of θ for which they occur.

(c) Sketch the graph of $y = 3\sin\theta - \sqrt{3}\cos\theta$ for $0 \leqslant \theta \leqslant 2\pi$.

(d) Solve the equation $3\sin\theta - \sqrt{3}\cos\theta = \sqrt{3}$ for $0 \leqslant \theta \leqslant 2\pi$.

8 **(a)** Express $2\sin2\theta + 3\cos2\theta$ in the form $r\sin(2\theta + \alpha)$,where $r > 0$ and $0 < \alpha < 90°$.

(b) State the maximum and minimum values of $2\sin2\theta + 3\cos2\theta$ and the smallest positive values of θ for which they occur.

(c) Sketch the graph of $y = 2\sin2\theta + 3\cos2\theta$ for $0 \leqslant \theta \leqslant 360°$

(d) Solve the equation $2\sin2\theta + 3\cos2\theta = 1$ for $0 \leqslant \theta \leqslant 360°$.

9 **(a)** Express $\cos\theta + \sqrt{2}\sin\theta$ in the form $r\cos(\theta - \alpha)$, where $r > 0$ and $0 < \alpha < 90°$.

(b) State the maximum and minimum values of $\cos\theta + \sqrt{2}\sin\theta$ and the smallest positive values of θ for which they occur.

(c) Sketch the graph of $y = \cos\theta + \sqrt{2}\sin\theta$ for $0 \leqslant \theta \leqslant 360°$.

(d) State the maximum and minimum values of

$$\frac{1}{3 + \cos\theta + \sqrt{2}\sin\theta}$$

and the smallest positive values for which they occur.

10 The diagram shows a table jammed in a corridor. The table is 120 cm long and 80 cm wide, and the width of the corridor is 130 cm.

(a) Show that $12\sin\theta + 8\cos\theta = 13$.

(b) Hence find the angle θ (there are two answers) by:

(i) expressing $12\sin\theta + 8\cos\theta$ in the form $r\sin(\theta + \alpha)$

(ii) using the t-formulae.

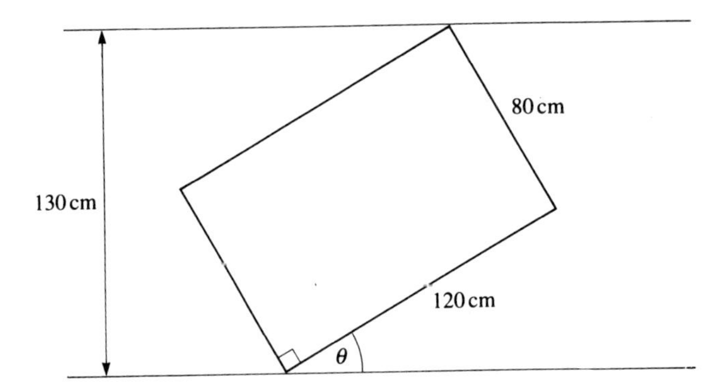

THE FACTOR FORMULAE

In algebra, the term 'factorising' means writing expressions as products. For example, 'factorise $x^2 - 3x + 2$' means write $x^2 - 3x + 2$ as $(x - 1)(x - 2)$. The same idea of factorising applies in trigonometry: you write sums or differences of trigonometric functions as products.

The factor formulae are derived from the compound-angle formulae.

Start with the compound-angle formulae for $\sin(\theta + \phi)$ and $\sin(\theta - \phi)$:

$$\sin(\theta + \phi) = \sin\theta\cos\phi + \cos\theta\sin\phi \qquad ①$$
$$\sin(\theta - \phi) = \sin\theta\cos\phi - \cos\theta\sin\phi \qquad ②$$

Adding ① and ② gives

$$\sin(\theta + \phi) + \sin(\theta - \phi) = 2\sin\theta\cos\phi. \qquad ③$$

At this point, it is helpful to change variables by writing

$$\theta + \phi = \alpha \quad \text{and} \quad \theta - \phi = \beta$$

so that $\theta = \frac{1}{2}(\alpha + \beta)$ and $\phi = \frac{1}{2}(\alpha - \beta)$.

Substituting for θ and ϕ in ③ gives

$$\sin\alpha + \sin\beta = 2\sin\frac{(\alpha + \beta)}{2}\cos\frac{(\alpha - \beta)}{2}.$$

The left-hand side is a sum, and the right-hand side is a product, so the expression has been factorised.

Similarly, subtracting ② from ① gives

$$\sin(\theta + \phi) - \sin(\theta - \phi) = 2\cos\theta\sin\phi \Rightarrow \sin\alpha - \sin\beta = 2\cos\frac{\alpha+\beta}{2}\sin\frac{\alpha-\beta}{2}.$$

Similarly, starting with

$$\cos(\theta + \alpha) = \cos\theta\cos\alpha - \sin\theta\sin\alpha$$

and $\cos(\theta - \alpha) = \cos\theta\cos\alpha + \sin\theta\sin\alpha$

gives the results

$$2\cos\theta\cos\phi = \cos(\theta + \phi) + \cos(\theta - \phi) \Rightarrow \cos\alpha + \cos\beta = 2\cos\left(\frac{\alpha+\beta}{2}\right)\cos\left(\frac{\alpha-\beta}{2}\right),$$

and

$$-2\sin\theta\sin\phi = \cos(\theta + \phi) - \cos(\theta - \phi) \Rightarrow \cos\alpha - \cos\beta = -2\sin\left(\frac{\alpha+\beta}{2}\right)\sin\left(\frac{\alpha-\beta}{2}\right).$$

The factor formulae are often useful in tidying up expressions and in solving equations as in the next example.

EXAMPLE 4.20

Solve $\sin3\theta + \sin\theta = 0$ for $0° \leqslant \theta \leqslant 360°$.

Solution Using

$$\sin\alpha + \sin\beta = 2\sin\left(\frac{\alpha+\beta}{2}\right)\cos\left(\frac{\alpha-\beta}{2}\right)$$

and putting $\alpha = 3\theta$ and $\beta = \theta$ gives

$$\sin3\theta + \sin\theta = 2\sin2\theta\cos\theta$$

so the equation becomes

$$2\sin2\theta\cos\theta = 0$$

$$\Rightarrow \quad \cos\theta = 0 \quad \text{or} \quad \sin2\theta = 0.$$

From the graphs for $y = \cos\theta$ and $y = \sin\theta$

$$\cos\theta = 0 \quad \text{gives} \quad \theta = 90° \text{ or } 270°$$

$$\sin2\theta = 0 \quad \text{gives} \quad 2\theta = 0°, 180°, 360°, 540° \text{ or } 720°$$

$$\text{so} \quad \theta = 0°, 90°, 180°, 270° \text{ or } 360°.$$

You should only list each root once in the final answer

The complete set of roots in the range given is $\theta = 0°, 90°, 180°, 270°, 360°$.

EXAMPLE 4.21

Prove that $\cos\frac{\pi}{4}\cos\frac{\pi}{12} = \frac{1+\sqrt{3}}{4}$.

Solution

$$\cos\frac{\pi}{4}\cos\frac{\pi}{12} = \frac{1}{2}\left(\cos\left(\frac{\pi}{4}+\frac{\pi}{12}\right) + \cos\left(\frac{\pi}{4}-\frac{\pi}{12}\right)\right)$$
$$= \frac{1}{2}\left(\cos\frac{\pi}{3} + \cos\frac{\pi}{6}\right)$$
$$= \frac{1}{2}\left(\frac{1}{2}+\frac{\sqrt{3}}{2}\right)$$
$$= \frac{1+\sqrt{3}}{4}.$$

EXERCISE 4F

1 Factorise the following expressions.

(a) $\sin 4\theta - \sin 2\theta$

(b) $\cos 5\theta + \cos\theta$

(c) $\cos 7\theta - \cos 3\theta$

(d) $\cos(\theta + 60°) + \cos(\theta - 60°)$

(e) $\sin(3\theta + 45°) + \sin(3\theta - 45°)$

2 Factorise $\cos 4\theta + \cos 2\theta$. Hence, for $0° < \theta < 180°$ solve

$$\cos 4\theta + \cos 2\theta = \cos\theta.$$

3 Simplify $\dfrac{\sin 5\theta + \sin 3\theta}{\sin 5\theta - \sin 3\theta}$.

4 Solve the equation $\sin 3\theta - \sin\theta = 0$ for $0 \leqslant \theta \leqslant 2\pi$.

5 Factorise $\sin(\theta + 73°) - \sin(\theta + 13°)$ and use your result to sketch the graph of $y = \sin(\theta + 73°) - \sin(\theta + 13°)$.

6 Prove that $\sin^2 A - \sin^2 B = \sin(A - B)\sin(A + B)$.

7 **(a)** Use a suitable factor formula to show that

$$\sin 3\theta + \sin\theta = 4\sin\theta\cos^2\theta.$$

(b) Hence show that $\sin 3\theta = 3\sin\theta - 4\sin^3\theta$.

8 Solve the following equations for $0° \leqslant \theta \leqslant 360°$.

(a) $\sin 5\theta + \sin\theta = 0$

(b) $\cos 4\theta - \cos 2\theta = 0$

(c) $\sin 3\theta = \sin\theta$

(d) $\cos 5\theta + \cos 3\theta + \cos\theta = 0$

9 Write the following as the sum, or difference, of two terms.

(a) $2\cos3\theta\cos\theta$

(b) $2\sin3\theta\sin5\theta$

(c) $4\cos6\theta\sin4\theta$

(d) $\sin3\theta\cos\theta$

10 Find the exact values of:

(a) $2\sin45°\cos15°$

(b) $\cos67.5°\cos22.5°$

(c) $\sin\frac{\pi}{4}\cos\frac{\pi}{12}$.

(d) $\cos\frac{\pi}{3}\sin\frac{\pi}{6}$.

EXERCISE 4G Examination-style questions

1 $7\cos x - 24\sin x \equiv R\cos(x + \alpha)$, where $R > 0$ and $0° \leqslant \alpha \leqslant 90°$.

(a) Find the exact value of R and the value of α to 1 decimal place.

(b) Solve $7\cos x - 24\sin x = 20$ giving your answers in the range $-180°$ to $180°$.

(c) State the maximum value of $7\cos x - 24\sin x$ and the value of x for which this occurs.

2 (a) Using the expansion for $\cos(A + B)$ find an expression for $\cos2x$ in terms of $\cos x$.

(b) Solve $\cos2x = \cos x$ for values of x in the interval 0 to 2π, giving your answers in terms of π.

(c) Solve $\cos x = \cos\frac{1}{2}x$ for values of x in the interval 0 to 2π, giving your answers in terms of π.

3 Express $\cos A\cos B$ in terms of $\cos(A + B)$ and $\cos(A - B)$.

(a) Find the exact value of $\cos105°\cos 45°$.

(b) Solve $\cos5x + \cos3x = 0$ for $0° \leqslant x \leqslant 360°$.

4 (a) On the same diagram, over the range $-360°$ to $360°$, sketch the graphs of $y = \cos x$ and $y = \sec x$.

(b) Solve $3\sec^2x = 10\tan x$ for $0° \leqslant x \leqslant 360°$.

(c) Prove that $\dfrac{\cos\theta}{\sec\theta - \tan\theta} \equiv 1 + \sin\theta$.

5 $f(x) \equiv 6\sin x - 8\cos x$.

Given that $f(x) \equiv R\sin(x - \alpha)$, where $R > 0$, $0 \leqslant \alpha \leqslant \frac{\pi}{2}$, and x and α are measured in radians,

(a) find R and the value of α to 2 decimal places.

Hence:

(b) find the minimum value of $f(x)$ and the value of x that gives this minimum.

(c) find the smallest angle x, in radians to 2 decimal places, for which

$$6\sin x - 8\cos x = 4.$$

6 Given that $y = \cos 2x + \sin x$, $0 < x < 2\pi$, and x is in radians, find, in terms of π, the values of x for which $y = 0$.

[Edexcel]

7 **(a)** Starting from the identity for $\cos(A + B)$ prove that

$$\cos 2x = 1 - 2\sin^2 x.$$

Find, in radians to 2 decimal places, the values of x in the interval $0 \leqslant x \leqslant 2\pi$ for which

(b) $2\cos 2x + 1 = \sin x$

(c) $2\cos x + 1 = \sin\frac{1}{2}x$.

[Edexcel]

8 **(a)** Solve the equation

$$2\cos^2 x + 5\sin x + 1 = 0, \qquad 0° \leqslant x \leqslant 360°$$

giving your answer in degrees.

(b) Using the half-angle formulae, or otherwise,

(i) show that $\dfrac{1 - \cos\theta}{\sin\theta} = \tan\frac{1}{2}\theta, \qquad 0 \leqslant \theta \leqslant \pi,$

(ii) solve $\dfrac{1 - \cos\theta}{\sin\theta} = \sqrt{3}\sin\theta, \qquad 0 < \theta < \pi,$

giving your answer in radians to 3 significant figures.

[Edexcel]

9 Given that

$$7\cos\theta + 24\sin\theta \equiv R\cos(\theta - \alpha), \text{ where } R > 0,\ 0° \leqslant \alpha \leqslant 90°,$$

(a) find the values of the constants R and α.

Hence find:

(b) the solutions of the equation $7\cos\theta + 24\sin\theta = 15$ in the range $0° \leqslant \theta \leqslant 360°$

(c) the range of the function $f(\theta)$ where

$$f(\theta) = \frac{1}{5 + (7\cos\theta + 24\sin\theta)^2},\ 0° \leqslant \theta \leqslant 360°.$$

[Edexcel, adapted]

10 $f(\theta) \equiv 9\sin\theta + 12\cos\theta$.

Given that $f(\theta) \equiv R\sin(\theta + \alpha)$, where $R > 0$, $0° \leqslant \alpha \leqslant 90°$

(a) find the values of the constants R and α.

(b) Hence find the values of θ, $0° \leqslant \theta \leqslant 360°$, for which

$$9\sin\theta + 12\cos\theta = -7.5,$$

giving your answers to the nearest tenth of a degree.

(c) Find, in radians in terms of π, the solutions to the equation

$$\sqrt{3}\sin(\theta - \tfrac{1}{6}\pi) = \sin\theta$$

in the interval $0 \leqslant \theta \leqslant 2\pi$.

[Edexcel, adapted]

KEY POINTS

1 Reciprocals

$$\operatorname{cosec}\theta = \frac{1}{\sin\theta} \qquad \sec\theta = \frac{1}{\cos\theta} \qquad \cot\theta = \frac{1}{\tan\theta}$$

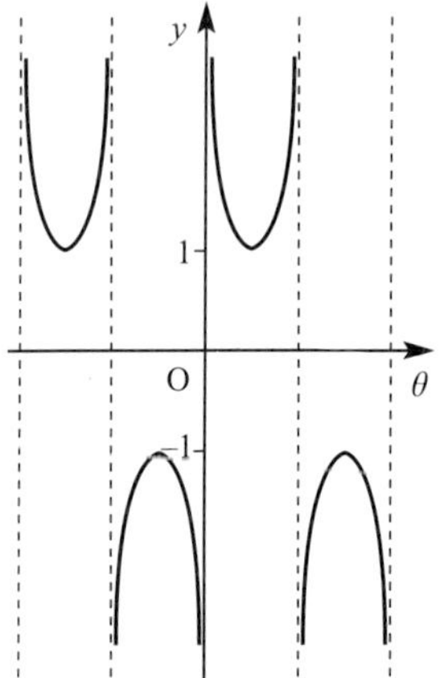

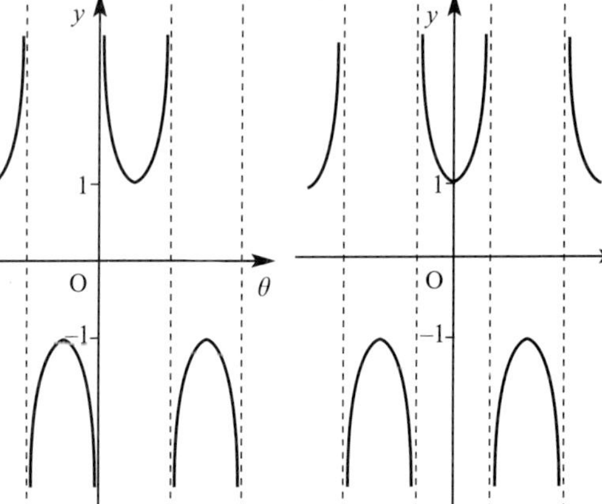

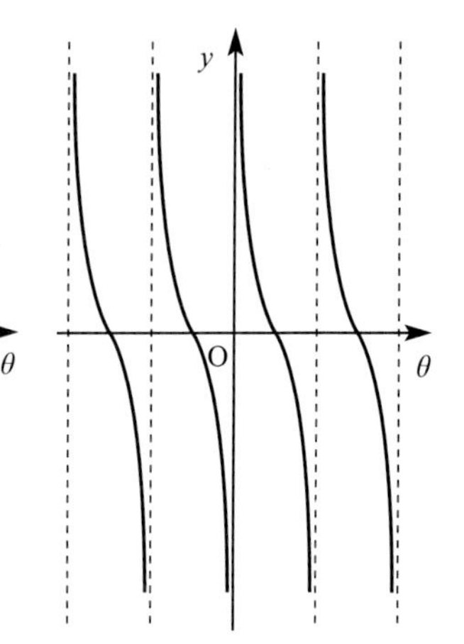

$$\tan\theta = \frac{\sin\theta}{\cos\theta} \qquad \cot\theta = \frac{\cos\theta}{\sin\theta}$$

2 Inverse functions

$\sin^{-1}\theta$ $\cos^{-1}\theta$ $\tan^{-1}\theta$

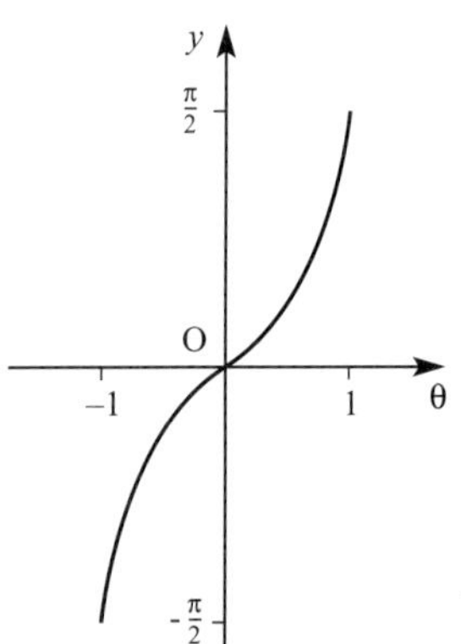

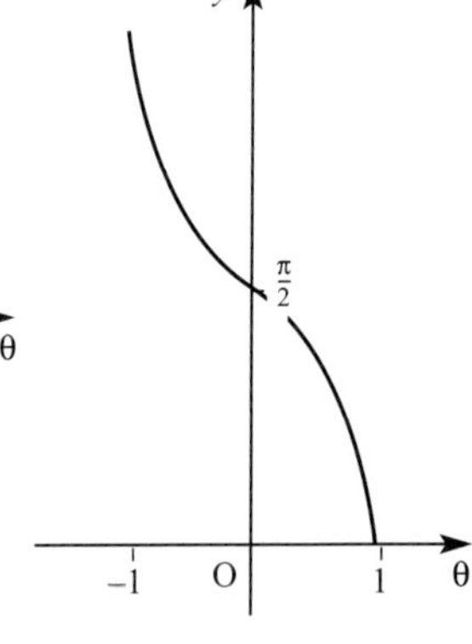

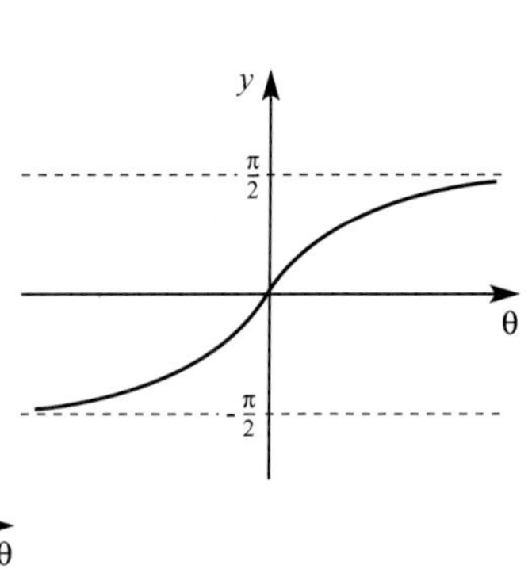

$-\frac{\pi}{2} \leqslant \theta \leqslant \frac{\pi}{2}$ $\qquad 0 \leqslant \theta \leqslant \pi$ $\qquad -\frac{\pi}{2} \leqslant \theta \leqslant \frac{\pi}{2}$

3 Fundamental trigonometric identity

$\sin^2\theta + \cos^2\theta = 1 \quad \tan^2\theta + 1 = \sec^2\theta \quad 1 + \cot^2\theta = \operatorname{cosec}^2\theta$

4 Compound-angle formulae:

$$\sin(A \pm B) = \sin A\cos B \pm \cos A\sin B$$

$$\cos(A \pm B) = \cos A\cos B \pm \sin A\sin B$$

$$\tan(A \pm B) = \frac{\tan A \pm \tan B}{1 \pm \tan A\tan B}$$

5 Double-angle formulae

$$\sin 2\theta = 2\sin\theta\cos\theta$$

$$\cos 2\theta = \cos^2\theta - \sin^2\theta = 2\cos^2\theta - 1 = 1 - 2\sin^2\theta$$

$$\tan 2\theta = \frac{2\tan\theta}{1 - \tan^2\theta}$$

6 *t*-formulae

$\sin\theta = \dfrac{2t}{1+t^2} \quad \cos\theta = \dfrac{1-t^2}{1+t^2} \quad \tan\theta = \dfrac{2t}{1-t^2}$ where $t = \tan\frac{1}{2}\theta$.

7 $a\cos\theta + b\sin\theta$

Let $a\cos\theta + b\sin\theta = r\cos(\theta \pm \alpha)$ or $a\cos\theta + b\sin\theta = r\sin(\theta \pm \alpha)$.

8 Factor formulae

$2\sin A\cos B = \sin(A+B) + \sin(A-B) \qquad \sin\alpha + \sin\beta = 2\sin\left(\frac{\alpha+\beta}{2}\right)\cos\left(\frac{\alpha-\beta}{2}\right)$

$2\cos A\sin B = \sin(A+B) - \sin(A-B) \qquad \sin\alpha + \sin\beta = 2\cos\left(\frac{\alpha+\beta}{2}\right)\sin\left(\frac{\alpha-\beta}{2}\right)$

$2\cos A\cos B = \cos(A+B) + \cos(A-B) \qquad \cos\alpha + \cos\beta = 2\cos\left(\frac{\alpha+\beta}{2}\right)\cos\left(\frac{\alpha-\beta}{2}\right)$

$-2\sin A\sin B = \cos(A+B) - \cos(A-B) \qquad \cos\alpha - \cos\beta = -2\sin\left(\frac{\alpha+\beta}{2}\right)\cos\left(\frac{\alpha-\beta}{2}\right).$

Chapter five

EXPONENTIALS AND LOGARITHMS

Prophesy upon these bones, and say unto them, O ye dry bones.

Bible

THE FUNCTION e^x

Think of a pond with weed growing on its surface. Initially let $1\,m^2$ of the surface be covered with weed and for the next few days, let the area double each day. Then after one day there will be $2\,m^2$ of weed, after two days there will be $4\,m^2$ of weed and so on. The pond weed graph would look like the solid curve shown in figure 5.1.

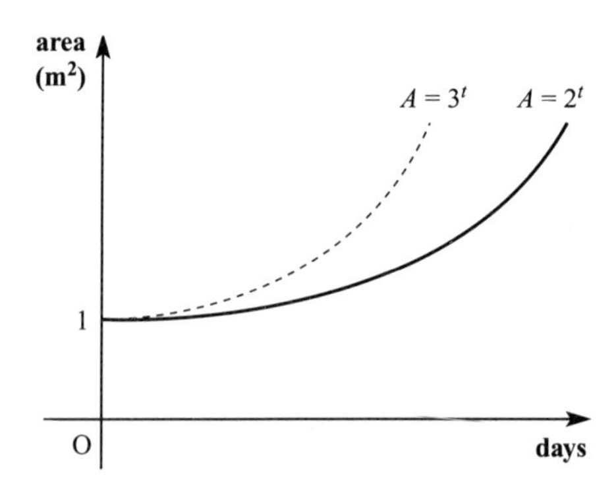

FIGURE 5.1

If A is the area and t the number of days then the equation connecting A and t is

$$A = 2^t.$$

t is the exponent of 2 and the expression is an *exponential equation*.

If the weed had trebled in area each day then the equation would have been $A = 3^t$ and the curve would be like the dotted one shown in figure 5.1.

Both of these functions grow at an ever-increasing rate and this is describe as *exponential growth*.

The functions are both examples of the general one

$y = a^x$ with $a > 1$.

The graphs of these functions can be extended for negative values of x. Curves of this type all have the shape shown in figure 5.2

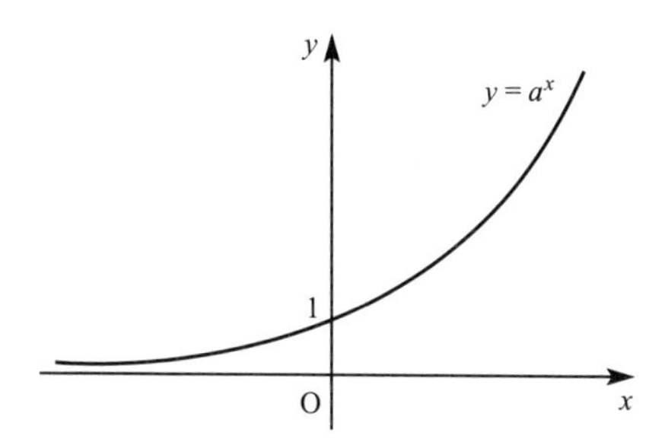

FIGURE 5.2

There is a particulary interesting graph of this type. At every point on its curve the gradient of the graph is equal to the value of y. (You will recall that to find the gradient you draw a tangent to the curve.) From *Pure Mathematics 1* the gradient is the derivative of the function.

Writing the equation of this particular graph in terms of x and y gives the exponential function $y = e^x$ and its graph looks like that in figure 5.3.

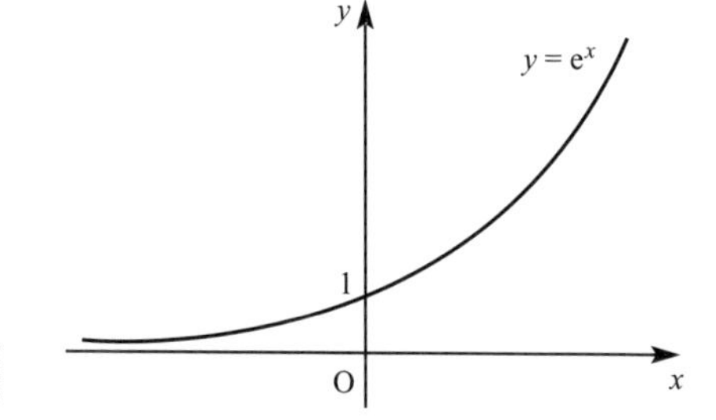

FIGURE 5.3

Note that the curve cuts the y axis at 1

Since its gradient equals its y value then $\frac{dy}{dx} = e^x$ (more of which you will read about in Chapter 6).

Note In some books e^x is written as $\exp(x)$.

If we reflect the graph of $y = e^x$ in the y axis it looks like that shown in figure 5.4.

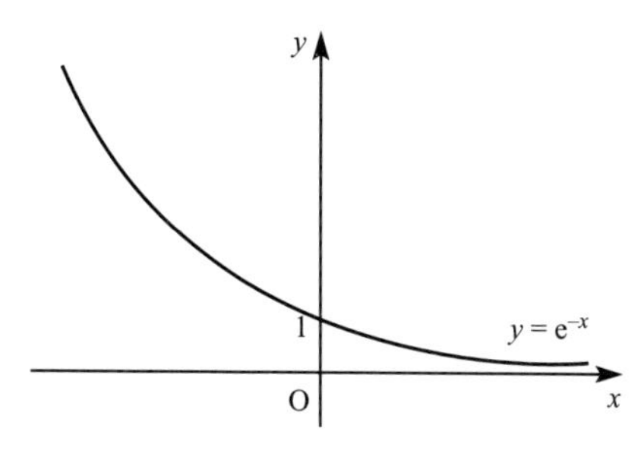

FIGURE 5.4

The equation for this graph is $y = e^{-x}$ and it describes *exponential decay*. You may have studied the half life of radioactive substances which is an example of exponential decay.

Note If $0 < a < 1$ the graph of $y = a^x$ will take the same shape as $y = e^{-x}$ and you may wish to work out why. Also, think about what the graph would look like if $a = 1$.

Historical Note The symbol e is used after Leonhard Euler (1707–83). Euler is often thought of as a hero by mathematicians. He was highly productive throughout the eighteenth century. Although he was Swiss by birth he worked at St Peterburg's Academy in Russia. Despite becoming blind in 1766 he continued to work by dictation and he was noted for his phenomenal memory. Euler established the use of the symbols π, ∞, i (for $\sqrt{-1}$) and e. He combined some of these in the statement

$$e^{i\pi} + 1 = 0$$

which must be one of the most exciting equations imaginable.

e is rather like π in that it is an irrational number. It is therefore a never-ending decimal and you will see from your calculator that e^1 has the value

$$e = 2.182\ 818\ 28.$$

(On most calculators you will have to press the keys e^x then 1. Do not use the power key.)

You will notice that the graphs of $y = e^x$ and $y = e^{-x}$ are both asymptotic to the x axis (that is they get closer and closer to the axis but do not quite reach it). Consequently if you translate the graph in the y direction you should draw a dotted line to indicate the position of the asymptote.

EXAMPLE 5.1

Sketch the graph of $y = e^x - 1$.

Solution This is a translation of $y = e^x$ by $\begin{pmatrix} 0 \\ -1 \end{pmatrix}$. Since $y = e^x$ cuts the y axis at 1 this graph will go through the origin when it is moved down one unit (see figure 5.5).

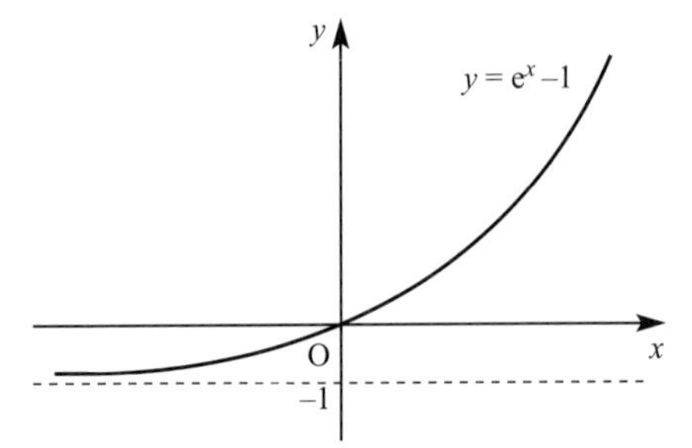

FIGURE 5.5

EXAMPLE 5.2

Sketch the graph of $y = 2 - e^{-x}$.

Solution From Chapter 2 you will recognise that this is a reflection of $y = e^{-x}$ in the x axis followed by a translation $\begin{pmatrix} 0 \\ 2 \end{pmatrix}$. See figure 5.6.

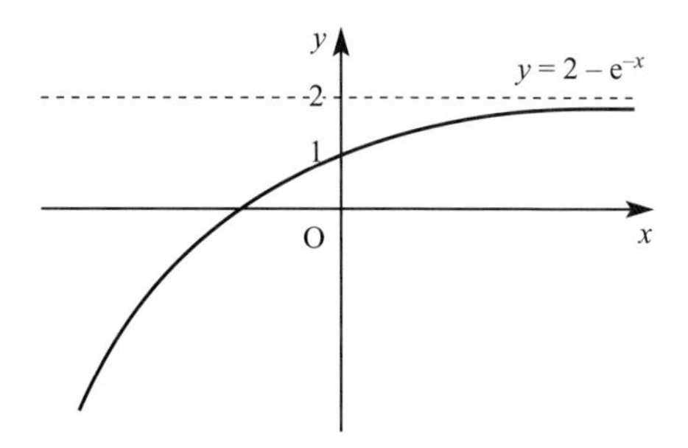

FIGURE 5.6

THE FUNCTION ln$x$

If the graph of $y = e^x$ is reflected in the line $y = x$ you obtain its inverse function as shown in figure 5.7.

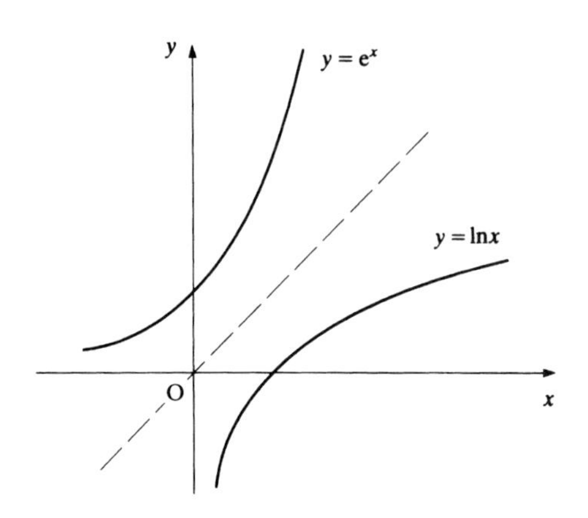

FIGURE 5.7

The inverse function is the natural logarithm $y = \ln x$. Since the domain of $y = e^x$ was $x \in \mathbb{R}$ and the range was $y > 0$ then the domain of $y = \ln x$ is $x > 0$ with a range $y \in \mathbb{R}$.

Note On most calculators the functions e^x and $\ln x$ are obtained using the same key.

ln is an abbreviation for the logarithm of x to the base e. This is written as

$$\ln x = \log_e x.$$

You will see the importance of natural logarithms in Chapters 6 and 7.

You can find the inverse of $y = a^x$ in exactly the same way to give the function

$$y = \log_a x.$$

This is said as 'the log of x to the base a'. The most familiar and frequently used logarithm is the inverse of $y = 10^x$ which is

$$y = \log_{10} x.$$

This is abbreviated to $y = \log x$ and you will find a key for this on your calculator labelled 'log'.

Whatever the value of a (the base of the logarithm), provided $a > 1$, the graph of $y = \log_a x$ has the same general shape (see figure 5.8)

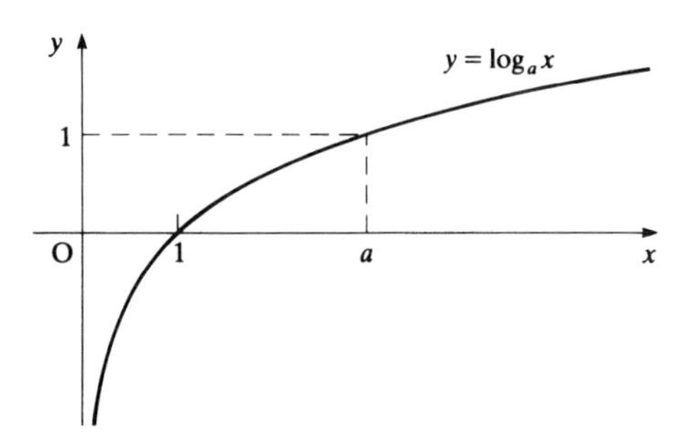

FIGURE 5.8

Note that:

- the curve crosses the x axis at 1
- it passes through the point $(a, 1)$
- it only exists for $x > 0$
- the y axis is an asymptote
- there is no limit to the height of the curve for large values of x though the gradient progressively decreases.

The function $y = \log_a x$ is also the inverse of $y = a^x$.

EXAMPLE 5.3

Sketch the graph of $y = \ln(x - 3)$.

Solution This is a translation of $y = \ln x$ by $\begin{pmatrix} 3 \\ 0 \end{pmatrix}$. The graph is shown in figure 5.9.

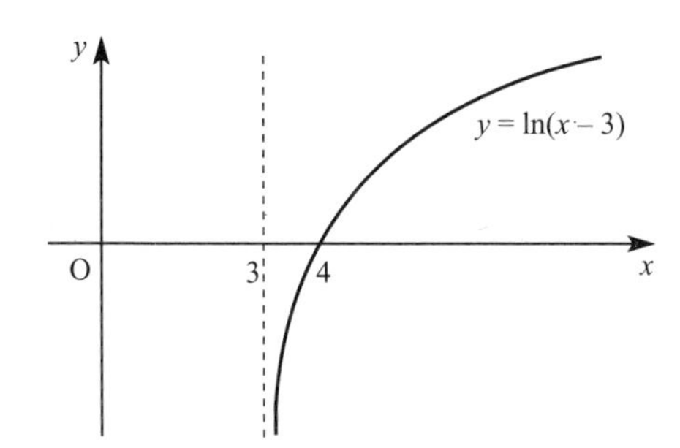

FIGURE 5.9

EXAMPLE 5.4

The function f(x) is defined by

$$f(x) \mapsto e^x + 2 \quad x \in \mathbb{R}$$

(a) Sketch the graph of $y = f(x)$.
(b) Sketch the graph of $y = f^{-1}(x)$.
(c) Find the inverse function $y = f^{-1}(x)$.
(d) State the domain of $f^{-1}(x)$.

Solution

(a)
(b)

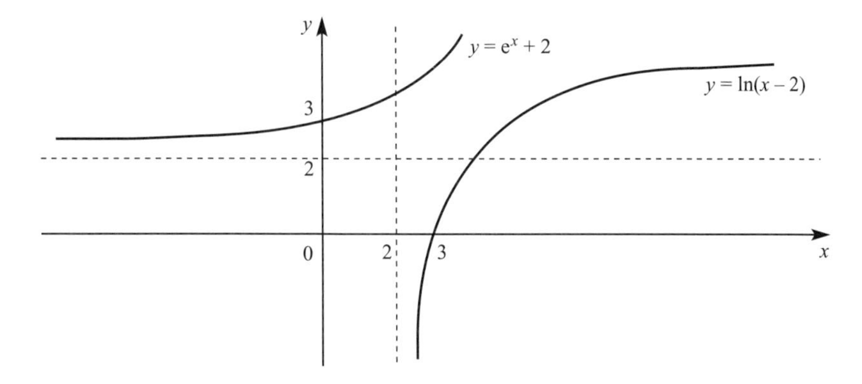

FIGURE 5.10

(c) The inverse function is $f^{-1}(x) = \ln(x - 2)$.
(d) The domain of $f^{-1}(x)$ is $x > 2$.

Historical note

Logarithms were discovered independently by John Napier (1550–1617), who lived at Merchiston Castle in Edinburgh, and Jolst Bürgi (1552–1632) from Switzerland. It is generally believed that Napier had the idea first, and so he is credited with their discovery. Natural logarithms are also called Naperian logarithms but there is no basis for this since Napier's logarithms were definitely not the same as natural logarithms. Napier was deeply involved in the political and religious events of his day and mathematics and science were little more than hobbies for him. He was a man of remarkable ingenuity and imagination and also drew plans for war chariots that look very like modern tanks, and for submarines.

EXERCISE 5A

The questions in this exercise refer to sketching curves of exponential and logarithmic functions. Use your knowledge of tranformations from Chapter 2 to make your drawings. You may then wish to check your answers using a graphical calculator or a computer drawing package.

1 On separate diagrams sketch

(a) $y = e^x$ **(b)** $y = -e^x$

(c) $y = e^{-x}$ **(d)** $y = -e^{-x}$.

2 On separate diagrams sketch

(a) $y = \ln x$ **(b)** $y = \ln(-x)$

(c) $y = -\ln x$ **(d)** $y = -\ln(-x)$.

3 On separate diagrams sketch the following curves showing asymptotes by drawing dotted lines

(a) $y = e^x + 1$ **(b)** $y = -e^x + 1$

(c) $y = 2e^x - 1$ **(d)** $y = e^{x+1}$.

4 On separate diagrams sketch

(a) $y = \ln(x - 2)$ **(b)** $y = \ln(x + 3)$

(c) $y = \ln(l - x)$ **(d)** $y = 2 + \ln x$.

5 **(a)** On the same diagram sketch $y = 2^x$ and $y = 4^x$.

(b) Describe a transformation that maps $y = 2^x$ onto $y = 4^x$.

6 The function $f(x) = 10^x$.

(a) On the same diagram sketch the graph of $y = f(x)$ and the graph of $y = f^{-1}(x)$.

(b) Describe the geometrical relationship between $f(x)$ and $f^{-1}(x)$.

(c) Write down the function $f^{-1}(x)$.

7 Draw a sketch of $y = \log_3 x$. On the same diagram sketch the inverse of this function and write down its equation.

8 The function $f(x)$ is given by

$$f: x \mapsto \ln(x - 3), \quad x > 3.$$

(a) Sketch the graph of $y = f(x)$.

(b) Find $f^{-1}(x)$ and sketch the graph of $y = f^{-1}(x)$.

(c) State the range of $f^{-1}(x)$.

EQUATIONS OF THE FORM $e^x = p$ AND $\ln x = q$

You have seen that the functions $y = e^x$ and $y = \ln x$ are inverses of each other. So you use this fact to solve equations of the form $e^x = p$ and $\ln x = q$.

The proof is that:

since $f(f^{-1}(x)) = x$ then $\ln(e^x) = x$ and $e^{\ln x} = x$

and so
if $e^x = p$ then taking natural logarithms of both sides gives

$$\ln(e^x) = \ln p$$
$$\therefore x = \ln p$$

and if $\ln x = q$ taking the exponential of both sides gives

$$e^{\ln x} = e^q$$
$$\therefore x = e^q.$$

EXAMPLE 5.5

Solve $e^x = 5$.

Solution Taking natural logarithms of both sides of $e^x = 5$ gives

$$x = \ln 5$$

and the answer may be left in this exact form or, using a calculator, could be given to, say, 4 significant figures as

$$x = 1.609.$$

EXAMPLE 5.6

Solve $\ln x = 5$.

Solution Taking the exponential of both sides of $\ln x = 5$ gives

$$x = e^5$$

and this answer may be left in this form or approximated using your calculator, again to 4 significant figures as

$$x = 148.4.$$

Reminder: key in e^x followed by 5. Do not use the power key

EXAMPLE 5.7

The number, N, of insects in a colony is given by $N = 2000e^{0.1t}$ where t is the number of days after observations have begun.

(a) Sketch the graph of N against t.

(b) What is the population of the colony after 20 days?

(c) How long does it take the colony to reach a population of 10 000?

Solution **(a)**

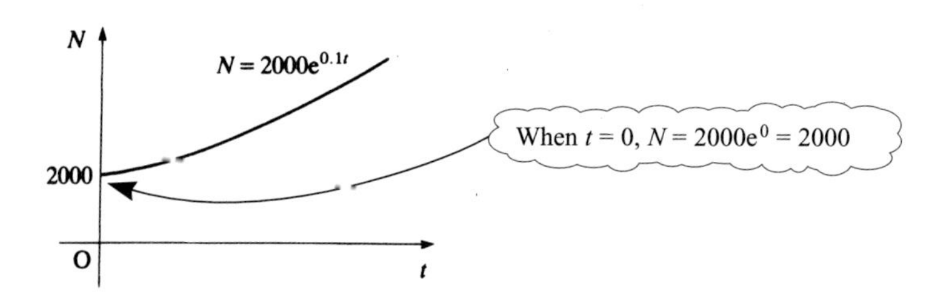

FIGURE 5.11

(b) When $t = 20$, $N = 2000e^{0.1\times20}$

$$= 14\,778.$$

The population is 14 778 insects.

(c) When $N = 10\,000$, $10\,000 = 2000e^{0.1t}$

$$5 = e^{0.1t}.$$

Taking natural logarithms of both sides,

$$\ln 5 = \ln(e^{0.1t})$$

$$= 0.1t$$

Remember $\ln(e^x) = x$

and so $t = 10\ln 5$

$$t = 16.09\ldots$$

It takes just over 16 days for the population to reach 10 000.

EXAMPLE 5.8

The radioactive mass, M g in a lump of material is given by $M = 25e^{-0.0012t}$ where t is the time in seconds since the first observation.

(a) Sketch the graphs of M against t.

(b) What is the initial size of the mass?

(c) What is the mass after 1 hour?

(d) The half-life of a radioactive substance is the time it takes to decay to half of its mass. What is the half-life of this material?

Solution **(a)**

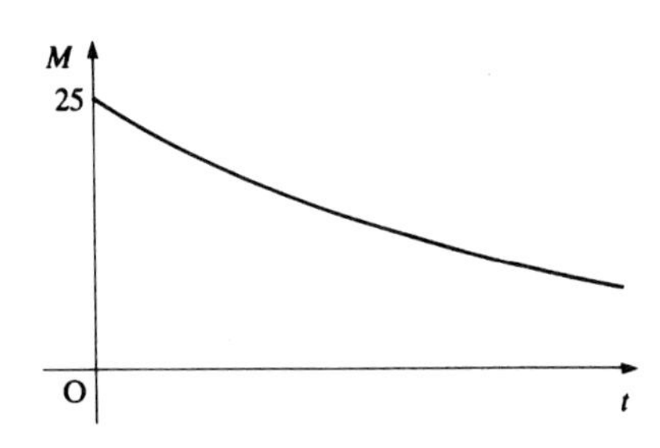

FIGURE 5.12

(b) When $t = 0$, $M = 25e^0 = 25$.

The initial mass is 25g.

(c) After one hour, $t = 3600$ $M = 25e^{-0.0012\times3600}$

The mass after one hour is 0.33g (to 2 decimal places).

(d) The initial mass is 25g, so after one half-life

$$M = \tfrac{1}{2} \times 25 = 12.5\text{g}.$$

At this point the value of t is given by

$$12.5 = 25e^{-0.0012t}.$$

Dividing both sides by 25 gives

$$0.5 = e^{-0.0012t}.$$

Taking logarithms of both sides

$$\ln 0.5 = \ln e^{-0.0012t} = -0.0012t$$

$$\Rightarrow \quad t = \frac{\ln 0.5}{-0.0012} = 557.6 \text{ (to 1 decimal place).}$$

The half-life is 577.6 seconds. (This is just under 10 minutes, so the substance is highly radioactive.)

EXERCISE 5B

1 Solve the following equations giving your answers to 3 significant figures:

(a) $e^x = 2$ **(b)** $e^{(x-1)} = 8$

(c) $e^{(x+3)} = 100$ **(d)** $50e^x = 1$

(e) $100e^x = 2 \times 10^6$ **(f)** $0.5e^{-x} = 10$

(g) $e^{(2-x)} = 0.5$ **(h)** $2.5e^{(3-2x)} = 4$

(i) $0.1e^{0.2x} = 7$ **(j)** $e^{5x} = 2e^{2x}$

2 Solve the following equations giving your answers to 3 significant figures:

(a) $\ln x = 2$ **(b)** $\ln(x - 3) = 0.3$

(c) $\ln(x + 5) = 5$ **(d)** $\ln(2x - 1) = 1.5$

(e) $\ln x = 0.04$ **(f)** $\ln(1 - x) = 0.1$

(g) $25\ln x = 8$ **(h)** $2\ln x = 0.75$

(i) $4\ln(0.2x - 1) = 9$ **(j)** $3\ln(2 - 3x) = 0.5$

3 Make p the subject of $\ln\left(\frac{p}{25}\right) = -0.02t$.

4 Make x the subject of $y - 5 = (y_0 - 5)e^x$.

5 A colony of humans settles on a previously uninhabited planet. After t years, their population, P, is given by

$$P = 100e^{0.05t}.$$

(a) Sketch the graph of P against t.

(b) How many settlers land on the planet initially?

(c) What is the population after 50 years?

(d) How long does it take the population to reach 1 million?

6 Ela sits on a swing. Her father pulls it back and then releases it. The swing returns to its maximum backwards displacement once every 5 seconds, but the maximum displacement, $\theta°$, becomes progressively smaller because of friction. At time t seconds, θ is given by

$$\theta = 25e^{-0.03t} \ (t = 0, 5, 10, 15, \ldots).$$

(a) Plot the values of θ for $0 \leqslant t \leqslant 30$ on graph paper.

(b) To what angle did Ela's father pull the swing?

(c) What is the value of θ after one minute?

(d) After how many swings is the angle θ less than 1°?

7 Alexander lives 800 metres from school. One morning he sets out at 8.00 am and t minutes later the distance s m, which he has walked is given by

$$s = 800\,(1 - e^{-0.1t}).$$

(a) Sketch the graph of s against t.

(b) How far has Alexander walked by 8.15 am?

(c) What time is it when Alexander is half-way to school?

(d) When does Alexander get to school?

8 A parachutist jumps out of an aircraft and some time later opens the parachute. His speed at time t seconds from when the parachute opens is v ms^{-1}. It is given by

$$v = 8 + 22e^{-0.07t}$$

(a) Sketch the graph of v against t.

(b) State the speed of the parachutist when the parachute opens, and the final speed that he would attain if he jumped from a very great height.

(c) Find the value of v as the parachutist lands, 60 seconds later.

(d) Find the value of t when the parachutist is travelling at 20 ms^{-1}.

9 A bacterium *mathematicus estus funius* grows such that its population P at time t is given by

$$P = 250e^{0.2t}$$

(a) What is the initial population (i.e. when $t = 0$)?

(b) Find the value of t when the population is four times the initial population.

(c) This is an example of exponential growth. Explain what is meant by exponential decay and draw a graph to illustrate your answer.

10 The manufacturers of the cream Acno claim that the bacteria *spoticus youthus* has its population halved within five days of its use. In his laboratory Professor Smiff finds that the number of bacteria N is connected by the number of days t by the model

$$N = 5000e^{-0.15t}.$$

Are the manufacturers accurate in their claim?

According to Professor Smiff's model

(a) How many *spoticus youthus* bacteria will there be after 10 days?

(b) After how many days will there only be 100 *spoticus youthus* bacteria left?

THE LAWS OF LOGARITHMS

You have seen that $f(x) = \log_a x$ and $g(x) = a^x$ are inverse functions so it follows that

$$n = a^x \Leftrightarrow x = \log_a n.$$

The proof is

$$\text{let } n = a^x$$

taking logarithms to the base a of both sides gives $\log_a n = \log_a a^x = x$.

EXAMPLE 5.9

Express $64 = 4^3$ in logarithmic form.

Solution Taking the logarithm to the base 4 of both sides gives $\log_4 64 = \log_4 4^3 = 3$.

EXAMPLE 5.10

Express $\log_a p = q$ in exponential form.

Solution Writing both sides as powers of a gives $a^{\log_a p} = p = a^q$.

You are now in a position to develop the rules of logarithms.

Let $x = a^p$ so that $p = \log_a x$ and let $y = a^q$ so that $q = \log_a y$. Then

multiplying: $$xy = a^p \times a^q = a^{p+q}$$

$$\therefore \log_a xy = \log_a a^{p+q} = p + q = \log_a x + \log_a y$$

dividing: $$\frac{x}{y} = \frac{a^p}{a^q} = a^{p-q}$$

$$\therefore \log_a \frac{x}{y} = p - q = \log_a x - \log_a y$$

powers: $$x^n = (a^p)^n = a^{pn}$$

$$\therefore \log_a x^n = pn = n\log_a x.$$

Further, since $a^0 = 1$ $(a \neq 0)$ and taking logarithms to the base a

$$\log_a 1 = 0$$

which was a result seen on the graph of $y = \log_a x$.

Also $a^1 = a$ which, after taking logarithms, gives the result

$$\log_a a = 1.$$

The laws of logarithms are summarised in the following box:

$$n = a^x \Leftrightarrow x = \log_a n$$
$$\log_a xy \equiv \log_a x + \log_a y$$
$$\log_a \frac{x}{y} \equiv \log_a x - \log_a y$$
$$\log_a x^n \equiv n\log_a x$$
$$\log_a 1 = 0$$
$$\log_a a = 1$$

EXAMPLE 5.11

Write $\log_3(x + 1) - \log_3 x$ as a single logarithm, hence solve

$$\log_3(x + 1) - \log_3 x = 4.$$

Solution When subtracting logarithms you divide:

$$\log_3(x + 1) - \log_3 x = \log_3\left(\frac{x + 1}{x}\right)$$

then

$$\log_3\left(\frac{x + 1}{x}\right) = 4$$

$$\therefore \frac{x + 1}{x} = 3^4 = 81$$

re-arranging and solving gives

$$x + 1 = 81x$$

$$\therefore x = \frac{1}{80}.$$

EXAMPLE 5.12

Write down the values of

(a) $\log_5 125$ **(b)** $\log_4(\frac{1}{16})$

without using a calculator.

Solution Express the numbers as powers of the base:

(a) $\log_5 125 = \log_5 5^3$

$= 3 \log_5 5$ using $\log_a x^n \equiv n\log_a x$

$= 3$ using $\log_a a \equiv 1$.

(b) $\log_4(\frac{1}{16}) = \log_4 4^{-2}$

$= -2.$

EQUATIONS OF THE FORM $a^x = b$

Equations of this type are solved using the logarithm rule that

$$\log_a x^n \equiv n\log_a x$$

and the method is illustrated in the following examples.

EXAMPLE 5.13

Solve the equation $2^n = 1000$.

Solution Taking logarithms to the base 10 of both sides (since these can be found on a calculator),

$$\log(2^n) = \log 1000$$

$$n\log 2 = \log 1000$$

$$n = \frac{\log 1000}{\log 2}$$

$$= 9.97 \text{ to 3 significant figures.}$$

Remember $\log_{10} \equiv \log$

Alternatively you could take the natural logarithms of both sides. You can check that the same answer is obtained whether you use log or ln.

EXAMPLE 5.14

Solve the equation $3^{2x} - 14 \times 3^x + 45 = 0$.

Solution You should recognise this as a quadratic in 3^x, so let $y = 3^x$ then $y^2 = 3^{2x}$ and the equation becomes

$$y^2 - 14y + 45 = 0$$

factorising $(y - 9)(y - 5) = 0$

solving $y = 3^x = 9$ or $y = 3^x = 5$.

By inspection, the first solution is $x = 2$ but to find the second solution you have to take logarithms:

$$\log 3^x = \log 5$$

$$x\log 3 = \log 5$$

$$\therefore x = \frac{\log 5}{\log 3} = 1.465 \text{ to 3 decimal places.}$$

EXERCISE 5C

1 Write the following in logarithmic form.

(a) $81 = 3^4$ **(b)** $256 = 2^8$

(c) $l = a^b$ **(d)** $x = y^z$

2 Write the following in exponential form.

(a) $\log_{10}10\,000 = 4$ **(b)** $\log_6 216 = 3$

(c) $q = \log_c b$ **(d)** $x = \log_n y$

3 Write the following in terms of $\log p$, $\log q$ and $\log r$.

(a) $\log\frac{pq}{r}$ **(b)** $\log\frac{p}{qr}$

(c) $\log p^2 q^3$ **(d)** $\log\frac{p \times \sqrt[3]{r}}{\sqrt{q}}$

4 Write the following expressions in the form $\log x$ where x is a number.

(a) $\log 5 + \log 2$ **(b)** $\log 6 - \log 3$

(c) $2\log 6$ **(d)** $-\log 7$

(e) $\frac{1}{2}\log 9$ **(f)** $\frac{1}{4}\log 16 + \log 2$

(g) $\log 5 + 3\log 2 - \log 10$ **(h)** $\log 12 - 2\log 2 - \log 9$

(i) $\frac{1}{2}\log\sqrt{16} + 2\log\left(\frac{1}{2}\right)$ **(j)** $2\log 4 + \log 9 - \frac{1}{2}\log 144$

5 Write down the values of the following without using a calculator.
Use your calculator to check your answers for those questions which use base 10.

(a) $\log_{10}1000$ **(b)** $\log_{10}\left(\frac{1}{10\,000}\right)$

(c) $\log_{10}\sqrt{10}$ **(d)** $\log_{10}1$

(e) $\log_3 81$ **(f)** $\log_3\left(\frac{1}{81}\right)$

(g) $\log_3\sqrt{27}$ **(h)** $\log_3\sqrt[4]{3}$

(i) $\log_4 2$ **(j)** $\log_5\left(\frac{1}{125}\right)$

6 Use logarithms to the base 10 to solve the following equations.

(a) $2^x = 1\,000\,000$ **(b)** $2^x = 0.001$

(c) $1.08^x = 2$ **(d)** $1.1^x = 100$

(e) $0.99^x = 0.000\,001$ **(f)** $2^{x+1} = 50$

(g) $5^{x-3} = 150$ **(h)** $3^{-0.2x} = 0.1$

(i) $5^x = 1\,000\,000$ **(j)** $5000 \times 4^{0.1x} = 2500$

7 Solve

(a) $4 \times 2^{2x} - 13 \times 2^x + 3 = 0$

(b) $5^{2x} - 23 \times 5^x - 50 = 0.$

8 Express $\log_3(x + 1) - \log_3(2x)$ as a single logarithm.
Hence find the exact value of x if $\log_3(x + 1) - \log_3(2x) = 2$.

9 Given that $f(x) = 6^x - 5$:

(a) sketch $y = |f(x)|$

(b) solve $|f(x)| < 2$.

10 At time t the population P is given by

$$t = 50(\ln P_0 - \ln P).$$

(a) Find an expression for P in terms of t.

(b) Find the value of P_0 given that when $t = 0$ the population is 1000.

(c) Show that when the value of t is 110 the population P is approximately 110.

EXERCISE 5D

Examination-style questions

1 **(a)** Express $\log_2(x + 4) - \log_2(x - 1)$ as a single logarithm.

(b) Solve $\log_2(x + 4) - \log_2(x - 1) = 4$.

(c) Solve $\log_2(x + 4) - \log_2(x - 1) = 2\log_2 3$.

2 **(a)** Solve $5^x = 10$ giving your answer to 3 significant figures.

(b) By making a suitable substitution, or otherwise, solve

$$5^{2x} - 25 \times 5^x + 150 = 0$$

giving your answers to 3 significant figures.

3 The population S of a sample of buglettes is given by the model

$$S = 6000\,e^{0.01t}$$

where t is the time in hours for which they have been growing.

(a) What is the initial population?

(b) Calculate the time to 3 significant figures when the population is three times its initial size.

A different strain of buglette is found to double its population P every 15 hours. If its initial population is P_0 find an equation connecting P, P_0 and t (the time in hours). Hence, or otherwise, find the number of hours it takes for its population to increase by factor of 100.

4 **(a)** On the same diagram draw the graphs of

$$y = \ln x \quad \text{and} \quad y = \ln(x - 3).$$

Describe the transformation that maps the first graph onto the second.

(b) $y = e^x$ is translated by moving it 2 units to the left. Write down the equation of the image.

State an alternative transformation which maps $y = e^x$ onto the image.

5 The number N of sproglets at time t (in weeks) is given by the equation

$$N = \frac{1300}{5 + 8e^{-0.2t}}.$$

(a) Explain why the graph of this function (shown below) indicates that the growth of sproglets is not exponential?

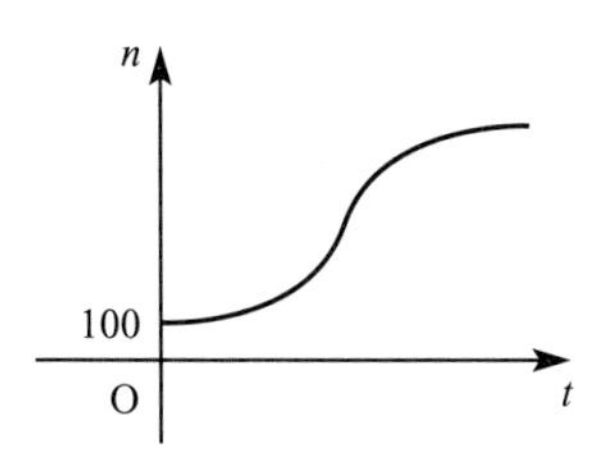

(b) How many sproglets are there after 5 weeks?

(c) After how many weeks are there 200 sproglets?

(d) What is the largest the population of sproglets can be?

6 Given that $y = 10^x$, show that:

(a) $y^2 = 100^x$

(b) $\frac{y}{10} = 10^{x-1}$.

(c) Using the results from parts (a) and (b) write the equation

$$100^x - 10\,001(10^{x-1}) + 100 = 0$$

as an equation in y.

(d) By first solving the equation in y, find the values of x which satisfy the given equation in x.

[Edexcel]

7 The function f is defined by $\text{f} : x \mapsto e^x + k$, $x \in \mathbb{R}$ and k is a positive constant.

(a) State the range of f.

(b) Find $\text{f}(\ln k)$, simplifying your answer.

(c) Find f^{-1}, the inverse function of f, in the form $\text{f}^{-1} : x \mapsto \ldots$, stating its domain.

(d) On the same axes, sketch the curves with equations $y = \text{f}(x)$, and $y = \text{f}^{-1}(x)$, giving the coordinates of all points where the graphs cut the axes.

[Edexcel]

8 The points P and Q lie on the curve with equation $y = e^{\frac{1}{2}x}$. The x coordinates of P and Q are $\ln 4$ and $\ln 16$ respectively.

(a) Find an equation of the line PQ.

(b) Show that this line passes through the origin O.

(c) Calculate the length, to 3 significant figures, of the line segment PQ.

[Edexcel]

9 A formula used to calculate the power gain of an amplifier has the form

$$G = h\ln\left(\frac{p_2}{p_1}\right)$$

Given that $G = 16$, $h = 4.3$ and $p_1 = 8$,

(a) calculate, to the nearest whole number, the value of p_2.

Given that the values of G and p_1 are exact but that the value of h has been given to 1 decimal place,

(b) find the range of possible values of p_2.

[Edexcel]

10 Given that $p = \log_b a$, $q = \log_c b$ and $r = \log_a c$ then

(a) Write $p = \log_b a$ in exponential form.

(b) Show that $p = \dfrac{\log a}{\log b}$.

(c) Using the results from parts (a) and (b) prove that

$$pqr = 1.$$

(d) Use the results of parts (a) and (b) to evaluate $\log_3 5$ to 3 significant figures.

KEY POINTS

1 $y = e^x$

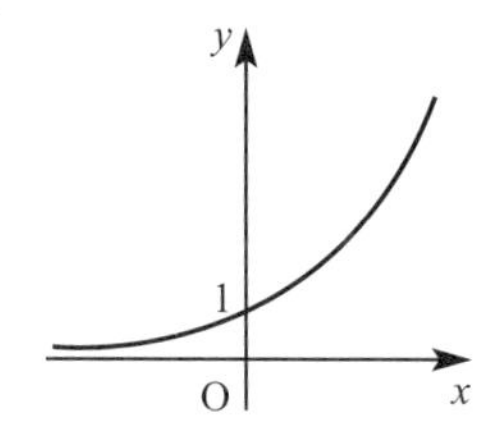

$y = e^{-x}$

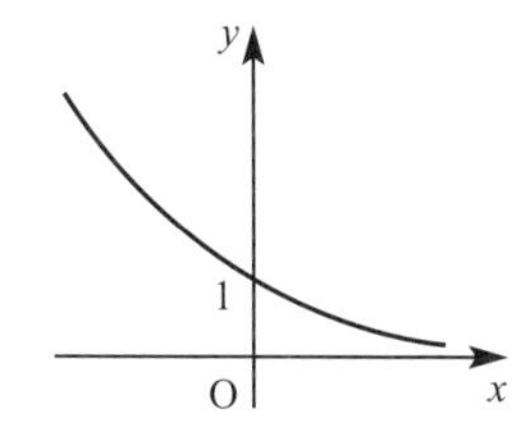

$y = \ln x$

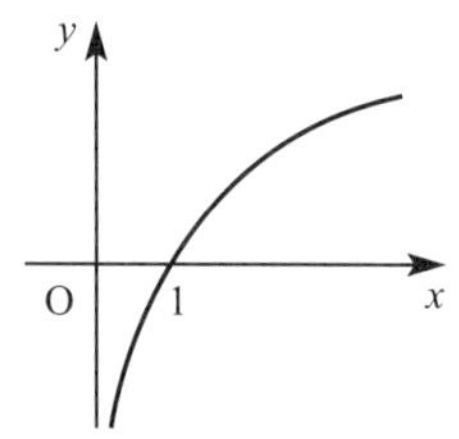

e^x and $\ln x$ are inverse functions

$e^{\ln x} = x$

$\ln(e^x) = x$

2 Laws of logarithms

$n = a^x \Leftrightarrow x = \log_a n$

$\log_a xy \equiv \log_a x + \log_a y$

$\log_a \dfrac{x}{y} \equiv \log_a x - \log_a y$

$\log_a x^n \equiv n\log_a x$

$\log_a 1 = 0$

$\log_a a = 1$

Chapter six

DIFFERENTIATION

The note I wanted, that of the strange and sinister, embroidered on the very type of the normal and easy.

Henry James

DERIVATIVE OF e^x

In Chapter 5 we defined the curve $y = e^x$ to be one for which the gradient at any point on the curve, say (x_1, y_1), is equal to the y value at that point. The graph of $y = e^x$ is shown in figure 6.1 with a tangent drawn at the point (x_1, y_1).

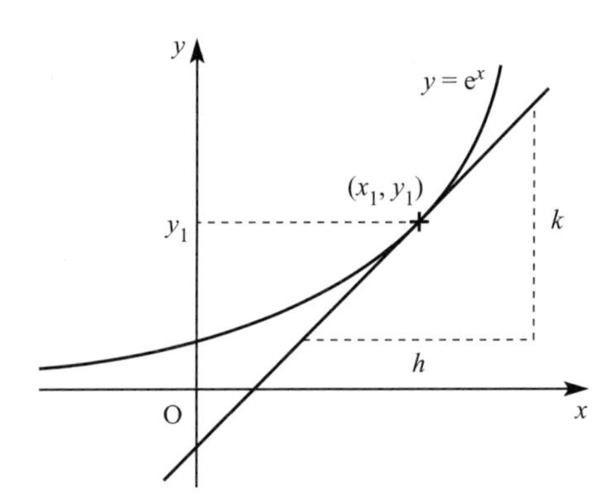

FIGURE 6.1

From figure 6.1 you can see that the gradient of the tangent is $\frac{k}{h} = y_1 = e^{x_1}$ and, in general, the gradient or derivative of $y = e^x$ is $\frac{dy}{dx} = e^x$, i.e.,

$$y = e^x \Rightarrow \frac{dy}{dx} = e^x.$$

If the curve is stretched by a constant scale factor k parallel to the y axis, with the x axis invariant, then both the y value and the gradient at each point will be multiplied by a factor of k, so that

$$y = ke^x \Rightarrow \frac{dy}{dx} = ke^x.$$

Note The derivative of $y = e^{kx}$ and the derivative of $y = e^{ax+b}$ will be studied in *Pure Mathematics 3*.

EXAMPLE 6.1

Find the derivative of $y = 5e^x$.

Solution Using $y = ke^x \Rightarrow \frac{dy}{dx} = ke^x$ means that

if $y = 5e^x$

then $\frac{dy}{dx} = 5e^x$.

EXAMPLE 6.2

Show that the gradient of $e^x + \sqrt{x}$ at the point where $x = 1$ is $\frac{1}{2}(2e + 1)$.

Solution Let $y = e^x + \sqrt{x}$

then $y = e^x + x^{\frac{1}{2}}$.

Differentiating $\frac{dy}{dx} = e^x + \frac{1}{2}x^{-\frac{1}{2}} = e^x + \frac{1}{2}\frac{1}{\sqrt{x}}$.

At the point where $x = 1$

$$\frac{dy}{dx} = e^1 + \frac{1}{2}\frac{1}{\sqrt{1}} = e + \frac{1}{2} = \frac{2e}{2} + \frac{1}{2}$$

$$= \frac{1}{2}(2e + 1).$$

DERIVATIVE OF ln$x$

You will recall from Chapter 5 that ln$x$ is the inverse of e^x. So if you take the graph of $y = e^x$ and reflect it in the line $y = x$ you will have the graph of $y = \ln x$ as shown on the left in figure 6.2. The tangent to the curve at the general point (x_1, y_1) is also shown on this graph.

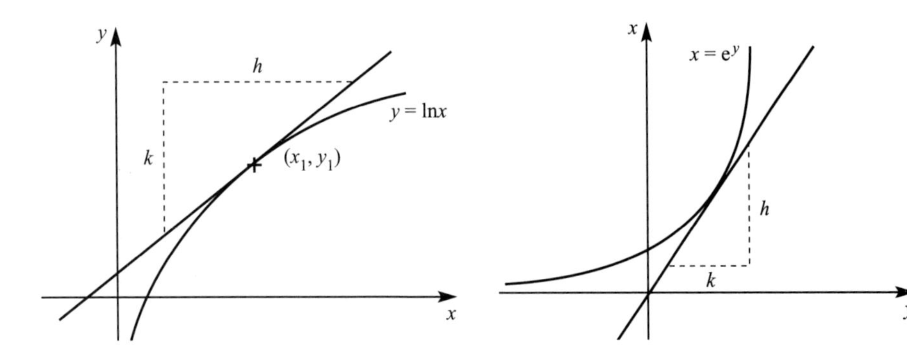

FIGURE 6.2

From figure 6.2 the gradient of the tangent to $y = \ln x$ is $\frac{k}{h}$. If you reflect $y = \ln x$ in $y = x$ you obtain the graph shown on the right in figure 6.2. As the x and y axes have been reflected the equation of this graph is $x = e^y$. The gradient of the tangent to $x = e^y$ is $\frac{h}{k} = e^y = x$. Combining these results gives the gradient of $y = \ln x$ as $\frac{k}{h} = \frac{1}{x}$, leading to the general result

$$y = \ln x \Rightarrow \frac{dy}{dx} = \frac{1}{x}.$$

If $y = \ln(kx)$, where k is a constant, we use the laws of logarithm to write this expression as $y = \ln k + \ln x$. Since k is a constant the logarithm $\ln k$ is also a constant and when it is differentiated the result is 0. Hence

$$y = \ln(kx) \Rightarrow \frac{dy}{dx} = \frac{1}{x}.$$

You can also use the laws of logarithms to differentiate $y = \ln x^n$:

$$y = \ln x^n = n\ln x \Rightarrow \frac{dy}{dx} = n \times \frac{1}{x} = \frac{n}{x}.$$

Note You will learn how to differentiate functions of the type $\ln(ax + b)$ in *Pure Mathematics 3*.

EXAMPLE 6.3

Differentiate $y = \ln(5x)$ with respect to x.

Solution Using the laws of logarithms

$$y = \ln(5x) = \ln 5 + \ln x$$

then

$$\frac{dy}{dx} = \frac{1}{x}.$$

ln5 is a constant so its deriative is 0

EXAMPLE 6.4

Find the gradient of $\ln x^2$ at the point where $x = \frac{1}{4}$.

Solution Using the laws of logarithms

$$y = \ln x^2 = 2\ln x$$

which gives

$$\frac{dy}{dx} = 2 \times \frac{1}{x} = \frac{2}{x}.$$

At the point where $x = \frac{1}{4}$ the gradient is $\frac{dy}{dx} = \frac{2}{\frac{1}{4}} = 8$.

EXERCISE 6A

1 Differentiate the following with respect to x.

(a) e^x
(b) $2e^x$
(c) $4 - 3e^x$
(d) $2x + 8e^x$
(e) $0.1e^x - 0.5x^2$
(f) $\sqrt{x} - \frac{1}{2}e^x$

2 Differentiate the following with respect to x.

(a) $\ln x$
(b) $2\ln x$
(c) $\ln(4x)$
(d) $\ln x^4$
(e) $0.8\ln x^5$
(f) $\ln(\frac{1}{x})$
(g) $2\ln(\frac{3}{x})$
(h) $6\ln(\frac{3}{x})$
(i) $x - \ln x$
(j) $\frac{2}{x} - 4\ln(\frac{1}{x})$

3 Find the exact values of the gradients of the following at the given values of x.

(a) $e^x + \ln x$ $\quad x = 1$
(b) $3\ln x - 4e^x$ $\quad x = 2$
(c) $\ln x^3 + 2e^x + x$ $\quad x = 1$
(d) $\dfrac{2}{x} - 4\ln(\sqrt{x}) + 2e^x$ $\quad x = 4$
(e) $\ln\dfrac{3}{x} + \dfrac{4}{x^3} - 5e^x$ $\quad x = 2$
(f) $\ln(3xe^x)$ $\quad x = \frac{1}{2}$

4 Prove that if $y = \sqrt{x} + \ln(\sqrt{x})$ then $\dfrac{dy}{dx} = \dfrac{x + \sqrt{x}}{2x\sqrt{x}}$.

5 The population P of the agents of a virus is found to be proportional to e^t, where t is the time measured in years. The initial population is 1000.

(a) Find the population, to 3 significant figures, after 10 days.
(b) What was the initial rate of growth of the population?

TANGENTS AND NORMALS

In Chapter 6 of *Pure Mathematics 1* you found the equations of tangents and normals to curves. To recapitulate: a tangent is a straight line that touches a curve at a given point and the normal is the straight line perpendicular to the tangent going through the same point. These are illustrated in figure 6.3.

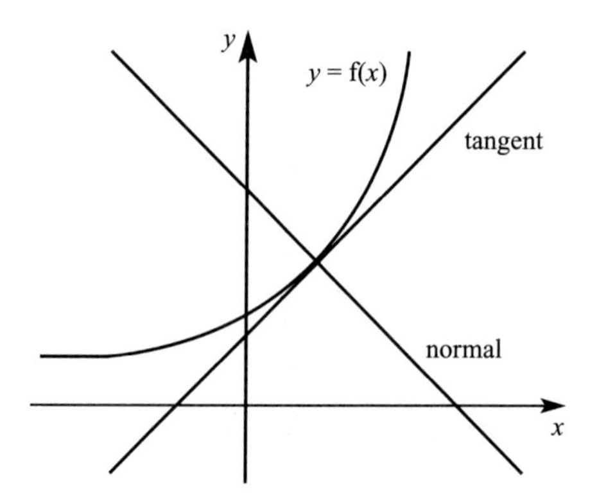

FIGURE 6.3

If the gradient of the tangent is m then the gradient of the normal is $-\frac{1}{m}$. To find the value of m you substitute the value of x into the expression for $\frac{dy}{dx}$. In addition to the derivatives of e^x and $\ln x$ you will recall from *Pure Mathematics 1* that

$$y = x^n \Rightarrow \frac{dy}{dx} = nx^{n-1}.$$

Further, the general equation of the straight line which is the most useful to solve questions involving tangents and normals is

$$y - y_1 = m(x - x_1).$$

EXAMPLE 6.5

Find the equation of the tangent and the normal to $y = 2x^2 - x + 3$ at the point where $x = 1$.

Solution When $x = 1$ the value of y is

$$y = 2 \times 1^2 - 1 + 3 = 4.$$

To find the gradient, first differentiate

$$\frac{dy}{dx} = 4x - 1.$$

At $x = 1$ the value of $\frac{dy}{dx}$ is the gradient of the tangent, so

$$m = 4 \times 1 - 1 = 3.$$

Tangent

Passes through (1, 4) with gradient 3, so its equation is

$$y - 4 = 3(x - 1)$$
$$y = 3x + 1.$$

Normal

Passes through (1, 4) with gradient $-\frac{1}{3}$, so its equation is

$$y - 4 = -\tfrac{1}{3}(x - 1)$$

multiply by 3 to avoid fractions

$$3y - 12 = -(x - 1)$$

and tidying up the equation gives an answer

$$x + 3y = 13.$$

EXAMPLE 6.6

Find the equation of the normal to the curve $y = 3x + 2\ln x$ at the point where $x = 2$.

Solution Where $x = 2$, $y = 6 + 2\ln 2$.

The gradient is

$$\frac{dy}{dx} = 3 + \frac{2}{x}$$

and when $x = 2$ the gradient $m = 4$.

The normal passes through $(2, 6 + 2\ln 2)$ and has gradient $-\frac{1}{4}$

$$y - (6 + 2\ln 2) = -\frac{1}{4}(x - 2)$$

$$4y - 24 - 8\ln 2 = -x + 2$$

$$x + 4y = 26 + 8\ln 2.$$

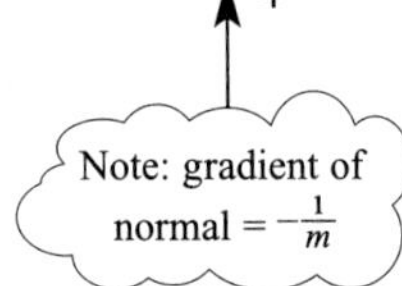

EXAMPLE 6.7

Find the equation of the tangent to $y = e^x - 2$ at the point where it crosses the x axis. Sketch the graph of the curve with its tangent.

Solution When it cuts the x axis the value of $y = 0$ so

$$e^x - 2 = 0$$

$$e^x = 2$$

$$x = \ln 2.$$

The gradient of the tangent at this point is given by

$$\frac{dy}{dx} = e^x = 2,$$

i.e. the tangent passes through the point $(\ln 2, 0)$ with gradient 2 so its equation is:

$$y - 0 = 2(x - \ln 2)$$

$$y = 2x - 2\ln 2.$$

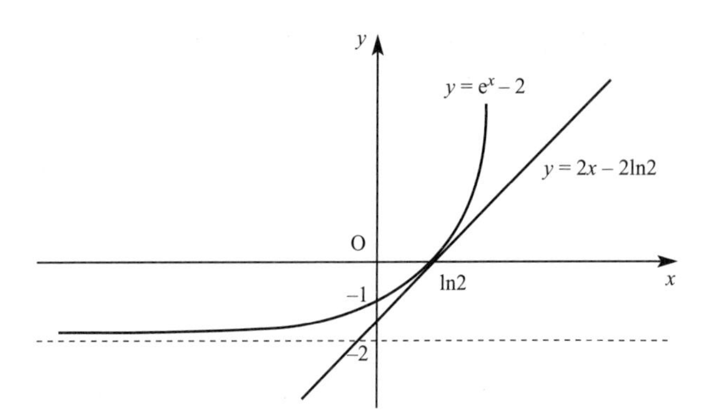

FIGURE 6.4

EXERCISE 6B

1 A curve has equation

$$y = \frac{1}{x}.$$

(a) Find the equation of the normal to the curve at the point where $x = 2$.

(b) Find where the normal cuts the curve again.

(c) Illustrate your solutions on a sketch of the curve.

2 Find the equation of the tangent to the curve

$$y = 2 + 4\ln x$$

at the point where $x = 1$.

3 Given that

$$y = e^x - 4$$

(a) Find the value of x at the point where $y = 0$.

(b) Find the gradient at this point.

(c) Find the equation of the tangent at this point.

4 **(a)** On the same diagram sketch $y = \ln x$ and $y = \ln(x - 2)$ stating the geometric relationship between the two curves.

(b) Using your sketches explain why

$$\frac{d(\ln(x-2))}{dx} = \frac{1}{x-2}.$$

(c) The equation of a curve is $y = \ln\frac{x-2}{x^2}$. Using the results given in part (b) and the laws of logarithms prove that

$$\frac{dy}{dx} = \frac{4-x}{x(x-2)}.$$

(d) Find the equation of the tangent at the point where $x = 3$.

5 Using the result that $a^{m+n} = a^m \times a^n$ prove that

$$\frac{d(e^{x+2})}{dx} = e^{x+2}.$$

Hence show that the equation of the normal to

$$y = e^{x+2} + x^2$$

at the point where $x = -1$ is

$$x + (e - 2)y = e^2 - e - 3.$$

6 Show that the tangent to the curve

$$y = 2 - \ln\left(\frac{x}{2}\right)$$

at the point where $x = 2$ cuts the x axis at the point $(6, 0)$.

7 A curve has equation

$$y = \sqrt{x} + \frac{1}{\sqrt{x}}.$$

(a) Show that $\dfrac{dy}{dx} = \dfrac{x - 1}{2x\sqrt{x}}$.

(b) Find the equation of the tangent at the point where $x = 4$.

8 Show that the area of the triangle enclosed between the x axis, the y axis and the tangent to $y = e^x + 1$ at the point where $x = -1$ is

$$\frac{e}{2} + 2 + \frac{2}{e}.$$

9 Find the equation of the normal to $y = 2\ln(5x) + e^x - 3x$ at the point where $x = 1$.

10 Show that the tangent to $y = e^x - \ln(3x)$ at the point where $x = 2$ cuts the y axis at $1 - e^2 - \ln 6$.

EXERCISE 6C

Examination-style questions

1 The curve with equation $y = e^x - 1$ meets the line $y = 3$ at the point $(h, 3)$.

(a) Find h, giving your answer in terms of natural logarithms.

(b) Find the equation of the tangent at $(h, 3)$ leaving natural logarithms in your answer.

2 It is given that $y = 1 - \dfrac{x^2}{2} + \ln\dfrac{x}{4}$.

(a) Show that $\dfrac{dy}{dx} = -\dfrac{15}{4}$ when $x = 4$.

(b) Obtain the equation of the normal to the curve at the point where $x = 4$.

3 Given that a curve has equation $y = 4x^2 - \ln(2x) + 3$

(a) Show that the gradient of the tangent at $x = \frac{1}{2}$ is 2.

(b) Find the equation of the tangent at the point where $x = \frac{1}{2}$.

(c) Show that the area of the triangle enclosed between the tangent and the axes is $\frac{9}{4}$.

4 A curve C has equation $y = \ln x - \frac{x^2}{8}$, where $x > 0$. Find, by differentiation, the x coordinate of the stationary point on C and determine whether this point is a minimum or a maximum. (You will recall from *Pure Mathematics 1* that at a stationary point the gradient of the function is 0.)

5 Find the equation of the tangent and the normal to the curve

$$y = e^x + 2\ln x + 3$$

at the point where $x = 1$. Leave your answer in terms of e.

6 A curve has equation $y = \ln x + \ln(2x)$, $x > 0$.

(a) P is the point $(h, \ln 18)$. Find h.

(b) Show that the gradient of the curve at P is $\frac{2}{3}$.

(c) Find the equation of the normal to the curve at the point P.

7 **(a)** Differentiate $y = 2e^x - 4x$.

(b) Find the x value of the stationary point on the curve.

(c) State the equation of the tangent at the stationary point.

(d) Determine whether the stationary point is a maximum or a minimum turning point.

8 Given that $y = \ln\frac{2}{x} - \frac{1}{x^2}$, $x > 0$.

(a) Find $\frac{dy}{dx}$.

(b) Show that when $x = 2$, $\frac{dy}{dx} = -\frac{1}{4}$.

(c) Find the equation of the normal at the point where $x = 2$

9 The diagram shows part of the curve $y = e^x - x^2$.

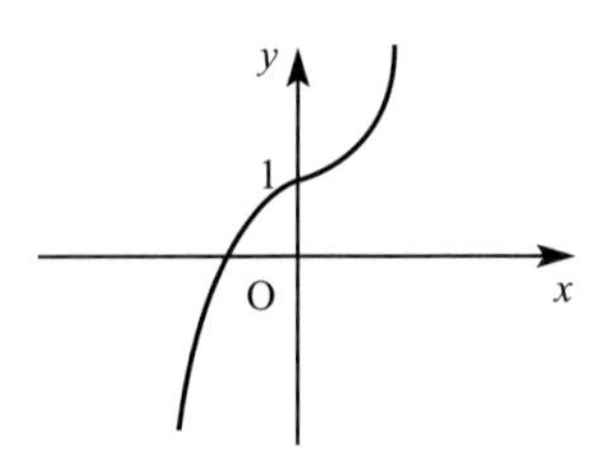

(a) Show that the gradient of the curve at the point $x = 2$ is $e^2 - 4$.

(b) Find the equation of the tangent to the curve at the point where $x = 2$, leaving your answer in terms of e.

(c) Verify that the tangent next touches the curve at a point close to $x = -4.2$.

10 The curve with equation $y = \frac{1}{2}e^x$ meets the y axis at the point A.

(a) Prove that the tangent at A to the curve has equation $2y = x + 1$.

The point B has x coordinate ln4 and lies on the curve. The normal at B to the curve meets the tangent at A to the curve at the point C.

(b) Prove that the x coordinate of C is $\frac{3}{2} + \ln 2$ and find the y coordinate of C.

[Edexcel]

KEY POINTS

1 Exponentials

$$y = e^x \Rightarrow \frac{dy}{dx} = e^x$$

$$y = ke^x \Rightarrow \frac{dy}{dx} = ke^x$$

2 Logarithms

$$y = \ln x \Rightarrow \frac{dy}{dx} = \frac{1}{x}$$

$$y = \ln(kx) \Rightarrow \frac{dy}{dx} = \frac{1}{x}$$

$$y = \ln x^n = n\ln^x \Rightarrow \frac{dy}{dx} = \frac{n}{x}$$

3 Rational powers

$$y = x^n \Rightarrow \frac{dy}{dx} = nx^{n-1}$$

4 Tangents and normals

Tangent passing through (x_1, y_1) with gradient m has equation

$$y - y_1 = m(x - x_1)$$

Normal will have gradient $-\frac{1}{m}$ and equation

$$y - y_1 = -\frac{1}{m}(x - x_1).$$

Chapter seven

INTEGRATION

I have been ever of opinion that revolutions are not to be evaded.

Benjamin Disraeli

INTEGRAL OF e^x

In Chapter 6 you learnt that if

$y = e^x$ then $\frac{dy}{dx} = e^x$.

You also know from *Pure Mathematics 1* that integration may be considered as the reverse process of differentiation, from which it follows that

$$\int e^x \, dx = e^x + c.$$

Remember: for indefinite integrals you need the constant c which disappears when you differentiate

EXAMPLE 7.1

Find $\int 2e^x \, dx$.

Solution

$$\int 2e^x \, dx = 2\int e^x \, dx$$
$$= 2e^x + c.$$

EXAMPLE 7.2

Work out $\int_0^1 5e^x \, dx$ leaving your answer in terms of e.

Solution

$$\int_0^1 5e^x \, dx = 5\int_0^1 e^x \, dx$$
$$= 5\left[e^x\right]_0^1$$
$$= 5(e^1 - e^0)$$
$$= 5(e - 1).$$

You will recall that the integral of a function with respect to x gives the area between the curve and the x axis between the two limits. Figure 7.1 shows the area defined by the integral in Example 7.2.

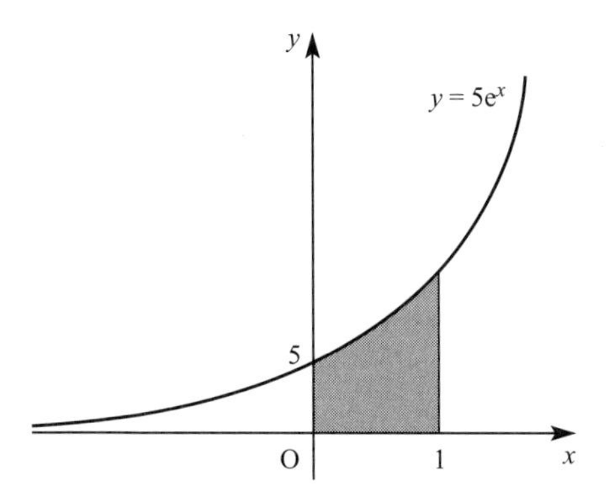

FIGURE 7.1

EXAMPLE 7.3

Find the area enclosed between the curve $y = e^x - 2$, the x axis and the lines $x = -1$ and $x = 1$. Leave your answer in terms of e and natural logarithms.

Solution First sketch the curve as shown in figure 7.2.

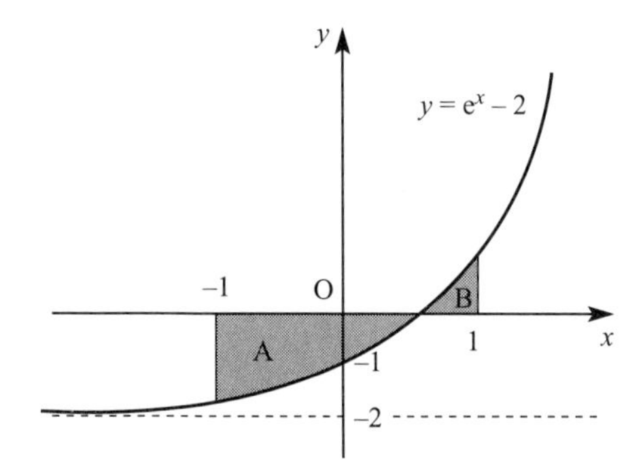

FIGURE 7.2

It cuts the x axis when $y = 0$,

i.e. when $e^x - 2 = 0$

so $e^x = 2$

and $x = \ln 2$.

To find the area enclosed consider the parts below and above the x axis separately.

$$\text{Area A} = \int_{-1}^{\ln 2} (e^x - 2)\,dx = [e^x - 2x]_{-1}^{\ln 2}$$

$$= e^{\ln 2} - 2\ln 2 - (e^{-1} + 2)$$

$$= -2\ln 2 - e^{-1}$$

Note that $e^{\ln 2} = 2$

Areas under the x axis are negative

Magnitude of the area A is $2\ln 2 + e^{-1}$.

$$\text{Area B} = \int_{\ln 2}^{1} (e^x - 2)\,dx = [e^x - 2x]_{\ln 2}^{1}$$
$$= e - 2 - (e^{\ln 2} - 2\ln 2)$$
$$= e + 2\ln 2 - 4.$$

So the total area enclosed is

$$e + e^{-1} + 4\ln 2 - 4.$$

EXERCISE 7A

1 Find the following indefinite integrals:

(a) $\int e^x \,dx$ **(b)** $\int 2e^x \,dx$

(c) $3\int e^x \,dx$ **(d)** $\int (e^x + 2)\,dx$

(e) $5\int (4 + 3e^x)\,dx$ **(f)** $\int (4 - e^x)\,dx$

(g) $\int (2x + 3e^x)\,dx$ **(h)** $\int (\tfrac{1}{2}e^x + \sqrt{x})\,dx$

(i) $\int \left(4e^x + \dfrac{2}{x^3}\right)dx$ **(j)** $\int e^{x+2}\,dx$ (Hint: $a^m \times a^n = a^{m+n}$.)

2 Find the following definite integrals, giving your answers to 3 significant figures. You may have a calculator that can compute these integrals. If so, use it to check your answers but ensure that you have shown sufficient working to prove that you can find the answers without such a facility.

(a) $\int_0^1 e^x \,dx$ **(b)** $\int_0^2 2e^x \,dx$

(c) $\int_{-1}^{0} (1 + e^x)\,dx$ **(d)** $\int_1^4 (\sqrt{x} - e^x)\,dx$

(e) $\int_{.5}^{.75} \left(e^x - \dfrac{1}{x^2}\right)dx$ **(f)** $\int_{-\infty}^{0} e^x \,dx$

(g) $4\int_1^2 (2e^x - \sqrt[3]{x})\,dx$ **(h)** $\int_3^5 (1 - 2e^x)\,dx$

(i) $\int_2^4 e^{x-2}\,dx$ **(j)** $\int_{-1}^{3} (e^x + x(1 + x))\,dx$
(Hint: multiply out the bracket.)

3 Prove that $\int_{-1}^{1} e^x \,dx = \dfrac{e^2 - 1}{e}$.

4 A curve has equation $y = e^x - 3$.

(a) Sketch the curve.

(b) Show that it cuts the x axis at $(\ln 3, 0)$.

(c) Show that the total area enclosed between the curve, the x axis and the lines $x = 0$ and $x = 2$ is

$$e^2 + 6\ln 3 - 11.$$

5 Prove that the area enclosed between $y = e^x$, the x axis and the lines $x = 1$ and $x = 2$ is $e(e - 1)$.

INTEGRAL OF $\frac{1}{x}$

In Chapter 6 you saw that $y = \ln x \Rightarrow \frac{dy}{dx} = \frac{1}{x}$, so it follows that when you reverse the process

$$\int \frac{1}{x}\,dx = \ln x + c.$$

The function $y = \ln x$ is defined for the domain $x > 0$ so it follows that the integral is also valid for $x > 0$.

If you have a multiplier, e.g. $y = \frac{k}{x}$, then take k out of the integral to give

$$\int \frac{k}{x}\,dx = k\int \frac{1}{x}\,dx = k\ln x + c.$$

EXAMPLE 7.4

Find the indefinite integral $\int \frac{1}{4x}\,dx$.

Solution

$$\int \frac{1}{4x}\,dx = \frac{1}{4}\int \frac{1}{x}\,dx$$

$$= \frac{1}{4}\ln x + c.$$

EXAMPLE 7.5

Find the exact value of $\int_1^3 \frac{2}{x}\,dx$.

Solution

$$\int_1^3 \frac{2}{x}\,dx = 2\int_1^3 \frac{1}{x}\,dx = 2[\ln x]_1^3$$

$$= 2(\ln 3 - \ln 1)$$

$\ln 1 = 0$

$$= 2\ln 3.$$

EXAMPLE 7.6

Evaluate $\int_2^4 \left(\frac{x+1}{x}\right) dx$.

Solution You must first remove the quotient by splitting it into two fractions and simplifying:

$$\int_2^4 \left(\frac{x+1}{x}\right) dx = \int_2^4 \left(\frac{x}{x} + \frac{1}{x}\right) dx$$

$$= \int_2^4 \left(1 + \frac{1}{x}\right) dx$$

$$= [x + \ln x]_2^4$$

$$= (4 + \ln 4) - (2 + \ln 2)$$
$$= 2 + \ln 4 - \ln 2$$
$$= 2 + \ln\tfrac{4}{2}$$
$$= 2 + \ln 2.$$

EXAMPLE 7.7

Sketch the graph of $y = \dfrac{1}{x}$ for all values of x.

Find the area between the curve, the x axis and

(a) the lines $x = 1$ and $x = 2$.

(b) the lines $x = -2$ and $x = -1$.

Solution

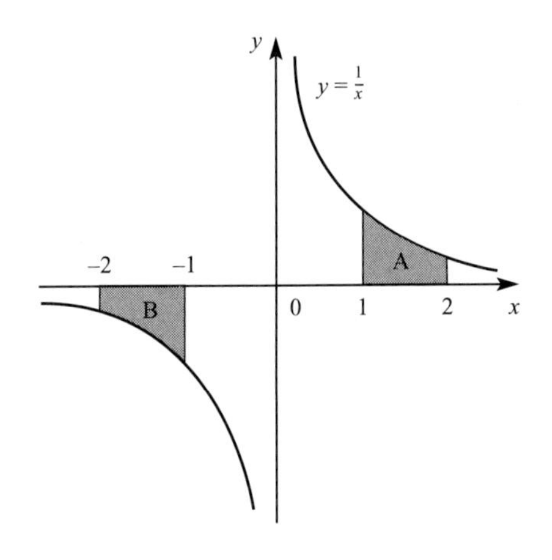

FIGURE 7.3

(a) Area A $= \displaystyle\int_1^2 \frac{1}{x}\,dx = [\ln x]_1^2 = \ln 2 - \ln 1 = \ln 2.$

(b) Area B $= \displaystyle\int_{-2}^{-1} \frac{1}{x}\,dx = [\ln x]_{-2}^{-1} = \ln(-1) - \ln(-2) = \ln\frac{-1}{-2} = \ln\frac{1}{2} = -\ln 2.$

But ln(−1) and ln(−2) do not exist – something very dubious here!

Example 7.7 illustrates a problem. It is clear from figure 7.3 that the areas A and B have equal magnitude and opposite sign. This allows us to drop the restriction $x > 0$ and to avoid dubious mathematics with logarithms of negative numbers by generalising the integral of $\frac{1}{x}$ to

$$\int \frac{1}{x}\,dx = \ln|x| + c \qquad x \neq 0$$

and

$$\int \frac{k}{x}\,dx = k\ln|x| + c \qquad x \neq 0.$$

Note

$x \neq 0$ since $\frac{1}{x}$ is undefined if $x = 0$.

EXAMPLE 7.8

Evaluate $\int_{-4}^{-2} \frac{1}{2x}\,dx$.

Solution $\int_{-4}^{-2} \frac{1}{2x}\,dx = \frac{1}{2}[\ln|x|]_{-4}^{-2} = \frac{1}{2}(\ln 2 - \ln 4) = \frac{1}{2}\ln\frac{1}{2}$.

EXERCISE 7B

1 Find the following indefinite integrals

(a) $\int \frac{2}{x}\,dx$ (b) $\int \frac{1}{2x}\,dx$

(c) $\int \left(\frac{3}{4x} + 5\right)dx$ (d) $\int \left(\frac{1}{x} + e^x\right)dx$

(e) $\int \left(\frac{1}{2x} + \frac{1}{2\sqrt{x}}\right)dx$ (f) $\int \left(\frac{1}{x^2} - \frac{1}{x}\right)dx$

(g) $\int \left(\frac{4 + x}{x}\right)dx$ (h) $\int \left(\frac{5}{x} + \frac{x}{5}\right)dx$

(i) $\int \left(\frac{(x + 1)^2}{x}\right)dx$ (j) $\int \left(\frac{x^2 - 1}{x}\right)dx$

2 Evaluate the following definite integrals

(a) $\int_1^2 \frac{1}{x}\,dx$ (b) $\int_1^4 \frac{3}{x}\,dx$

(c) $\int_1^4 \left(\frac{1}{2x} - \sqrt{x}\right)dx$ (d) $\int_{-6}^{-2} \frac{1}{x}\,dx$

(e) $\int_4^9 \left(\frac{1}{\sqrt{x}} + \frac{1}{x}\right)dx$ (f) $\int_2^3 \left(\frac{x - 2}{x}\right)dx$

(g) $2\int_1^2 \left(\frac{1}{x} + e^x\right)dx$ (h) $\int_1^2 \left(\frac{(1 + x)^3}{x}\right)dx$

(i) $\int_2^4 \left(1 + \frac{1}{x}\right)^2 dx$ (j) $\int_{-2}^{-1} \left(\frac{4}{x} - \frac{x^2}{4}\right)dx$

3 Sketch the graph of $y = \frac{1}{x} + 2$.

(a) Find the area between the graph, the x axis and the lines $x = 2$ and $x = 3$.

(b) Show that the area between the graph, the x axis and the lines $x = -3$ and $x = -2$ is $2 - \ln\left(\frac{3}{2}\right)$.

4 The diagram shows part of the curve $y = \dfrac{4x^2 + 1}{x}$.

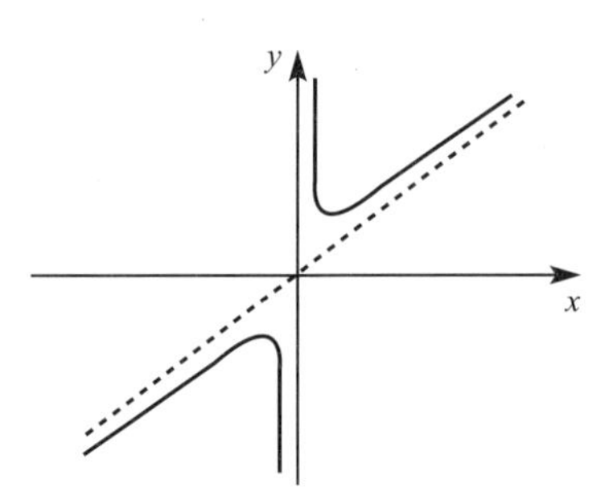

Show that the area between the curve, the x axis and the lines $x = 1$ and $x = \mathrm{e}$ is $2\mathrm{e}^2 - 1$.

5 Evaluate $\int_2^3 \left(\frac{1}{2x} - \frac{1}{4}\mathrm{e}^x + \sqrt[3]{x}\right) \mathrm{d}x$ correct to 4 significant figures.

FINDING VOLUMES BY INTEGRATION

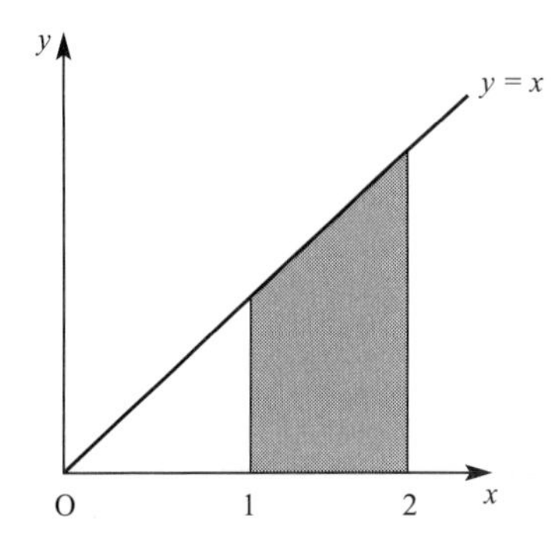

FIGURE 7.4

FIGURE 7.5

When the shaded region in figure 7.4 is rotated through 360° about the x axis, the solid obtained, illustrated in figure 7.5 is called a *solid of revolution*. In this particular case, the volume of the solid could be calculated as the difference between the volumes of two cones, i.e. using $V = \frac{1}{3}\pi r^2 h$ gives

$$\text{volume of solid of revolution} = \tfrac{1}{3}\pi \times 2^2 \times 2 - \tfrac{1}{3}\pi \times 1^2 \times 1$$

$$= \tfrac{7}{3}\pi \text{ cubic units.}$$

If the line $y = x$ in figure 7.4 was replaced by a curve, such a simple calculation would no longer be possible.

SOLIDS FORMED BY ROTATION ABOUT THE x AXIS

The shaded region in figure 7.6 is bounded by $y = f(x)$, the x axis and the lines $x = a$ and $x = b$.

Now look at the solid formed by rotating the shaded region through 360° about the x axis.

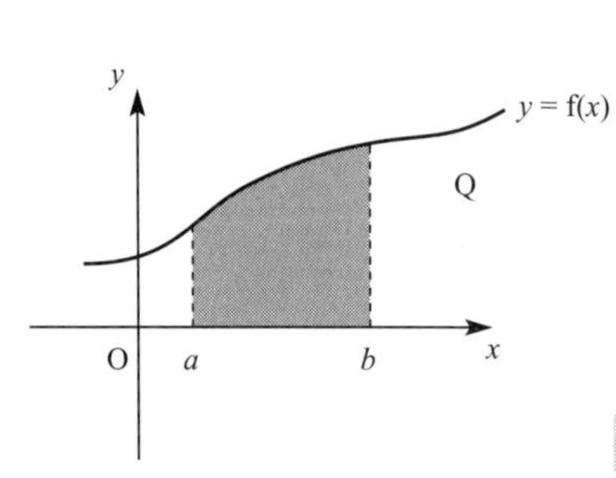

FIGURE 7.6

FIGURE 7.7

The volume of the solid of revolution (which is usually called the *volume of revolution*) can be found by imagining that the solid can be sliced into thin discs.

The disc shown in figure 7.7 is approximately cylindrical with radius y and thickness δx, so its volume is given by

$$\delta V = \pi y^2 \delta x.$$

The volume of the solid is the limit of the sum of all these elementary discs as $\delta x \to 0$,

i.e. the limit as $\delta x \to 0$ of $\sum_{\text{over all discs}} \delta V = \sum_{x=a}^{x=b} \pi y^2 \delta x.$

The limiting values of sums such as these are integrals, so

$$V = \int_a^b \pi y^2 \, dx$$

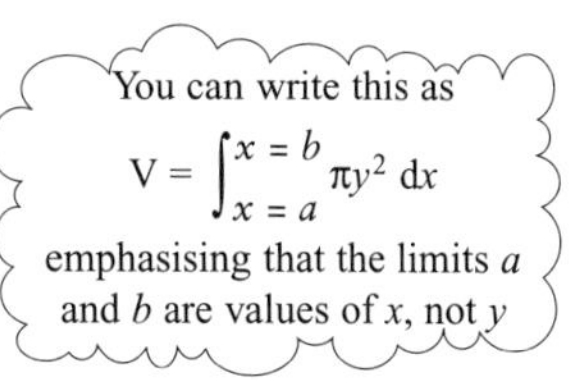

The limits are a and b because x takes values from a to b.

Taking π out of the integral the volume of revolution about the x axis from $x = a$ to $x = b$ is

$$V = \pi \int_a^b y^2 \, dx.$$

Note Since the integration is 'with respect to x', indicated by the dx and the fact that the limits a and b are values of x, it cannot be evaluated unless the function y is also written in terms of x.

EXAMPLE 7.9

The region between the curve $y = x^2$, the x axis and the lines $x = 1$ and $x = 3$ is rotated through 360° about the x axis. Find the volume of revolution which is formed.

Solution The region is shaded in the diagram.

Using $V = \pi \int_a^b y^2 \, dx$

volume $= \pi \int_1^3 (x^2)^2 \, dx$

$= \pi \int_1^3 x^4 \, dx$

$= \pi \left[\frac{x^5}{5} \right]_1^3$

$= \frac{\pi}{5}(243 - 1)$

$= \frac{242\pi}{5}.$

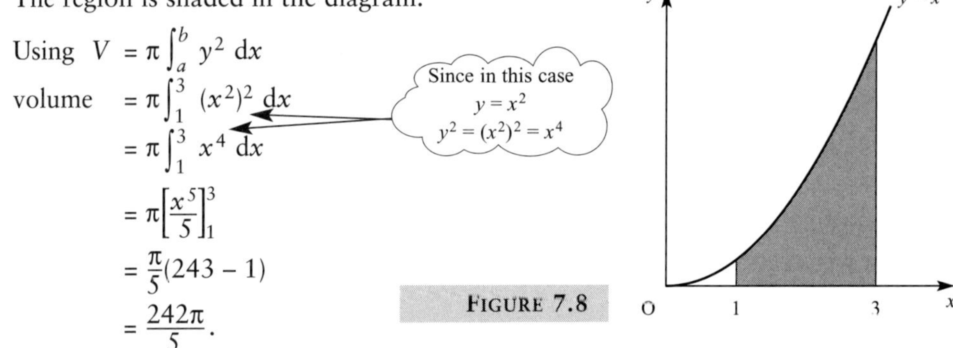

FIGURE 7.8

The volume is $\frac{242\pi}{5}$ cubic units or 152 cubic units (3 s.f).

Note Unless a decimal answer is required, it is usual to leave π in the answer, which is then exact.

EXAMPLE 7.10

(a) Find the volume of a spherical ball of radius 2 cm using integration.

(b) Verify your result using the formula for the volume of a sphere.

Solution **(a)** The volume is obtained by rotating the top half of the circle $x^2 + y^2 = 4$ through 360° about the x axis.

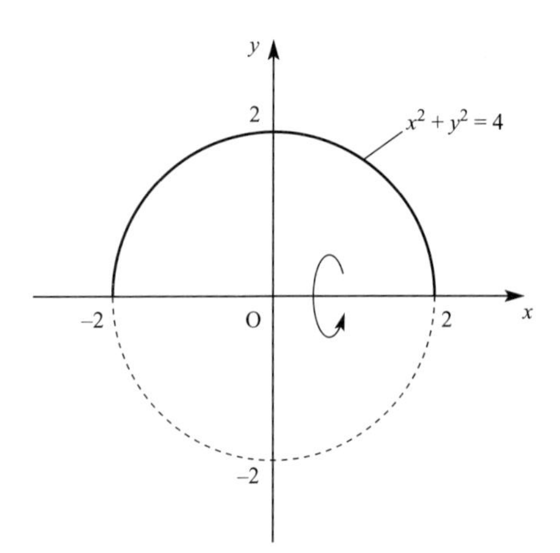

FIGURE 7.9

Using $V = \pi \int_a^b y^2 \, dx$ and $y^2 = 4 - x^2$ from the circle equation

$$\text{volume} = \pi \int_{-2}^{2} (4 - x^2) \, dx$$

$$= \pi \left[4x - \frac{x^3}{3} \right]_{-2}^{2}$$

$$= \pi \left[\left(8 - \frac{8}{3}\right) - \left(-8 + \frac{8}{3}\right) \right]$$

$$= \frac{32\pi}{3} \text{ cm}^3.$$

(b) Volume of a sphere

$$= \frac{4}{3}\pi r^3$$

$$= \frac{4}{3}\pi \times 2^3$$

$$= \frac{32\pi}{3} \text{ cm}^3$$

which verifies the result in part (a).

ROTATION ABOUT THE y AXIS

When a region is rotated about the y axis a very different solid is obtained.

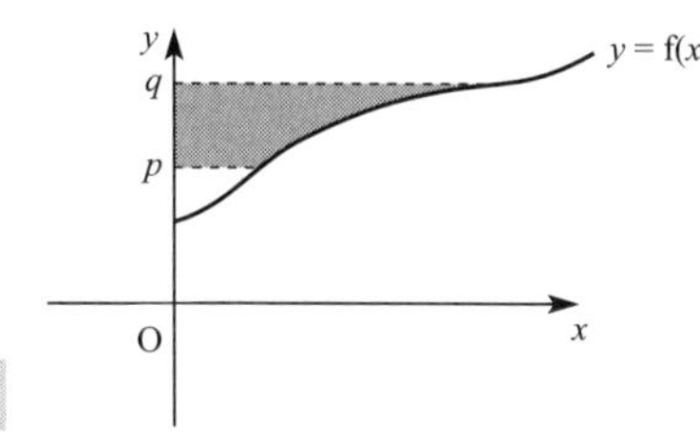

FIGURE 7.10

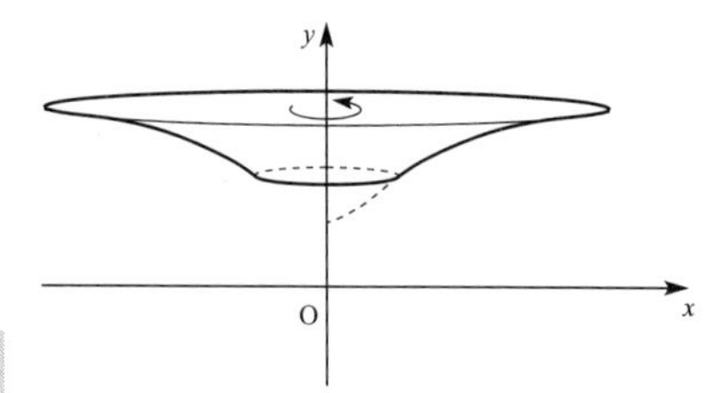

FIGURE 7.11

Notice the difference between the solid obtained in figure 7.11 and that in figure 7.7.

For rotation about the x axis you obtained the formula

$$V_{x \text{ axis}} = \pi \int_a^b y^2 \, dx.$$

In a similar way, the formula for rotation about the y axis from $y = p$ to $y = q$ can be obtained using

$$V_{y \text{ axis}} = \pi \int_p^q x^2 \, dy.$$

ote

In this case, because you are integrating with respect to y you will need to substitute for x^2 in terms of y.

EXAMPLE 7.11

The region between the curve $y = x^2$, the x axis and the lines $y = 2$ and $y = 5$ is rotated through 360° about the y axis. Find the volume of revolution which is formed.

Solution The region is shaded in figure 7.12.

Using $V = \pi\int_p^q x^2 \, dy$

$$\text{volume} = \pi\int_2^5 y \, dy \quad \text{since } x^2 = y$$

$$= \pi\left[\frac{y^2}{2}\right]_2^5$$

$$= \frac{\pi}{2}(25 - 4)$$

$$= \frac{21\pi}{2} \text{ cubic units.}$$

From the derivation of the formula for the volume you only need to consider $x \geqslant 0$

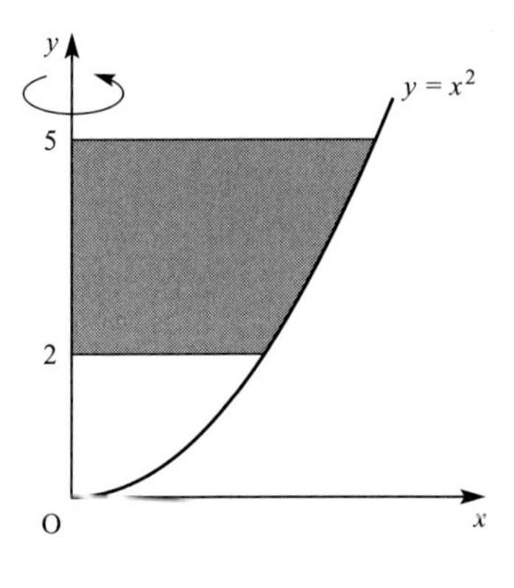

FIGURE 7.12

EXERCISE 7C

1 In each part of this question a region is defined in terms of the lines which form its boundaries. Draw a sketch of the region and find the volume of the solid obtained by rotating it through 360° about the x axis:

(a) $y = 2x$, the x axis and the lines $x = 1$ and $x = 3$;

(b) $y = x + 2$, the x axis, the y axis and the line $x = 2$;

(c) $y = x^2 + 1$, the x axis and the lines $x = 0$ and $x = 1$;

(d) $y = \sqrt{x}$, the x axis and the line $x = 4$.

(e) $y = \dfrac{1}{\sqrt{x}}$, the x axis and the lines $x = 1$ and $x = 4$;

(f) $y = e^{\frac{x}{2}}$, the x axis and the lines $x = 0$ and $x = 3$.

2 In each part of this question a region is defined in terms of the lines which form its boundaries. Draw a sketch of the region and find the volume of the solid obtained by rotating through 360° about the y axis.

(a) $y = 3x$, the y axis and the lines $y = 3$ and $y = 6$;

(b) $y = x - 3$, the y axis, the x axis and the line $y = 6$;

(c) $y = x^2 - 2$, the y axis and the line $y = 4$;

(d) $y = 2\ln x$, the y axis and the lines $y = 1$ and $y = 2$;

(e) $y = x^3$, the y axis and the line $y = 8$;

(f) $y = \dfrac{2}{x}$, the y axis and the lines $y = 2$ and $y = 4$.

3 **(a)** Find the coordinates of A and B, the points of intersection of the circle $x^2 + y^2 = 25$ and the line $y = 4$.

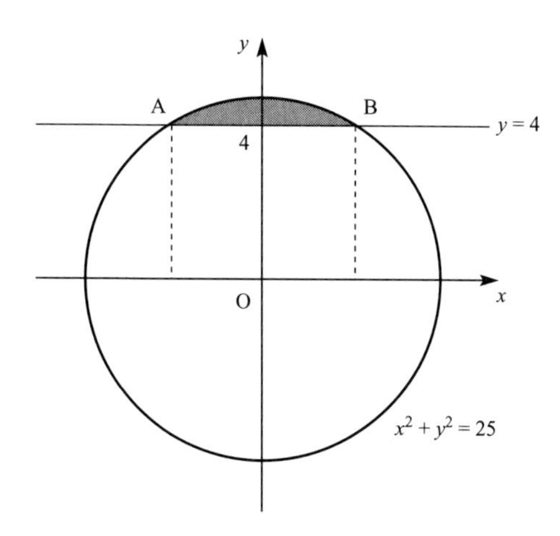

(b) A napkin ring is formed by rotating the shaded area through 360° about the x axis. By considering the shaded area as the difference between two areas, and hence the volume of the napkin ring as the difference between two volumes, find the volume of the napkin ring.

4 **(a)** Sketch the line $4y = 3x$ for $x \geqslant 0$.

(b) Identify the area between this line and the x axis which, when rotated through 360° about the x axis, would give a cone of base radius 3 and height 4.

(c) Calculate the volume of the cone using:

(i) integration;

(ii) a formula.

5 The diagram shows a sketch of part of the graph $y = \dfrac{2}{x^2}$.

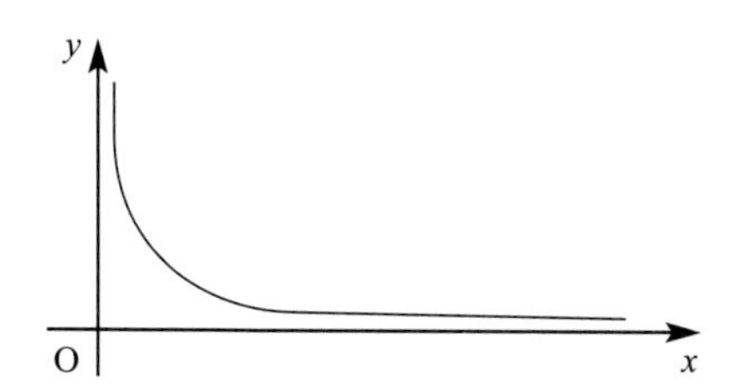

A potter designs two new sculptures that are based on this curve.

(a) For the first she rotates the curve about the x axis from $x = 1$ to $x = 10$. Find the volume of this sculpture.

(b) For the second she rotates the curve about the y axis from $y = 1$ to $y = 10$. Find the volume of this sculpture.

(c) Comment on the shapes of the two sculptures.

6 A hemispherical bowl is formed by rotating the bottom half of the circle $x^2 + y^2 = 100$ about the y axis as shown in the diagram (units are cm).

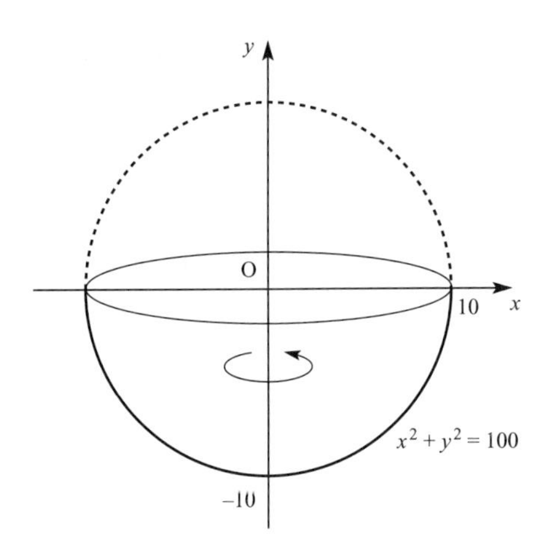

(a) Find the volume of the bowl.

(b) The bowl is filled with water to a depth of 8 cm. Find the volume of water in the bowl.

7 A mathematical model for a large garden pot is obtained by rotating through 360° about the y axis the part of the curve $y = 0.1x^2$ which is between $x = 10$ and $x = 25$ and then adding a flat base. Units are in centimetres.

(a) Draw a sketch of the curve and shade in the cross-section of the pot, indicating which line will form its base.

(b) Garden compost is sold in litres. How many litres will be required to fill the pot to a depth of 35 cm. (Ignore the thickness of the pot.)

8 **(a)** Sketch the graph of $y = (x - 2)^2$ for values of x between $x = -1$ and $x = +5$. Shade in the region under the curve, between $x = 0$ and $x = 2$.

(b) Calculate the area you have shaded.

(c) Show that $(x - 2)^4 = x^4 - 8x^3 + 24x^2 - 32x + 16$.

(d) The shaded region is rotated about the x axis to form a volume of revolution. Calculate this volume, using your answer to part (c) or otherwise.

[MEI]

9 The diagram shows the graph of $y = x^3$.

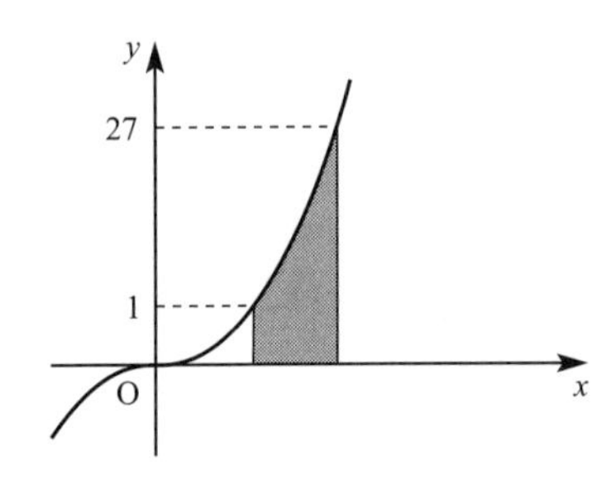

(a) Find the volume of revolution generated when the region bounded by the curve, the y axis and the lines $y = 1$ and $y = 27$ is rotated through 360° about the y axis.

(b) Find the values of x when $y = 1$ and $y = 27$.

(c) Show that when the shaded region shown in the diagram is rotated about the y axis the volume generated is

$$\frac{484}{5}\pi.$$

10 The diagram shows the curve $y = \dfrac{x + 1}{\sqrt{x}}$.

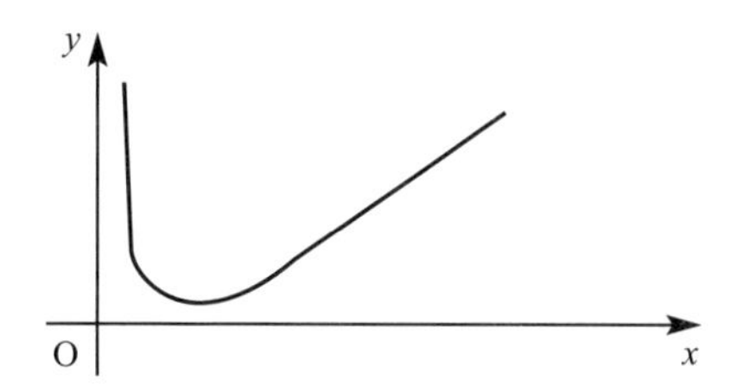

Show that the volume generated when the region enclosed between the curve, the x axis and the lines $x = 2$ and $x = 4$ is rotated around the x axis is $\pi(10 + \ln x)$.

EXERCISE 7D Examination-style questions

1 A finite region R is enclosed by the curve with equation $y = x + \dfrac{4}{x}$, the lines $x = 1$ and $x = \mathrm{e}$, and the x axis.

(a) Use integration to find the area of R, giving your answer in terms of e.

(b) The region R is rotated through 360° about the x axis. Find the volume of revolution, giving your answer to 3 significant figures.

2 It is given that $y = x^{\frac{3}{2}} + \dfrac{3}{2x}$, $x > 0$.

(a) Show that the curve has a minimum turning point at $(1, \frac{5}{2})$.

(b) The region R is bounded by the curve, the lines $x = 1$ and $x = 4$ and the x axis. Find, by integration, the area of R giving your answer in the form $p + q\ln r$ where the numbers p, q and r are to be found.

3 Prove that $\displaystyle\int_{-1}^{1} (\mathrm{e}^x + x)\,\mathrm{d}x = \frac{(\mathrm{e} + 1)(\mathrm{e} - 1)}{\mathrm{e}}$.

4 Find the general solution of $\dfrac{\mathrm{d}y}{\mathrm{d}x} = \sqrt{x} + \dfrac{1}{4x}$ (i.e. find an equation for y).

Given that $y = \frac{8}{3}$ when $x = 1$ find the value of y when $x = 9$, leaving your answer in terms of natural logarithms.

5 A curve has equation $y = 4e^x - 2$.

(a) Sketch the curve showing any asymptotes and stating the exact value of the point at which it cuts the x axis.

(b) Show that the finite area enclosed between the curve, the x axis and the line $x = 2$ is $4e^2 - \ln 4 - 6$.

6 The diagram shows part of the graph of $y = \frac{1}{2x}$.

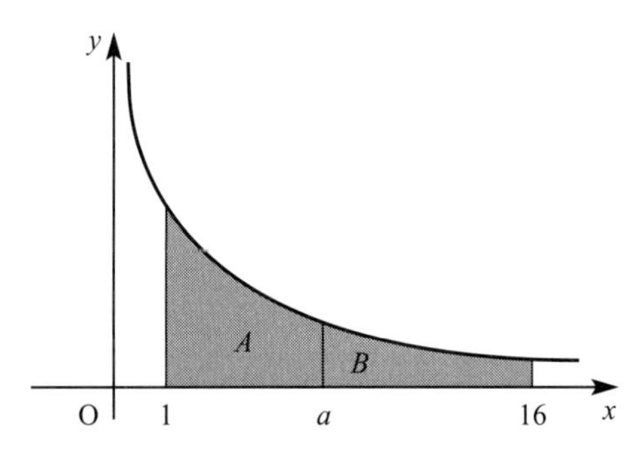

Find the value of a such that the area marked A is three times the area marked B.

7 The area R, as shown in the diagram, is bounded by the curve $y = 3e^x - 1$, the x and y axes, and the line $x = 2$. Find the area of R, leaving your answer in terms of e.

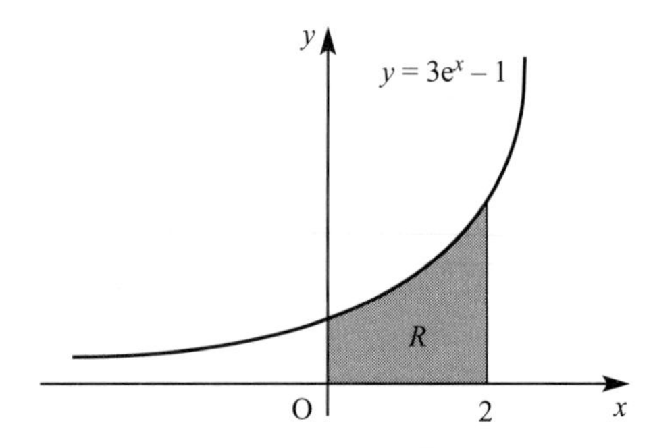

8 The region R is bounded by the curve $y = \sqrt{x} + \frac{1}{4x}$, the lines $x = 4$ and $x = 9$ and the x axis.

(a) Show that the area of R is $\frac{38}{3} + \frac{1}{2}\ln\frac{3}{2}$.

(b) Find the volume when R is rotated through 360° about the x axis, giving your answer to 3 significant figures.

9 It is given that $f(x) = \dfrac{px + q}{x}$, $x > 0$.

(a) Find the values of p and q if $\int_2^6 f(x)\,dx = 20 - \ln 9$.

(b) Show that the equation of the tangent at the point where $x = 1$ is $y = 2x + 1$.

10

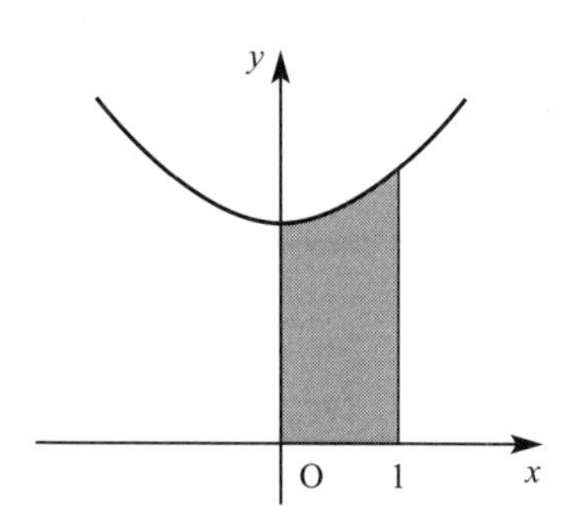

The figure shows the finite shaded region bounded by the curve with equation $y = x^2 + 3$, the lines $x = 1$, $x = 0$ and the x axis. This region is rotated through 360° about the y axis. Find the volume generated.

[Edexcel]

KEY POINTS

1 Integral of e^x

$$\int e^x\,dx = e^x + c$$

$$\int ke^x\,dx = ke^x + c$$

2 Integral of $\frac{1}{x}$

$$\int \frac{1}{x}\,dx = \ln|x| + c \qquad x \neq 0$$

$$\int \frac{k}{x}\,dx = k\ln|x| + c \qquad x \neq 0$$

3 Volumes of revolution

About the x axis

$$V = \pi\int_a^b y^2\,dx$$

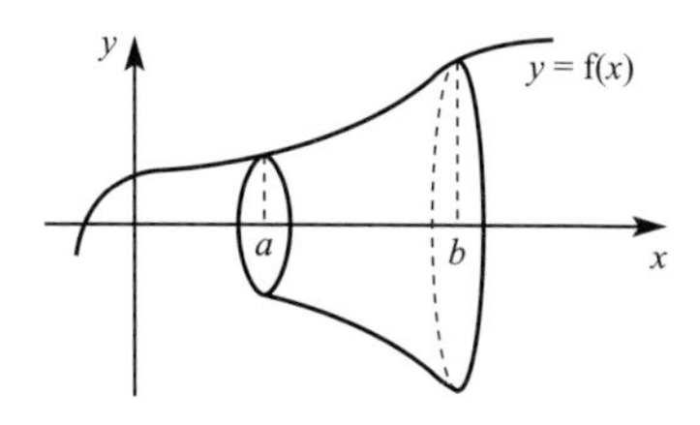

About the y axis

$$V = \pi\int_p^q x^2\,dy$$

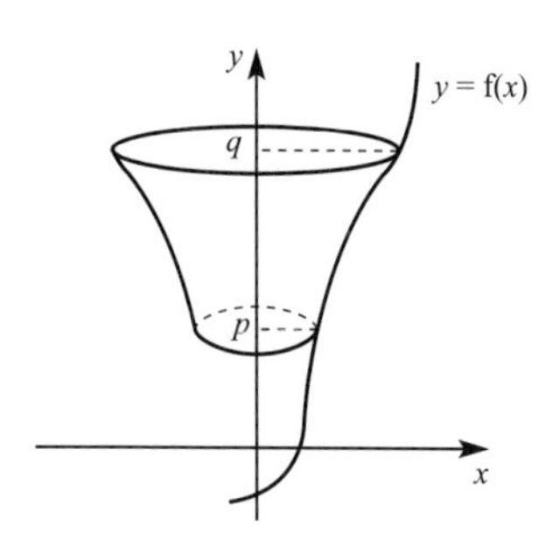

Chapter eight

NUMERICAL METHODS

It is the true nature of mankind to learn from his mistakes.

Fred Hoyle

NUMERICAL SOLUTIONS OF EQUATIONS

Which of the following equations can be solved algebraically, and which cannot? For each equation find a solution, accurate or approximate.

- $x^2 - 4x + 3 = 0$
- $x^2 + 10x + 8 = 0$
- $x^5 - 5x + 3 = 0$
- $x^3 - x = 0$
- $e^x = 4x$

You probably realised that the equations $x^5 - 5x + 3 = 0$ and $e^x = 4x$ cannot be solved algebraically. You may have decided to draw their graphs, either manually or using a graphics calculator or computer package (see figure 8.1).

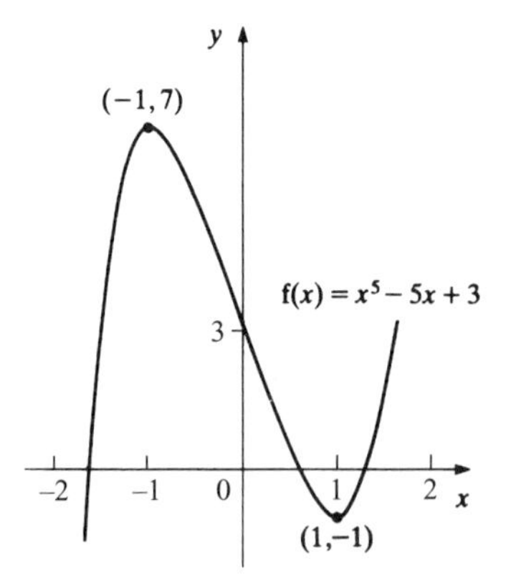

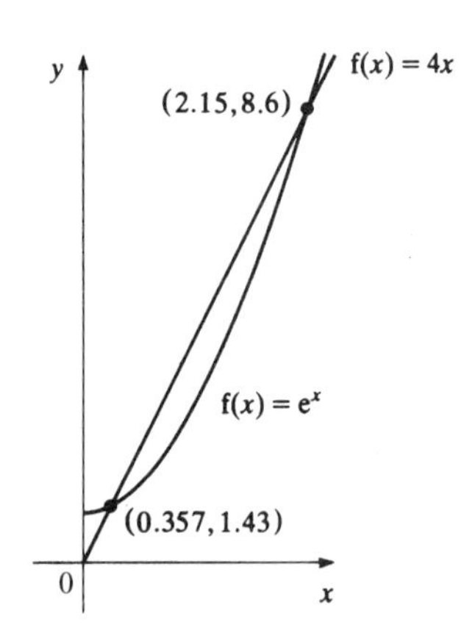

FIGURE 8.1

The graphs show you that:

- $x^5 - 5x + 3 = 0$ has three roots, lying in the intervals [–2, –1], [0, 1] and [1, 2].
- $e^x = 4x$ has two roots, lying in the intervals [0, 1] and [2, 3].

The problem now is how to find the roots to any required degree of accuracy, and as efficiently as possible.

In many real problems, equations are obtained for which solutions using algebraic or analytical methods are not possible, but for which you nonetheless want to know the answers. In this chapter you will be introduced to numerical methods for solving such equations. In applying these methods, keep the following points in mind.

- Only use numerical methods when algebraic ones are not available. If you can solve an equation algebraically (e.g. a quadratic equation), that is the right method to use.
- Before starting to use a calculator or computer program, always start by drawing a sketch graph of the function whose equation you are trying to solve. This will show you how many roots the equation has and their approximate positions. It will also warn you of possible difficulties with particular methods.
- Always give a statement about the accuracy of an answer (e.g. to 5 decimal places, or $\pm 0.000\,005$). An answer obtained by a numerical method is worthless without this; the fact that at some point in the procedure your calculator display reads, say, 1.676 470 588 2 does not mean that all these figures are valid.
- Your statement about the accuracy must be obtained from within the numerical method itself. Usually you find a sequence of estimates of ever-increasing accuracy.
- Remember that the most suitable method for one equation may not be that for another.

Note

An interval written as $[a, b]$ means the interval between a and b, including a and b. This notation is used in this chapter. If a and b are not included, the interval is written (a, b). You may also elsewhere meet the notation $]a, b[$, indicating that a and b are *not* included.

CHANGE OF SIGN METHOD

Assume that you are looking for the roots of the equation $f(x) = 0$. This means that you want the values of x for which the graph of $y = f(x)$ crosses the x axis. As the curve crosses the x axis, $f(x)$ changes sign, so provided that $f(x)$ is a continuous function (its graph has no asymptotes or other breaks in it), once you have located an interval in which $f(x)$ changes sign, you know that that interval must contain a root. In both of the graphs in figure 8.2, there is a root lying between a and b.

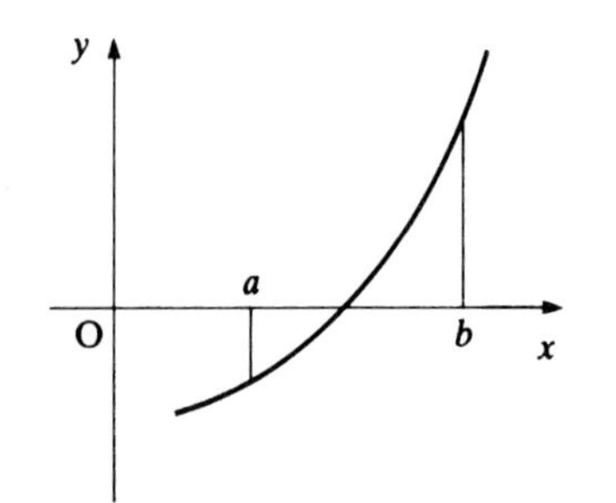

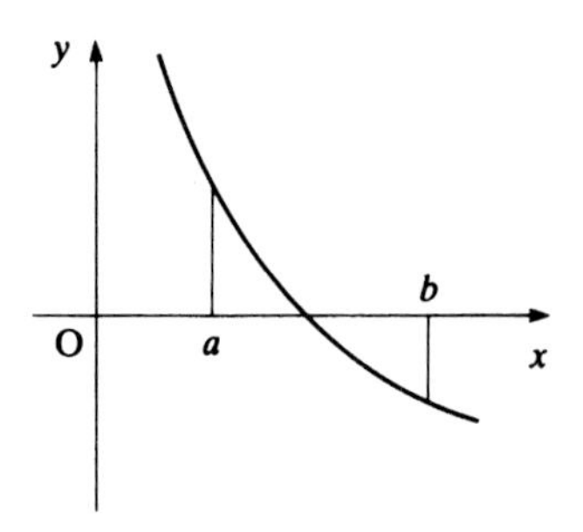

FIGURE 8.2

You have seen that $x^5 - 5x + 3 = 0$ has roots in the intervals [–2, –1], [0, 1] and [1, 2]. A decimal search will be used as an example to find the root in the interval [0, 1].

DECIMAL SEARCH

In this method you first take increments in x of size 0.1 within the interval [0, 1], working out the value of the function f(x), where $f(x) = x^5 - 5x + 3$ for each one. You do this until you find a change of sign.

x	0	0.1	0.2	0.3	0.4	0.5	0.6	0.7
f(x)	3.00	2.50	2.00	1.50	1.01	0.53	0.08	–0.33

There is a sign change, and therefore a root, in the interval [0.6, 0.7] since the function is continuous. Having narrowed down the interval, you can now continue with increments of 0.01 within the interval [0.6, 0.7].

x	0.60	0.61	0.62
f(x)	0.08	0.03	–0.01

This shows that the root lies in the interval [0.61, 0.62].

Alternative ways of expressing this information are:

(a) the root can be taken as 0.615 with a maximum error of ±0.005, or
(b) the root is 0.6 (to 1 decimal place).

This process can be continued by considering $x = 0.611$, $x = 0.612$, ... to obtain the root to any required number of decimal places.

When you use this procedure on a computer or calculator you should be aware that the machine is working in base 2, and that the conversion of many simple numbers from base 10 to base 2 introduces small rounding errors. This can lead to simple roots such as 2.7 being missed and only being found as 2.699 999.

EXAMPLE 8.1

Show that $2 + x^2 - x^3 = 0$ has a root between 1 and 2. By considering changes of sign find this root to 1 decimal place.

Solution Let $f(x) = 2 + x^2 - x^3$.

Using a graphical calculator the graph of $y = f(x)$ looks like that in figure 8.3.

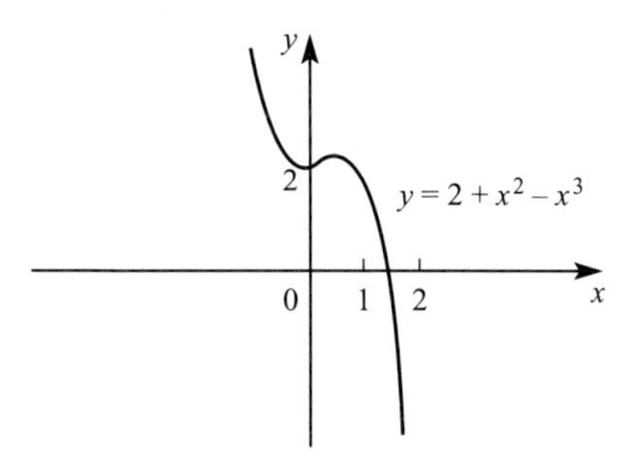

FIGURE 8.3

The graph is continuous over the interval [1, 2] and

$$f(1) = 2$$
$$f(2) = -2$$

so the root lies between $x = 1$ and $x = 2$.

Improving this root to 1 decimal place:

$$f(1.5) = 0.875$$
$$f(1.6) = 0.464$$
$$f(1.7) = -0.023$$ so the root lies in [1.6, 1.7].

Putting $x = 1.65$ confirms whether the root is nearer 1.6 or 1.7 (even when, as in this case, it appears obvious you should use this type of check).

$$f(1.65) = 0.23$$

So the root is close to 1.7.

EXAMPLE 8.2

Show that $\sin\theta + \cos\theta = 1$ has a root between 1 and 2 and find θ to 1 decimal place.

Solution First rewrite in the form $f(\theta) = 0$.

Let $f(\theta) = \sin\theta + \cos\theta - 1$.

(Note that when angles are used they should be measured in radians.)

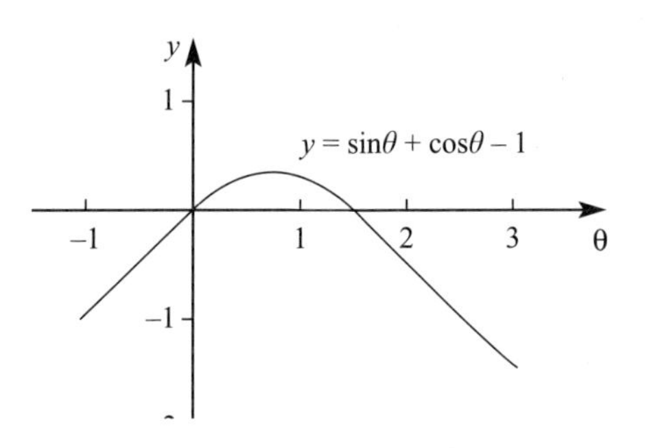

FIGURE 8.4

A calculator gives a sketch of $y = f(\theta)$ to be as in figure 8.4.
Using a decimal search

$$f(1) = 0.3818$$
$$f(2) = -0.5068$$

so the root lies between $\theta = 1$ and $\theta = 2$.

$$f(1.5) = 0.0682$$
$$f(1.6) = -0.0296$$
$$f(1.55) = 0.0206$$

so, to 1 decimal place, $\theta = 1.6$.

PROBLEMS WITH CHANGE OF SIGN METHODS

There are a number of situations which can cause problems for change of sign methods if they are applied blindly, for example by entering the equation into a computer program without prior thought. In all cases you can avoid problems by first drawing a sketch graph, provided that you know what dangers to look out for.

The curve touches the *x* axis

In this case there is no change of sign, so change of sign methods are doomed to failure (figure 8.5).

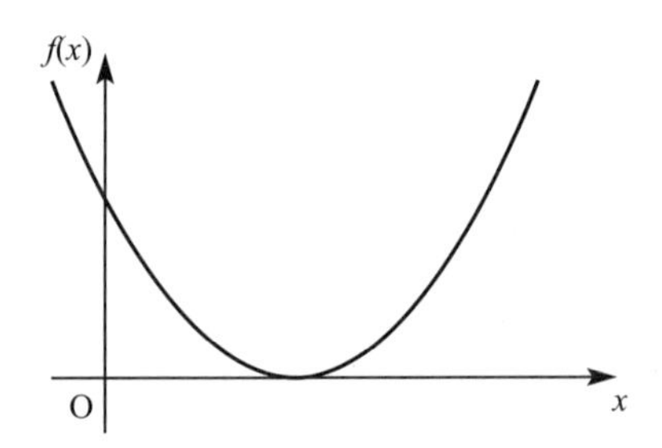

FIGURE 8.5

There are several roots close together

Where there are several roots close together, it is easy to miss a pair of them. The equation:

$$f(x) = x^3 - 1.9x^2 + 1.11x - 0.189 = 0$$

has roots at 0.3, 0.7 and 0.9. A sketch of the curve of f(x) is shown in figure 8.6.

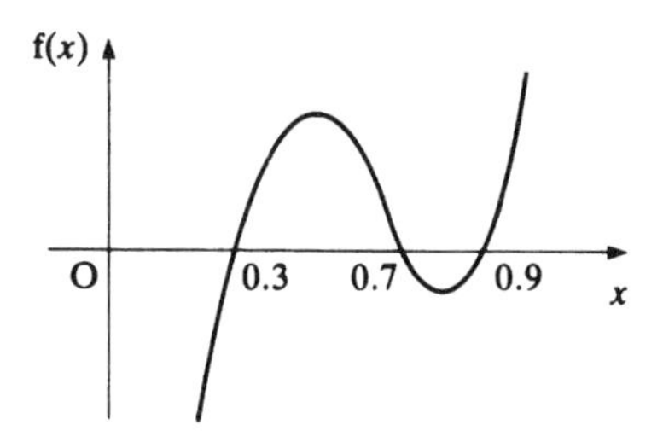

FIGURE 8.6

In this case $f(0) < 0$ and $f(1) > 0$, so you know there is a root between 0 and 1.

A decimal search would show that $f(0.3) = 0$, so that 0.3 is a root. You would be unlikely to search further in this interval.

There is a discontinuity in f(x)

The curve $y = \frac{1}{x - 2.7}$ has a discontinuity at $x = 2.7$, as shown in figure 8.7.

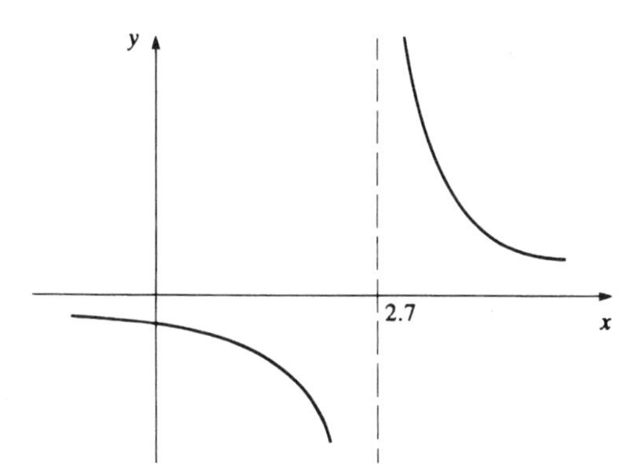

FIGURE 8.7

The equation $\frac{1}{x - 2.7} = 0$ has no root, but all change of sign methods will converge on a false root at $x = 2.7$.

None of these problems will arise if you start by drawing a sketch graph.

Note: Use of technology It is important that you understand how each method works and are able, if necessary, to perform the calculations using only a scientific calculator. However, these repeated operations lend themselves to the use of a spreadsheet or a programmable calculator and if possible you will benefit from using a variety of approaches when working through the following exercises.

EXERCISE 8A

1 Use a decimal search to find the roots of $x^5 - 5x + 3 = 0$ in the intervals $[-2, -1]$ and $[1, 2]$, correct to 2 decimal places.

2 (a) Use a systematic search for a change of sign, starting with $x = -2$, to locate intervals of unit length containing each of the three roots of $x^3 - 4x^2 - 3x + 8 = 0$.
(b) Sketch the graph of $f(x) = x^3 - 4x^2 - 3x + 8$.
(c) Find each of the roots correct to 2 decimal places.
(d) Use your last intervals to give each of the roots in the form $x = a$ with a maximum error of $(0.5)n$ stating your values of a and n.

3 The diagram shows a sketch of the graph of $f(x) = e^x - x^3$ without scales.

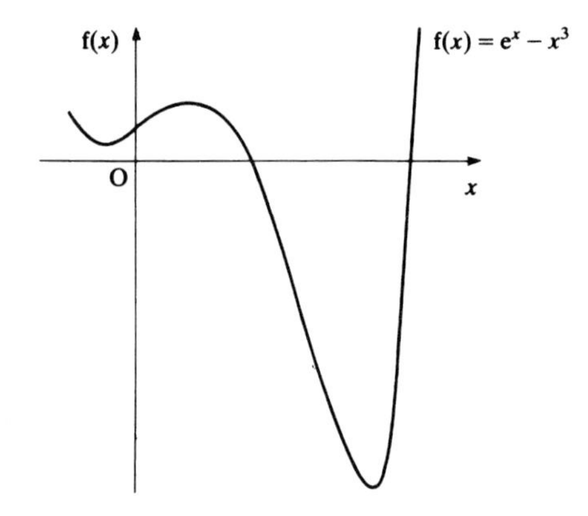

(a) Use a systematic search for a change of sign to locate intervals of unit length containing each of the roots.
(b) Find each of the roots correct to 3 decimal places.

4 (a) Show that the equation $x^3 + 3x - 5 = 0$ has no turning points.
(b) Show with the aid of a sketch that the equation can have only one root, and that this root must be positive.
(c) Find the root, correct to 3 decimal places.

5 (a) How many roots has the equation

$$e^x - 3x = 0.$$

(b) Find an interval of unit length containing each of the roots.
(c) Find each root correct to 2 decimal places.

6 **(a)** Sketch $y = 2^x$ and $y = x + 2$ on the same axes.
(b) Use your sketch to deduce the number of roots of the equation $2^x = x + 2$.
(c) Find each root, correct to 3 decimal places if appropriate.

7 Find all the roots of $x^3 - 3x + 1 = 0$, giving your answers correct to 2 decimal places.

8 For each of the equations below:
(i) sketch the curve;
(ii) write down any roots;
(iii) investigate what happens when you use a change of sign method with a starting interval of $[-0.3, 0.7]$.

(a) $y = \dfrac{1}{x}$ **(b)** $y = \dfrac{x}{x^2 + 1}$ **(c)** $y = \dfrac{x^2}{x^2 + 1}$

9 Given that $f(x) = x^3 + ax^2 + b$ and $f(1) = 2$ with $f(2) = 21$ find the value of a and show that $b = -3$.

$f(x) = 0$ has one positive root. By evaluating $f(x)$ for appropriate values of x find this root correct to 1 decimal place.

10 On a single diagram sketch, over $-\pi \leqslant x \leqslant \pi$, the graphs of

$y = \sin x$ and $y = \frac{1}{5}x + \frac{2}{5}$.

From your sketch state the approximate values of x for which

$\sin x = \frac{1}{5}x + \frac{2}{5}$ over $-\pi \leqslant x \leqslant \pi$.

This equation has a root α in the range 0 to $\frac{\pi}{2}$ radians. By rewriting the equation as $5\sin x - x - 2 = 0$ show that α lies betweeen 0.5 and 0.6 and find the value of α correct to 2 decimal places.

ITERATIVE METHODS TO SOLVE $f(x) = 0$

The method requires the equation $f(x) = 0$ to be rewritten in the form $x = g(x)$ and using an *iterative process* with this equation to generate a sequence of numbers by continued repetition of the same procedure. If the numbers obtained in this manner approach some limiting value, then they are said to *converge* to this value.

Figure 8.8 illustrates the fact that $f(x) = 0$ and its rearrangement $x = g(x)$ has the same root using the equations $f(x) = x^2 - x - 2$ and $x = g(x) = x^2 - 2$.

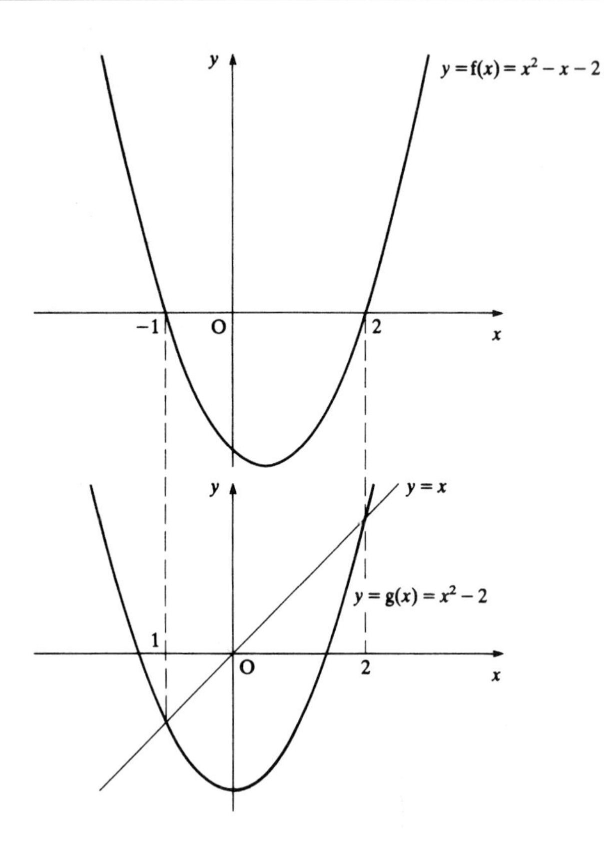

FIGURE 8.8

The equation $x^5 - 5x + 3 = 0$ which you met earlier can be re-written in a number of ways. One of these is $5x = x^5 + 3$, giving

$$x = g(x) = \frac{x^5 + 3}{5}$$

Figure 8.9 shows the graphs of $y = x$ and $y = g(x)$ in this case.

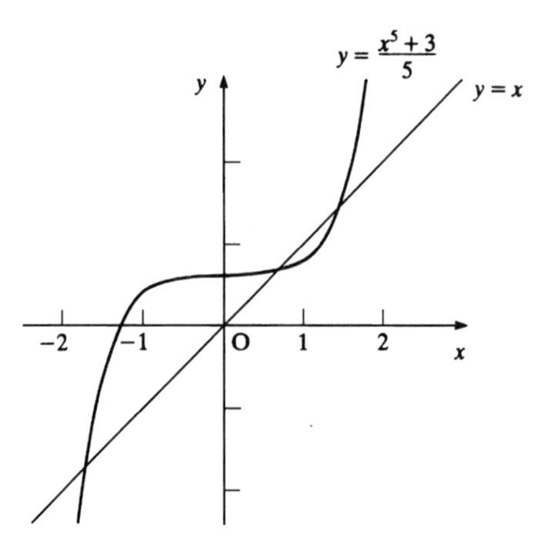

FIGURE 8.9

The rearrangement provides the basis for the iterative formula

$$x_{n+1} = \frac{x_n^{\,5} + 3}{5}.$$

It is known that there is a root in the interval [0, 1] so take $x_1 = 1$ as a starting point and use the iteration to find successive approximations:

$x_1 = 1$
$x_2 = 0.8$
$x_3 = 0.6655$
$x_4 = 0.6261$
$x_5 = 0.6192$
$x_6 = 0.6182$
$x_7 = 0.6181$
$x_8 = 0.6180$
$x_9 = 0.6180$

On your calculator if you have an ANS key, key in 1 then = followed by $(\text{ANS}^5 + 3) \div 5$, then repeatedly press the = key

In this case the iteration has converged quite rapidly to the root for which you were looking.

EXAMPLE 8.3

Show that $x^5 - 5x + 3 = 0$ can be rearranged to $x = \sqrt[5]{5x - 3}$. Use this to provide an iterative formula to find the root between 1 and 2 to 4 decimal places.

Solution Rearranging

$$x^5 - 5x + 3 = 0$$
$$x^5 = 5x - 3$$
$$x = \sqrt[5]{5x - 3}.$$

This becomes the iterative formula

$$x_{n+1} = \sqrt[5]{5x_n - 3}.$$

Starting with $x_1 = 1$

$x_2 = 1.148\,69$
$x_3 = 1.223\,65$
$x_4 = 1.255\,40$
$\vdots$
$x_{12} = 1.275\,67$
$x_{13} = 1.275\,68$
$x_{14} = 1.275\,68$

So the root between 1 and 2 is 1.2757 correct to 4 decimal places.

Note

To give an answer to a particular number of decimal places it is often sufficient to carry out the iteration until the next decimal place does not change, and then round off. Care is needed, particularly if convergence is slow, since the iteration may continue to change the rounded off value. The more iterations you can do the better!

The iteration process is easiest to understand if you consider the graph. Rewriting the equation $f(x) = 0$ in the form $x = g(x)$ means that instead of looking for points where the graph of $y = f(x)$ crosses the x axis, you are now finding the points of intersection of the curve $y = g(x)$ and the line $y = x$.

What you do	What it looks like on the graph
• Choose a value, x_1, of x	Take a starting point on the x axis
• Find the corresponding value of $g(x_1)$	Move vertically to the curve $y = g(x)$
• Take this value $g(x_1)$ as the new value of x, i.e. $x_2 = g(x_1)$	Move horizontally to the line $y = x$
• Find the value of $g(x_2)$ and so on.	Move vertically to the curve

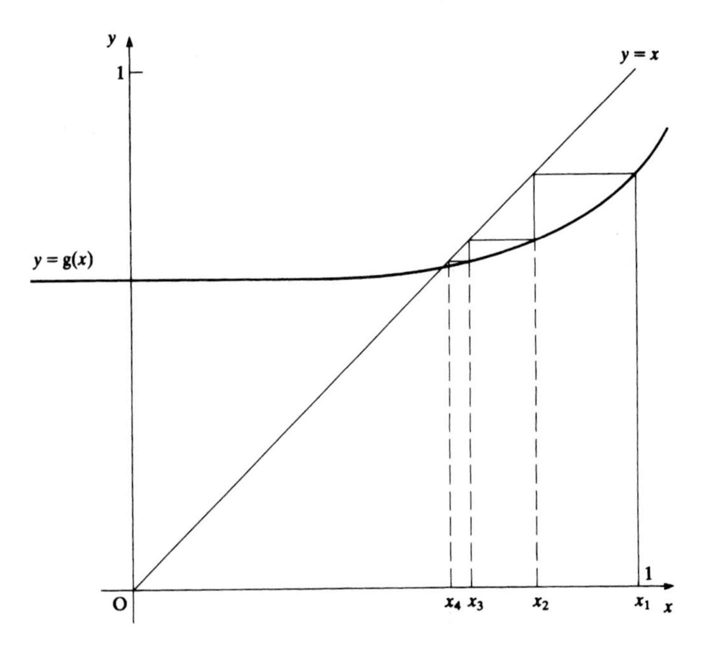

FIGURE 8.10

The effect of several repeats of this procedure is shown in figure 8.10. The successive steps look like a staircase approaching the root: this type of diagram is called a *staircase diagram*. In other examples, a *cobweb diagram* may be produced, as shown in figure 8.11.

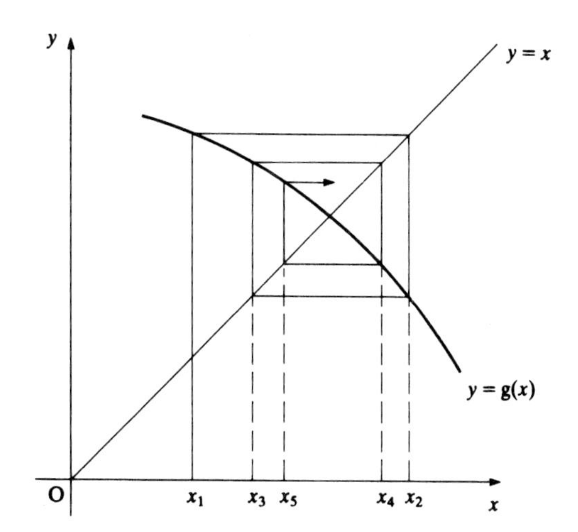

FIGURE 8.11

Successive approximations to the root are found by using the formula

$$x_{n+1} = g(x_n).$$

This is an example of an *iterative formula*. If the resulting values of x_n approach some limit, a, then $a = g(a)$, and so a is a *fixed point* of the iteration. It is also a root of the original equation $f(x) = 0$.

Note

In the staircase diagram, the values of x_n approach the root from one side, but in a cobweb diagram they oscillate about the root. From figures 8.10 and 8.11 it is clear that the error (the difference between a and x_n) is decreasing in both diagrams.

USING DIFFERENT ARRANGEMENTS OF THE EQUATION

So far we have used two possible arrangements of the equation $x^5 - 5x + 3 = 0$.

The first, $x = \frac{x^5 + 3}{5}$ converged to 0.618 and the second $x = \sqrt[5]{5x - 3}$ converged to 1.2757 when a starting point $x_1 = 1$ was used.

The processes have clearly converged. If instead you had taken $x_1 = 0$ as your starting point and applied the second formula, you would have obtained a sequence converging to the value –1.6180, the root in the interval [–2, –1]. The second formula does not appear to converge to the root in the interval [0, 1].

THE CHOICE OF g(x)

A particular rearrangement of the equation $f(x) = 0$ into the form $x = g(x)$ will allow convergence to a root a of the equation, provided that $-1 < g'(a) < 1$ for values of x close to the root.

Look again at the two rearrangements of $x^5 - 5x + 3 = 0$ which were suggested.

When you look at the graph of:

$$y = \mathrm{g}(x) = \sqrt[5]{5x - 3},$$

you can see that its gradient near A, the root you were seeking, is greater than 1 (figure 8.12). This makes

$$x_{n+1} = \sqrt[5]{5x_n - 3},$$

an unsuitable iterative formula for finding the root in the interval [0, 1], as you saw earlier.

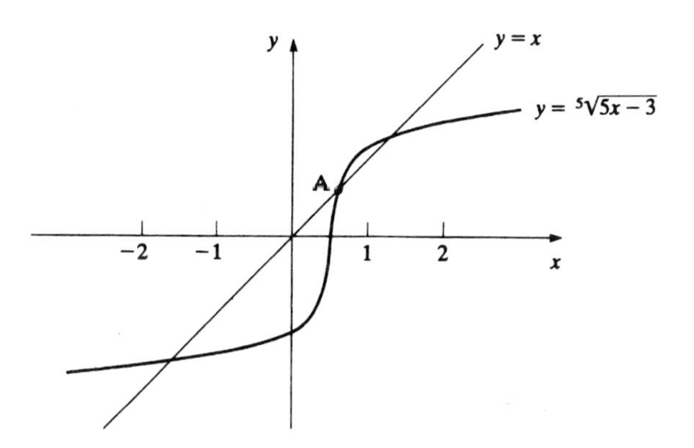

FIGURE 8.12

When an equation has two or more roots, a single rearrangement will not usually find all of them. This is demonstrated in figure 8.13.

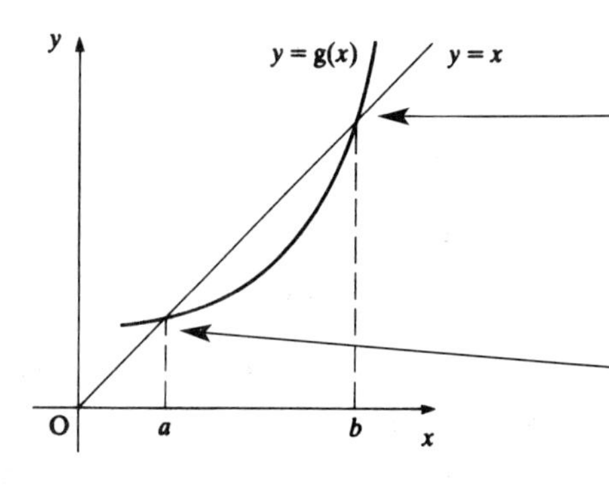

FIGURE 8.13

EXAMPLE 8.4

Show that $x_{n+1} = \sqrt{5x_n - 3}$ is an iterative formula which may be used to solve $x^2 - 5x + 3 = 0$.

Use the formula with $x_1 = 4$ to find one root of the equation correct to 3 decimal places.

Illustrate the convergence to the root on an appropriate diagram.

Solution Rearranging the equation

$$x^2 - 5x + 3 = 0$$

gives $x^2 = 5x - 3$

then $x = \sqrt{5x - 3}$

which, as an iterative formula is,

$$x_{n+1} = \sqrt{5x_n - 3}.$$

The iteration gives

$$x_1 = 4$$
$$x_2 = 4.123...$$
$$x_3 = 4.197...$$
$$x_4 = 4.240...$$

which eventually converges to 4.303 to 3 decimal places.

To illustrate the convergence sketch $y = x$ and $x = \sqrt{5x - 3}$ on the same diagram (see figure 8.14).

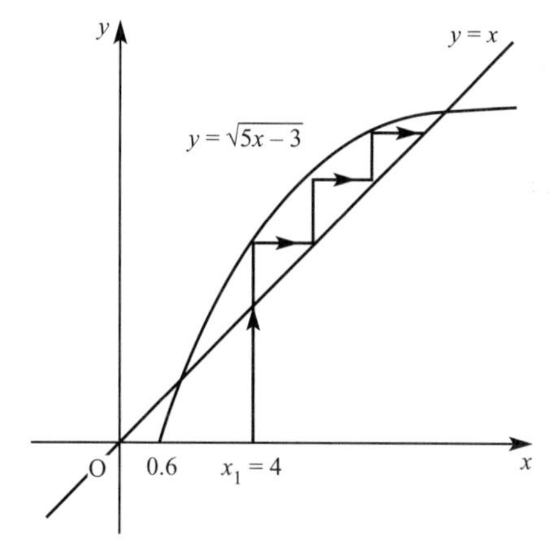

FIGURE 8.14

WHEN DOES THIS METHOD FAIL?

It is always possible to rearrange an equation $f(x) = 0$ into the form $x = g(x)$, but this only leads to a successful iteration if:

(a) successive iterations converge;

(b) they converge to the root for which you are looking.

EXERCISE 8B

1 **(a)** Show that the equation $x^3 - x - 2 = 0$ has a root between 1 and 2.

(b) The equation is rearranged into the form $x = g(x)$, where $g(x) = \sqrt[3]{x + 2}$. Sketch $y = g(x)$ and show that starting values of both 1 and 2 will converge to the root in the interval [1, 2].

(c) Use the iterative formula suggested by this rearrangement to find the value of the root to 3 decimal places.

2 **(a)** Show that the equation $e^{-x} - x + 2 = 0$ has a root in the interval [2, 3].

(b) The equation is rearranged into the form $x = g(x)$ where $g(x) = e^{-x} + 2$. Use the iterative formula suggested by this rearrangement to find the value of the root to 3 decimal places.

3 **(a)** By considering $f'(x)$, where $f(x) = x^3 + x - 3$, show that there is exactly one real root of the equation $x^3 + x - 3 = 0$.

(b) Show that the root lies in the interval [1, 2].

(c) Rearrange the equation to give the iterative formula $x_{n+1} = \sqrt[3]{3 - x_n}$.

(d) Hence find the root correct to 4 decimal places.

4 **(a)** Show that the equation $e^x + x - 6 = 0$ has a root in the interval [1, 2].

(b) Show that this equation may be written in the form $x = \ln(6 - x)$.

(c) Hence find the root correct to 3 decimal places.

5 **(a)** Sketch the curves $y = e^x$ and $y = x^2 + 2$ on the same graph.

(b) Use your sketch to explain why the equation $e^x - x^2 - 2 = 0$ has only one root.

(c) Rearrange this equation to give the iteration $x_{n+1} = \ln(x_n^2 + 2)$.

(d) Find the root correct to 3 decimal places

6 **(a)** On the same diagram sketch $y = x$ and $y = 3\ln(x + 1)$.

(b) State one solution of $x = 3\ln(x + 1)$ and find the second, to 3 decimal places, by an iterative process.

(c) Show on your diagram how the iterative process has converged to the root.

7 **(a)** Sketch the graphs of $y = x$ and $y = \cos x$ on the same axes, for

$$0 \leqslant x \leqslant \frac{\pi}{2}.$$

(b) Find the solution of the equation $x = \cos x$ to 5 decimal places.

8 Given that $x_{n+1} = \dfrac{(x_n^2 + 1)}{3}$.

(a) Find the equation that this iteration solves.

(b) Use the iteration with a starting value of 1 to find a solution of the equation correct to 3 significant figures.

(c) By drawing $y = x$ and $y = \dfrac{(x^2 + 1)}{3}$ explain the convergence of your solutions.

(d) Find an alternative rearrangement that gives a convergence to the other root, and find this root to 3 significant figures.

NUMERICAL INTEGRATION

There are times when you need to find the area under a graph but cannot do this by the integration methods you have met already.

- The function may be one that cannot be integrated algebraically. (There are many such functions.)
- The function may be one that can be integrated algebraically but which requires a technique with which you are unfamiliar.
- It may be that you do not know the function in algebraic form, but just have a set of points (perhaps derived from an experiment).

In these circumstances you can always find an approximate answer using a numerical method, but you must:

(a) have a clear picture in your mind of the graph of the function, and how your method estimates the area beneath it;

(b) understand that a numerical answer without any estimate of its accuracy, or error bounds, is valueless.

THE TRAPEZIUM RULE

In this chapter just one numerical method of integration is introduced, namely the *trapezium rule*. As an illustration of the rule, we shall use it to find the area under the curve $y = \sqrt{5x - x^2}$ for values of x between 0 and 4.

It is in fact possible to integrate this function algebraically, but not using the techniques that you have met so far.

Note You should not use a numerical method when an algebraic (sometimes called analytic) technique is available to you. Numerical methods should be used only when other methods fail.

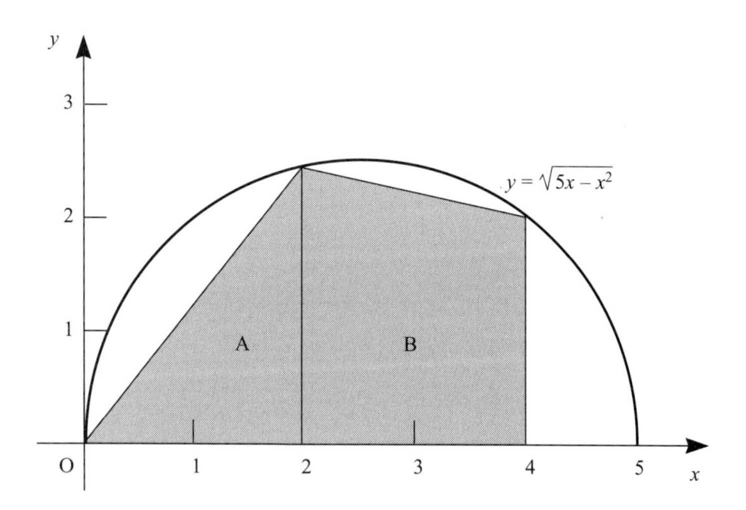

FIGURE 8.15

Figure 8.15 shows the area approximated by two trapezia of equal width.

Remember the formula for the area of a trapezium

$$\text{Area} = \tfrac{1}{2}h(a + b)$$

where a and b are the lengths of the parallel sides and h the distance between them.

In the cases of the trapezia A and B, the parallel sides are vertical. The left-hand side of trapezium A has 0 height, and so the trapezium is also a triangle.

When $x = 0 \Rightarrow y = \sqrt{0} = 0$

when $x = 2 \Rightarrow y = \sqrt{6} = 2.4495$ (to 4 dp)

when $x = 4, \Rightarrow y = \sqrt{4} = 2.$

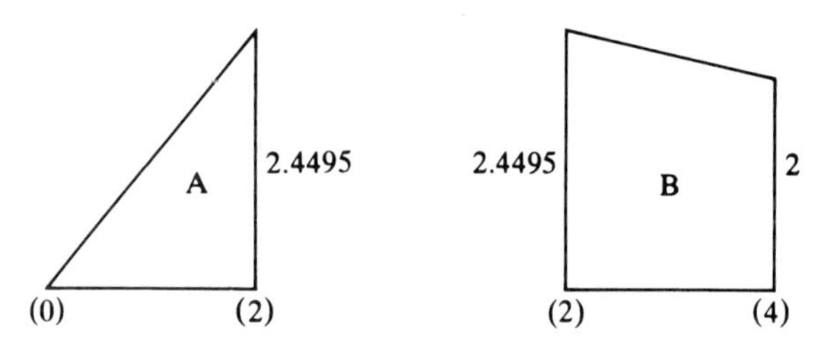

FIGURE 8.16

The area of trapezium A = $\frac{1}{2} \times 2 \times (0 + 2.4495) = 2.4495$
The area of trapezium B = $\frac{1}{2} \times 2 \times (2.4495 + 2) = 4.4495$

Total 6.8990

For greater accuracy we can use four trapezia, P, Q, R and S, each of width 1 unit as in figure 8.17. The area is estimated in just the same way.

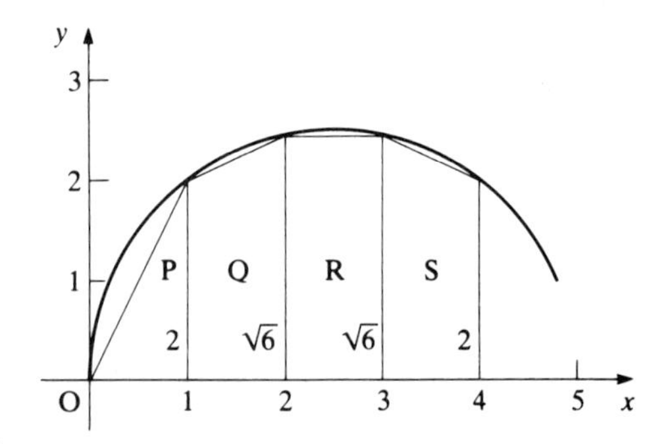

FIGURE 8.17

Trapezium P: $\frac{1}{2} \times 1 \times (0 + 2)$ = 1.0000
Trapezium Q: $\frac{1}{2} \times 1 \times (2 + 2.4495)$ = 2.2247
Trapezium R: $\frac{1}{2} \times 1 \times (2.4495 + 2.4495)$ = 2.4495
Trapezium S: $\frac{1}{2} \times 1 \times (2.4495 + 2)$ = 2.2247

Total 7.8990

These figures are given to 4 decimal places but the calculation has been done to more places on a calculator

Accuracy

In this example, the first two estimates are 6.8989... and 7.8989... . You can see from figure 8.17 that the trapezia all lie underneath the curve, and so in this case the trapezium rule estimate of 7.8989... must be too small. You cannot, however, say by how much. To find that out you will need to take progressively more strips and see how the estimate homes in. Using 8 strips gives an estimate of 8.2407..., and 16 strips gives 8.3578... . The first figure, 8, looks reasonably certain but it is still not clear whether the second is 3, 4 or even 5. You need to take even more strips to be able to decide. In this example, the convergence is unusually slow because of the high curvature of the curve.

In general, the area that the integral $\int_a^b f(x)\,dx$ represents is divided into trapezia. The width of each trapezium is h. This is illustrated in figure 8.18.

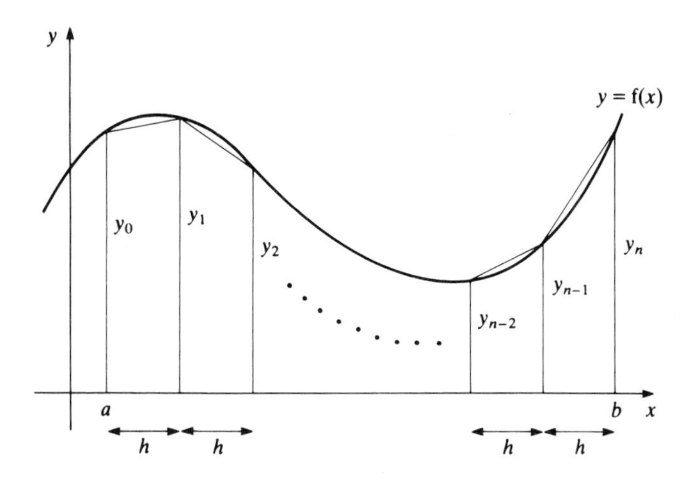

FIGURE 8.18

If y_0 is the value of the function at the left-hand limit $x = a$, y_1 is the value at $x = a + h$, y_2 at $a + 2h$ up to y_n at $x = b$ then $\int_a^b f(x)\,dx$ is approximated by the sum of the areas A of all the trapezia,
i.e.

$$A \approx \tfrac{1}{2}(y_0 + y_1)h + \tfrac{1}{2}(y_1 + y_2)h + \tfrac{1}{2}(y_2 + y_3)h + \ldots + \tfrac{1}{2}(y_{n-1} + y_n)h$$

factorising gives

$$A \approx \tfrac{1}{2}h(y_0 + y_1 + y_1 + y_2 + y_2 + y_3 + \ldots\ y_{n+1} + y_n)$$

and this simplifies to the trapezium rule:

$$A \approx \tfrac{1}{2}h\{y_0 + y_n + 2(y_1 + y_2 + y_3 + \ldots + y_{n-1})\}.$$

This is often said in words as

$$A \approx \tfrac{1}{2} \times \text{strip width} \times \{\text{first} + \text{last} + 2 \times \text{rest}\}.$$

The proof of the derivation of the trapezium rule should be known.

You have to decide from the shape of the graph of $y = f(x)$ as to whether the trapezium rule gives an overestimate, an underestimate or whether you cannot tell.

The following diagrams illustrate different situations:

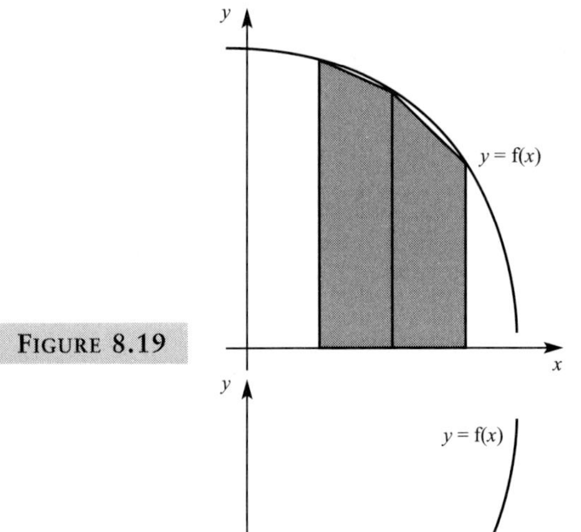

FIGURE 8.19

The trapezium rule will give an underestimate.

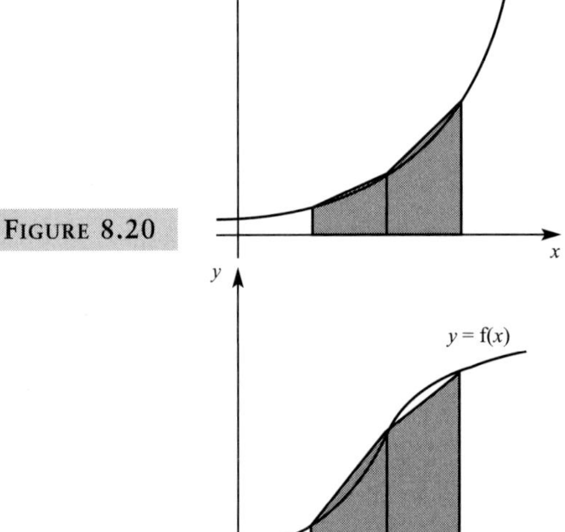

FIGURE 8.20

The trapezium rule will give an overestimate.

FIGURE 8.21

You cannot tell whether the trapezium rule is an underestimate or an overestimate.

EXAMPLE 8.5

Use the trapezium rule with 5 ordinates to evaluate $\int_1^2 \frac{1}{x^2}\,dx$, correct to 4 significant figures.

By drawing a sketch state whether the trapezium rule gives an overestimate or an underestimate to the exact value.

Calculate the percentage error between the trapezium rule value and the exact value.

Solution The word 'ordinate' means y value, so 5 ordinates means 4 strips. As you wish to find the integral from $x = 1$ to $x = 2$ the width of each strip, h, is 0.25. Tabulating the values gives:

x	$\frac{1}{x^2}$	
1	1	
1.25	0.64	
1.5	0.444 44	} 1.410 975
1.75	0.032 65	
2	0.25	

and these are used in the trapezium rule to give

$$\int_1^2 \frac{1}{x^2}\,dx \approx \tfrac{1}{2} \times 0.25 \times \{1 + 0.25 + 2 \times 1.410\,975\} = 0.508\,993\ldots$$

$$\approx 0.5090 \text{ (4 significant figures).}$$

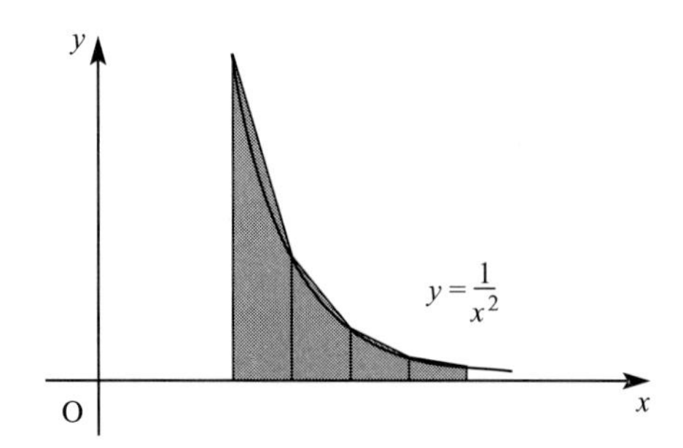

FIGURE 8.22

From figure 8.22 you can see that the trapezium rule gives an overestimate of the integral.

The exact value is

$$\int_1^2 \frac{1}{x^2}\,dx = \left[-\frac{1}{x}\right]_1^2 = -\frac{1}{2} - (-1) = 0.5.$$

So the percentage error is

$$\frac{0.5090 - 0.5}{0.5} \times 100 = 1.8\%.$$

EXERCISE 8C

In questions 1–3, use the trapezium rule to estimate the areas shown.

Start with 2 strips, then 4 and then 8 in each case.

State, with reasons, whether your final estimate is an overestimate or an underestimate.

1 $y = \dfrac{3}{(x+1)}, x > -1$

2 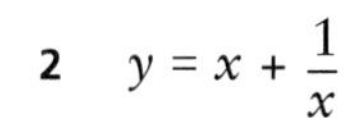 $y = x + \dfrac{1}{x}$

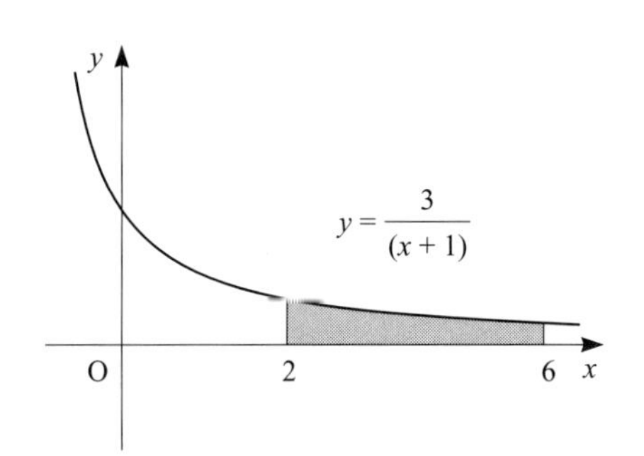

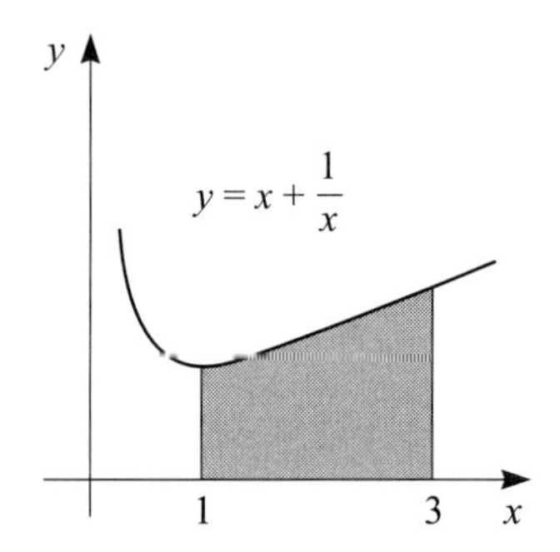

3 $y = \dfrac{1}{x^2}, x > 0$

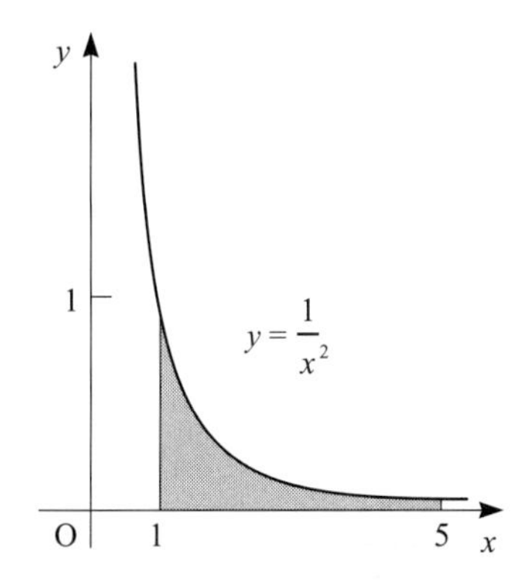

4 Evaluate $\int_0^\pi e^{\sin x}\, dx$ using the trapezium rule with four strips.

5 **(a)** Draw the graph of $y = \frac{1}{x}$ for values of x between 1 and 10. Shade the region A bounded by the curve, the lines $x = 1$ and $x = 2$ and the x axis, and the region B bounded by the curve, the lines $x = 4$ and $x = 8$ and the x axis.

(b) For each of the regions A and B, find estimates of the area using the trapezium rule with **(i)** 2 **(ii)** 4 **(iii)** 8 strips.

(c) Which region has the greater area?

6 The speed v in ms^{-1} of a train is given at time t seconds in the following table.

t	0	10	20	30	40	50	60
v	0	5.0	6.7	8.2	9.5	10.6	11.6

The distance the train has travelled is given by the area under the graph of the speed (vertical axis) against time (horizontal axis).

(a) Estimate the distance the train travels in this 1-minute period.

(b) Give two reasons why your method cannot give a very accurate answer.

7 The definite integral $\int_0^1 \frac{1}{1 + x^2}\,dx$ is known to equal $\frac{\pi}{4}$.

(a) Using the trapezium rule for 5 strips, find an approximation for π.

(b) Repeat your calculation with 10 and 20 strips to obtain closer estimates.

(c) If you did not know the value of π, what value would you give it with confidence on the basis of your estimates in parts (a) and (b)?

8 The table below gives the values of a function $f(x)$ for different values of x.

x	0	0.5	1.0	1.5	2.0	2.5	3.0
$f(x)$	1.000	1.225	1.732	2.345	3.000	3.674	4.359

(a) Apply the trapezium rule to the values in this table to obtain an approximation for $\int_0^3 f(x)\,dx$.

(b) By considering the shape of the curve $y = f(x)$, explain whether the approximation calculated in part (a) is likely to be an overestimate or an underestimate of the true area under the curve $y = f(x)$ between $x = 0$ and $x = 3$.

[MEI]

9

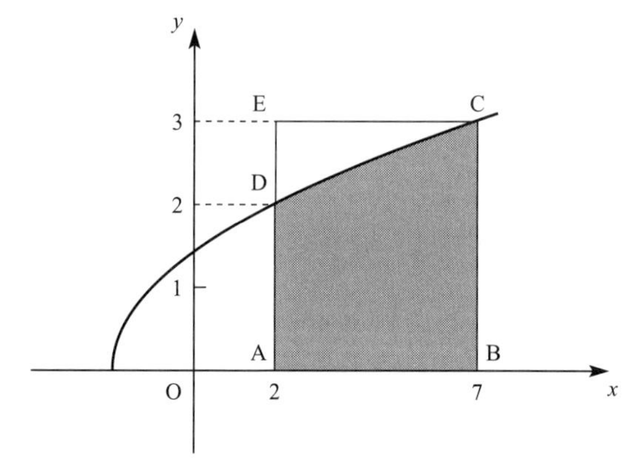

The graph of the function $y = \sqrt{2 + x}$ is given in the diagram. The area of the shaded region ABCD is to be found.

(a) Make a table of values for y, for integer values of x from $x = 2$ to $x = 7$, giving each value of y correct to 4 decimal places.

(b) Use the trapezium rule with 5 strips, each 1 unit wide, to calculate an estimate for the area ABCD. State, giving a reason, whether your estimate is too large or too small.

Another method is to consider the area ABCD as the area of the rectangle ABCE minus the area of the region CDE.

(c) Show that the area CDE is given by $\int_2^3 (y^2 - 4)\,dy$. Calculate the exact value of this integral.

(d) Find the exact value of the area ABCD. Hence find, as a percentage, the relative error in using the trapezium rule.

[MEI]

10 (a) The table gives the values of the function $(1 + x^2)^5$ for $x = 0$, 0.2 and 0.4.

x	0	0.1	0.2	0.3	0.4
$(1 + x^2)^5$	1.00000		1.21665		2.10034

Complete the table for $x = 0.1$ and 0.3.

Use the trapezium rule with 4 strips to estimate the value of

$$\int_0^{0.4} (1 + x^2)^5\,dx.$$

(b) Use the binomial theorem to expand $(1 + x^2)^5$ as a polynomial in ascending powers of x.

The integral $\int_0^{0.4} (1 + 5x^2 + 10x^4)\,dx$ is to be used to estimate the integral in part (a).

Evaluate $\int_0^{0.4} (1 + 5x^2 + 10x^4)\,dx$.

(c) The diagram below shows a sketch of the graph of $y = (1 + x^2)^5$. State whether the estimate in part (a) is too high or too low and use the graph to explain your answer.

State also whether the estimate in part (b) is too high or too low and explain your answer.

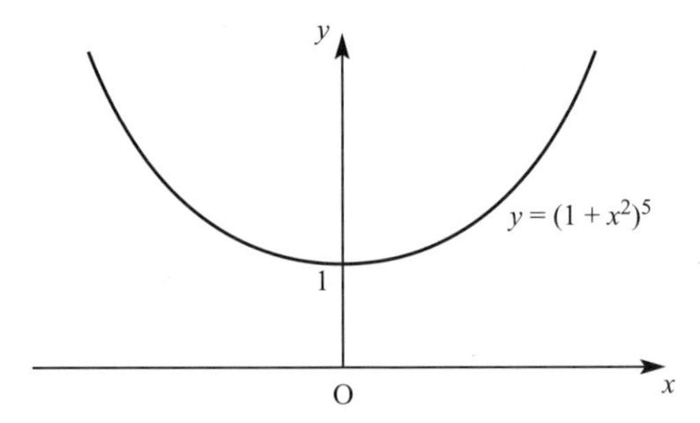

(d) How could the methods for estimating $\int_0^{0.4} (1 + x^2)^5\,dx$ used in parts (a) and (b) be improved to give greater accuracy?

[MEI]

EXERCISE 8D

Examination-style questions

1 A function f(x) is given by

$$f(x) = e^x - 2x - 6.$$

(a) Show that $f(x) = 0$ has a root in the interval [2, 3].

(b) Find the value of the integer N, such that the root found in part (a) lies in the interval $\left[\frac{N}{10}, \frac{N+1}{10}\right]$.

(c) Find between which two integers the negative root of $f(x) = 0$ lies.

2 A function f is given by

$$f(x) = x^3 - 2x^2 - 8x + 8.$$

(a) By a change of sign method show that $f(x) = 0$ has a root in the interval [0, 1].

(b) Find the value of the negative integer, N, for which the negative root of $f(x) = 0$ lies in the interval $[N, N + 1]$. Show sufficient working to justify your answer.

(c) How many other roots does $f(x) = 0$ have?

3 A circle of radius r has a sector OAB making an angle of θ radians at the centre O.

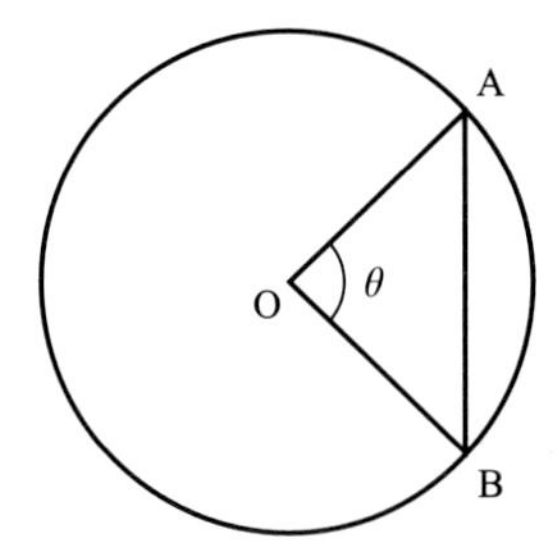

(a) If the area of the sector is twice the area of the triangle OAB show that

$$\sin\theta = \tfrac{1}{2}\theta.$$

(b) On the same diagram sketch $y = \sin\theta$ and $y = \frac{1}{2}\theta$ for $0 \leqslant \theta \leqslant \pi$.

(c) From your diagram give an approximate value for the non-zero solution to $\sin\theta = \frac{1}{2}\theta$ and, by a change of sign method, find the range of numbers within which this value lies correct to 2 decimal places.

4 A curve has equation $x^2 - 2x - 5 = 0$.

(a) Show that the equation can be rearranged to give the iteration

$$x_{n+1} = \sqrt{ax_n + b}$$

stating the values of a and b. Use the iteration with $x_1 = 3$ to find the positive solution to the quadratic equation correct to 3 decimal places, showing the result of each iteration.

(b) Prove that an alternative rearrangement gives the iteration

$$x_{n+1} = \frac{5}{x_n - 2}$$

and use this iteration to find the negative root to 3 decimal places.

5 Show that the equation $x^5 - 5x - 6 = 0$ has a root in the interval [1, 2]. Stating the values of the constants p, q and r, use an iteration of the form

$$x_{n+1} = (px_n + q)^{\frac{1}{r}}$$

the appropriate number of times to calculate this root of the equation $x^5 - 5x - 6 = 0$ correct to 3 decimal places. Show sufficient working to justify your final answer.

[Edexcel]

6 **(a)** By sketching the curves with equations $y = 4 - x^2$ and $y = e^x$ show that the equation $x^2 + e^x - 4 = 0$ has one negative root and one positive root.

(b) Use the iteration formula

$$x_{n+1} = -(4 - e^{x_n})^{\frac{1}{2}}$$

with $x_0 = -2$ to find in turn x_1, x_2, x_3 and x_4 and hence write down an approximation to the negative root of the equation, giving your answer to 4 decimal places.

An attempt to evaluate the positive root of the equation is made using the iteration formula

$$x_{n+1} = (4 - e^{x_n})^{\frac{1}{2}}$$

with $x_0 = 1.3$.

(c) Describe the result of such an attempt.

[Edexcel]

7 **(a)** Rearrange the cubic equation $x^3 - 6x - 2 = 0$ into the form

$$x = \pm\sqrt{a + \frac{b}{x}}.$$

State the values of a and b.

(b) Use the iterative formula

$$x_{n+1} = \sqrt{a + \frac{b}{x_n}}$$

with $x_0 = 2$ and your values of a and b to find the approximate solution x_4 of the equation, to an appropriate degree of accuracy. Show all your intermediate answers.

[Edexcel]

8 The diagram shows the region R bounded by part of the curve $y = (1 + x^2)^{\frac{3}{2}}$ and the lines $x = 1$ and $x = 2$.

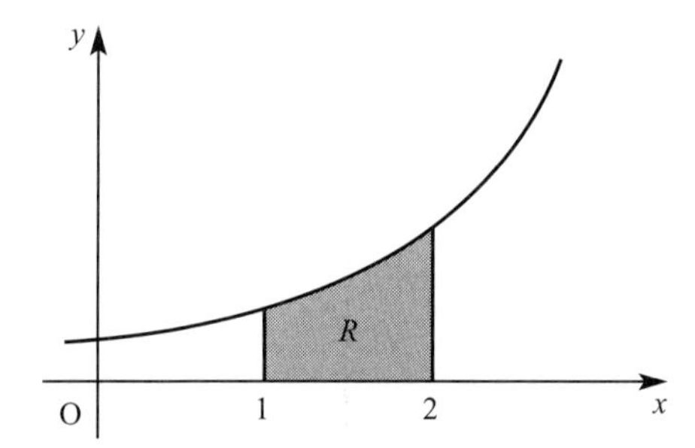

Use the trapezium rule with five strips to calculate an approximation for the area of R, showing your working and giving your answer to a suitable degree of accuracy.

Explain with the aid of a sketch whether the the approximation is an overestimate or an underestimate.

9 Use the trapezium rule with six ordinates to find an approximation to

$$\int_0^{0.5} \frac{1}{\sqrt{1 - x^2}}\, dx$$

giving your answer to 4 significant figures.

It is known that $\int_0^{0.5} \frac{1}{\sqrt{1 - x^2}}\, dx = \frac{\pi}{6}$. Use your answer to find an approximate value for π.

10 A function f is given by

$$f(x) = e^x + \frac{1}{x}.$$

(a) Find $\int_1^2 f(x)\, dx$ using the trapezium rule with five strips.

(b) Find the exact value of $\int_1^2 f(x)\, dx$.

(c) Find the percentage error between the trapezium rule value and the exact value.

KEY POINTS

1 Location of roots by change of sign

If $y = \mathrm{f}(x)$ is continuous over $[a, b]$ then $\mathrm{f}(x) = 0$ has a root α in $[a, b]$ if $\mathrm{f}(a)$ and $\mathrm{f}(b)$ are of opposite sign. For example:

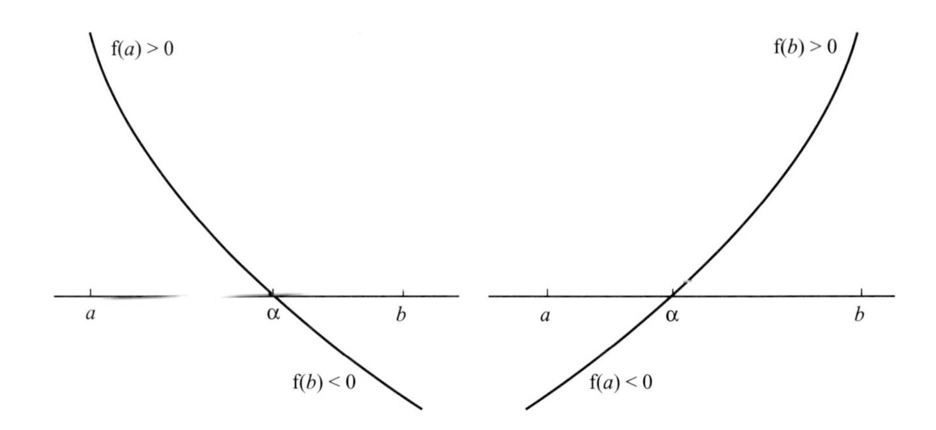

2 Iterative methods

To solve $\mathrm{f}(x) = 0$ re-write the equation as $x = \mathrm{g}(x)$ to give the iterative formula

$$x_{n+1} = \mathrm{g}(x_n).$$

The iteration will converge if $|\mathrm{g}'(x)| < 1$ close to the root.

3 Trapezium rule

$\int_a^b \mathrm{f}(x)\,\mathrm{d}x$ is approximated by the area A under the curve where

$$A \approx \tfrac{1}{2}h\{y_0 + y_n + 2(y_1 + y_2 + y_3 + \ldots + y_{n-1})\}$$

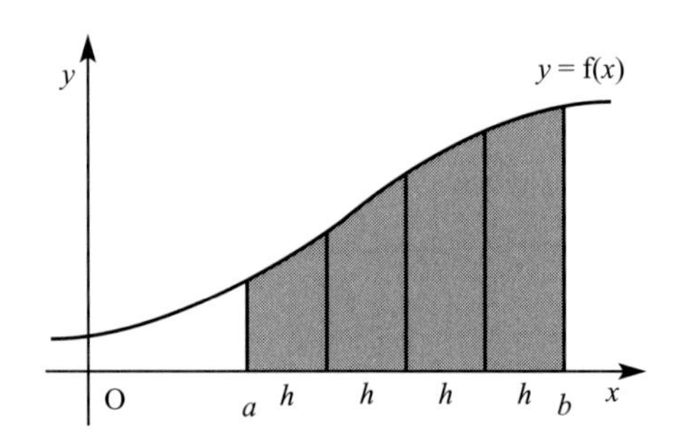

Appendix

PROOF

The Edexcel subject criteria states that: 'methods of proof, including proof by contradiction and disproof by counter example, are required. At least one paper will require the use of proof.' Throughout this book you have seen a number of examples of proof. The purpose of this appendix is to illustrate the three different methods and to give an exercise which will not only test ideas of proof but will serve to help with revision.

EXAMPLE A.1

Prove that if $f(x) = \dfrac{x-1}{3}$ then $f^{-1}(x) = 3x + 1$.

Solution Let $y = \dfrac{x-1}{3}$

Interchange x and y (recalling that the inverse is the reflection of the function in the line $y = x$),

so $\quad x = \dfrac{y-1}{3}$.

Rearrange to make y the subject:

$$3x = y - 1$$
$$y = 3x + 1\ .$$

Hence the inverse function is

$$f^{-1}(x) = 3x + 1.$$

EXAMPLE A.2

Prove that $\sqrt{2}$ is irrational.

Solution We shall prove this result by *contradiction*.

First assume that $\sqrt{2}$ is rational.

Let $\sqrt{2} = \frac{a}{b}$ where a and b are co-prime (that is, thay have no common factors).

So, squaring both sides and re-writing

$$a^2 = 2b^2.$$

$2b^2$ is even, so a^2 is even, hence a must be even.

Let $a = 2c$ from which

$$2b^2 = (2c)^2 = 4c^2 \text{ so that } b^2 = 2c^2.$$

But this implies that b is even, so a and b cannot be co-prime. This contradicts the initial condition, so the assumption that $\sqrt{2}$ is rational is false.

Therefore $\sqrt{2}$ is irrational.

EXAMPLE A.3

Prove that $\cos(\theta + 30°)$ is not the same as $\cos\theta + \cos 30°$ by putting $\theta = 60°$.

Solution When $\theta = 60°$,

$$\cos(\theta + 30°) = \cos 90° = 0$$

$$\begin{aligned}\cos\theta + \cos 30° &= \cos 60° + \cos 30° \\ &= \frac{1}{2} + \frac{\sqrt{3}}{2} \\ &= \frac{1 + \sqrt{3}}{2} \neq 0.\end{aligned}$$

This example illustrates a *disproof by counter example*.

EXERCISE AA

1 Prove that $\dfrac{2x + 3}{6x^2 + 5x - 6} \equiv \dfrac{1}{3x - 2}$.

2 Give an example of a quadratic equation which does not have real roots.

State the condition that proves that a quadratic equation does not have real roots.

3 Prove by counter example that if $u_n = 4^n + 1$ for $n \geqslant 0$ than u_n is not always prime.

4 Prove that $\tan(45° + 30°) \equiv \dfrac{\sqrt{3} + 1}{\sqrt{3} - 1}$.

5 Prove by counter example that $\sec(A + B)$ is equivalent to $\sec A + \sec B$, for all values of A and B.

6 Prove that $f(x) = \sin(x^2)$ is an even function but $f(x) = \sin(x^3)$ is not.

7 A student rearranges the expression $2x^2 + 2x - 5 = 0$ to obtain an iteration as follows:

$$2x^2 + 2x - 5 = 0$$

$$2x^2 = 5 - 2x \qquad ①$$

$$x^2 = \frac{5 - 2x}{2} \qquad ②$$

$$x^2 = 5 - x \qquad ③$$

$$x_{n+1} = \sqrt{5 - x_n} \qquad ④$$

In which line is there an error in the student's working?

8 A Mersenne number is a number of the form $2^n - 1$, where n is prime. Prove by counter example that not all Mersenne numbers are prime numbers.

9 Prove that the tangent to the curve $y = 2e^x + \ln x$ at the point where $x = 1$ is $y = (2e + 1)x - 1$.

10 Prove that if n is odd then n^2 is odd.

11 Show that $\cos 60° + \cos 30°$ is not equal to $\cos 90°$.

12 Prove that $\log_a xy \equiv \log_a x + \log_a y$.

13 Sketch the graph of $y = \frac{h}{r}x$.

The section of this line from $y = 0$ to $y = h$ is rotated about the y axis.

Show that the volume of revolution is $\frac{1}{3}r^2h$.

14 A curve has equation $y = \dfrac{1}{x - 1}$.

Sketch this curve for the domain $-3 \leqslant x \leqslant 5$.

From the graph explain why $x \neq 1$.

15 Prove by contradiction that $\sqrt{5}$ is irrational.

16 Show that the solution of $3^x = 5$ to 4 significant figures is 1.465.

Explain, with the aid of a sketch, why there is no solution to $3^x = -5$.

17 What is wrong with

$$\int_1^2 \frac{2x + 1}{2x}\,dx = \left[\frac{x^2 + x}{x^2}\right]_1^2$$

$$= \left[1 + \frac{1}{x}\right]_1^2$$

$$= \left(1 + \frac{1}{2}\right) - (1 + 1)$$

$$= -\frac{1}{2}?$$

Find the exact value of the integral.

18 Prove that

$$\sin 3x + \sin x \equiv 4\sin x - 4\sin^3 x.$$

19 Prove by counter example that $|\sin x|$ is not the same as $\sin|x|$ when x is measured in radians.

20 Prove by contradiction that there is not a largest prime number (that is, there is an infinite number of primes).

ANSWERS

CHAPTER 1

EXERCISE 1A (Page 4)

1 $\frac{2a^2}{3b^3}$

2 $\frac{1}{9y}$

3 $\frac{x+3}{x-6}$

4 $\frac{x+3}{x+1}$

5 $\frac{2x-5}{2x+5}$

6 $\frac{3(a+4)}{20}$

7 $\frac{x(2x+3)}{(x+1)}$

8 $\frac{2}{5(p-2)}$

9 $\frac{a-b}{2a-b}$

10 $\frac{(x+4)(x-1)}{x(x+3)}$

11 $\frac{9}{20x}$

12 $\frac{x-3}{12}$

13 $\frac{a^2+1}{a^2-1}$

14 $\frac{5x-13}{(x-3)(x-2)}$

15 $\frac{2}{(x+2)(x-2)}$

16 $\frac{2p^2}{(p^2-1)(p^2+1)}$

17 $\frac{a^2-a+2}{(a+1)(a^2+1)}$

18 $-\frac{2(y^2+4y+8)}{(y+2)^2(y+4)}$

19 $\frac{x^2+x+1}{x+1}$

20 $-\frac{(3b+1)}{(b+1)^2}$

21 $\frac{13x-5}{6(x-1)(x+1)}$

22 $\frac{4(3-x)}{5(x+2)^2}$

23 $\frac{3a-4}{(a+2)(2a-3)}$

24 $\frac{3x^2-4}{x(x-2)(x+2)}$

EXERCISE 1B (Page 8)

1 (a) 84
(b) 4
(c) –2
(d) 5.24 or 0.76
(e) 3 or $\frac{1}{3}$
(f) 0 or 3
(g) 1.71 or 0.29

2 (a) $\frac{600}{x}$
(b) $\frac{600}{x-1}$
(c) $x^2 - x - 600 = 0$, $x = 25$

3 (a) $\frac{270}{x}, \frac{270}{x-10}$
(b) $x^2 - 10x - 9000 = 0$, $x = 100$
(c) Arrive 1 pm

4 Cost = £16, 16 staff left

5 12 thick slices

6 (a) 1.714 ohms
(b) 4 ohms
(c) Equivalent to half

EXERCISE 1C (Page 9)

1 $(x+1)(x^2-x+1)$
$\frac{x^2-x+1}{(x+2)}$

2 –2, 3

3 $\frac{1}{(x-1)(x-2)}$

4 $\frac{x}{x+4}$

6 $\frac{3x-7}{(x-1)(x-5)}$

7 –2, $\frac{1}{2}$

8 $\frac{x^2+x+1}{x-1}$

10 $\frac{2(x+1)}{x+3}$

CHAPTER 2

EXERCISE 2A (Page 16)

1 **(a)** one-to-one function
(b) many-to-one function
(c) one-to-one function
(d) many-to-one function
(e) many-to-many
(f) one-to-one function
(g) one-to-one function
(h) many-to-one function

2 **(a)** one-to-one $x \in \mathbb{R}$ $y \in \mathbb{R}$
(b) many-to-one $x \neq 0$ $y > 0$
(c) one-to-many
(d) one-to-one $x \in \mathbb{R}$ $y \in \mathbb{R}$
(e) many-to-many
(f) one-to-one $x \neq 0$ $y \neq 0$
(g) many-to-one $x \in \mathbb{R}$ $y \geqslant -4$
(h) many-to-one $x \in \mathbb{R}$ $y \in \mathbb{R}$
(i) one-to-one $x \in \mathbb{R}$ $y > 0$
(j) many-to-many

3 **(a)** **(i)** -5
(ii) 9
(iii) -11
(b) **(i)** 3
(ii) 5
(iii) 10
(c) **(i)** 32
(ii) 82.4
(iii) 14
(iv) -40

4 **(a)** $f(x) \leqslant 2$
(b) $0 \leqslant f(\theta) \leqslant 1$
(c) $y \in \{2, 3, 6, 11, 18\}$
(d) $y \in \mathbb{R}^+$
(e) $\mathbb{R}$
(f) $\{\frac{1}{2}, 1, 2, 4\}$
(g) $0 \leqslant y \leqslant 1$
(h) $f(\theta) \geqslant 1$ or $f(\theta) \leqslant -1$
(i) $0 < f(x) \leqslant 1$
(j) $f(x) \geqslant 3$

5 For f, every value of x (including $x = 3$) gives a unique output, whereas g(2) can equal either 4 or 6.

EXERCISE 2B (Page 21)

1 **(a)** $8x^3$
(b) $2x^3$
(c) $(x + 2)^3$
(d) $x^3 + 2$
(e) $8(x + 2)^3$
(f) $2(x^3 + 2)$
(g) $4x$
(h) $\{(x + 2)^3 + 2\}^3$
(i) $x + 4$

2 **(a)** $x^2 + 8x + 14$
domain $x \in \mathbb{R}$
range $y \geqslant -2$
(b) $x = \pm 5$

3 **(a)** $\sqrt{x + 1}$
domain $x \geqslant -1$
range $y \geqslant 0$
(b) $x = 8$
(c) $x = 4$

4 **(a)** $gf(x) = \dfrac{1}{2(x + 1)}$
domain $x \neq -1$
(b) 0.229, -0.729

5 **(a)** $gh = \dfrac{3}{x - 3}$
(b) $fgh = \dfrac{6}{x - 3} + 3$
(c) f and h

6 **(a)** fg
(b) g^2
(c) fg^2
(d) gf

7 **(a)** $gf(x) = x^2 - 4x + 7$
(c) $y \geqslant 3$
(d) many-to-one
(e) Yes

8 **(b)** $3p + q = 7$
(c) $p = 3$ with $q = -2$ or $p = 5$ with $q = -8$

9 **(a)** 3
(b) 2
(c) 1
(d) 61

10 **(a)** $fg(x) = 2x^2 - 2x + 3$
(b) $fg(x) = 2(x - \frac{1}{2})^2 + \frac{5}{2}$
$p = 2, q = -\frac{1}{2}, r = \frac{5}{2}$
(c) $fg(x) \geqslant \frac{5}{2}$

EXERCISE 2C (Page 28)

1 **(a)** $f^{-1}(x) = \dfrac{x - 7}{2}$
(b) $f^{-1}(x) = 4 - x$
(c) $f^{-1}(x) = \dfrac{2x - 4}{x}$
(d) $f^{-1}(x) = \sqrt[x]{x + 3}, x \geqslant -3$

2 **(a), (b)**

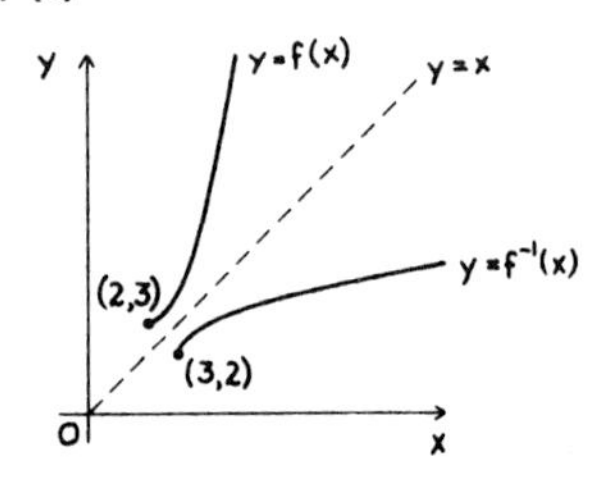

3 **(a)** $f(x)$ not defined for $x = 4$;
$g(x)$ is defined for all x;
$h(x)$ not defined for $x > 2$
(b) $f^{-1}(x) = \dfrac{4x + 3}{x}$;
$h^{-1}(x) = 2 - x^2, x \geqslant 0$
(c) $g(x)$ is not one-to-one.
(d) Suitable domain: $x \geqslant 0$
(e) No: $fg(x) = \dfrac{3}{x^2 - 4}$, not defined for $x = \pm 2$;
$gf(x) = \left(\dfrac{3}{x - 4}\right)^2$, not defined for $x = 4$.

4 **(a)** x **(b)** $\frac{1}{x}$ **(c)** $\frac{1}{x}$ **(d)** $\frac{1}{x}$

5 **(a)** a = 3
(b)

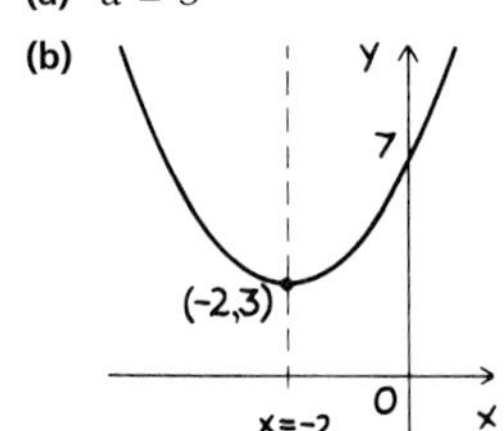

(c) $f(x) \geqslant 3$
(d) many-to-one function; possible domain is $x \geqslant -2$

6 (a) $f^{-1}: x \mapsto \sqrt[3]{\frac{x-3}{4}}, x \in \mathbb{R}$.

The graphs are reflections of each other in the line $y = x$.

7 (a) $f^{-1}: x \mapsto \frac{x+2}{4}$

$g^{-1}: x \mapsto \sqrt{x}$

(b)

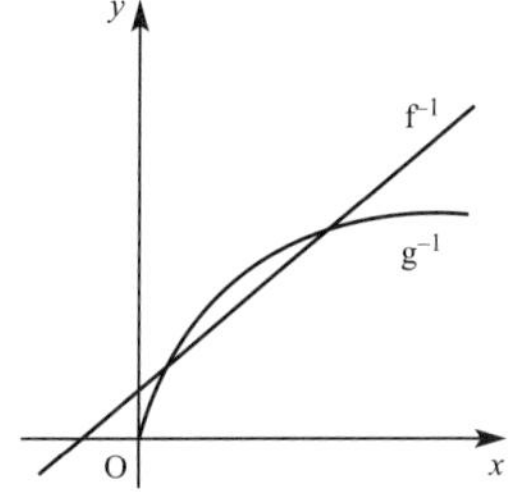

(c) 2

(d) $6 \pm 4\sqrt{2}$

8 (a) $a = -3, b = 1$

(b) (3, 1)

(c)

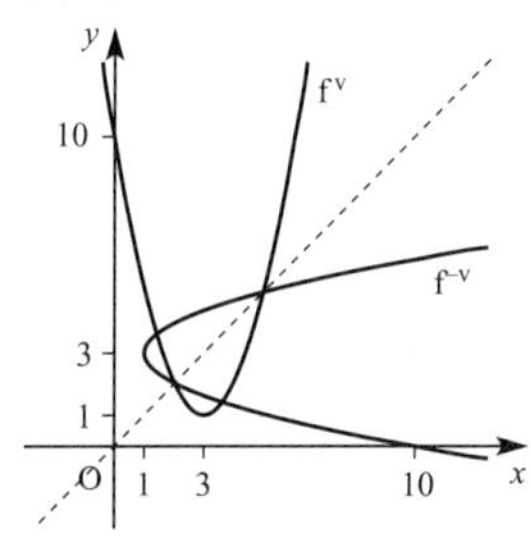

(d) $x \geqslant 3, y \geqslant 1$

(e) see graph; reflection in $y = x$

9 (b) $y \leqslant 4$

(c) $f^{-1}: x \mapsto \sqrt{4-x}, x \leqslant 4$

(d) $x = -5$

10 (b) $x > 1$

(c) $g^{-1}: x \mapsto \sqrt{x-1}$

(d)

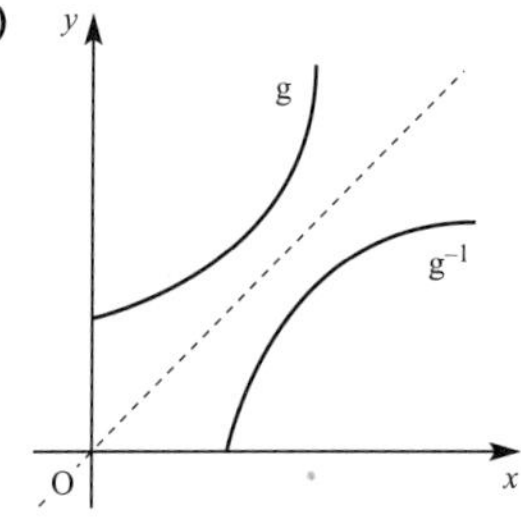

EXERCISE 2D (Page 34)

1 (a) Even

(b) Odd

(c) Neither

(d) Neither

(e) Odd

(f) Even

2 (a) Even

(b) Odd, periodic; $\frac{2}{3}\pi$

(c) None

(d) Odd

(e) Periodic; 2π

(f) Odd, periodic; π

3 (a)

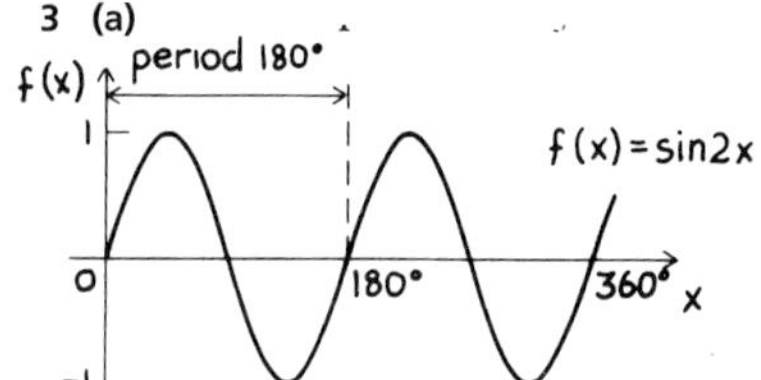

(b) Half the period of $\sin x$

(c) (i) 90°

(ii) 120°

(iii) 720°

4

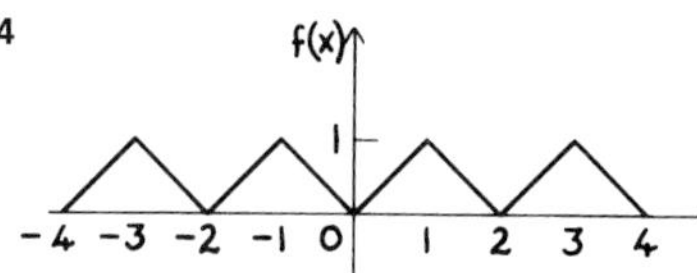

5 (a) even

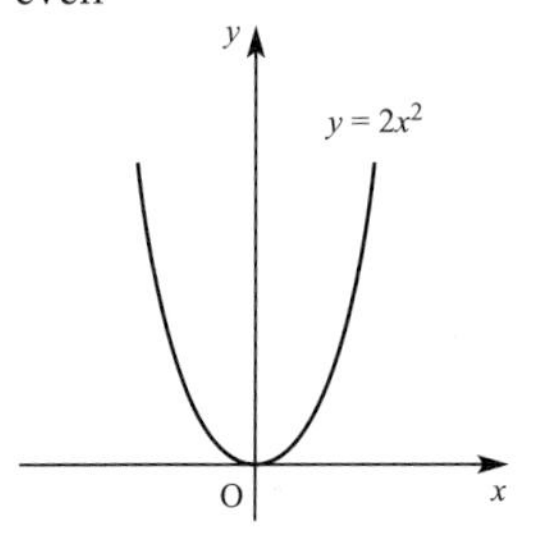

(b) odd

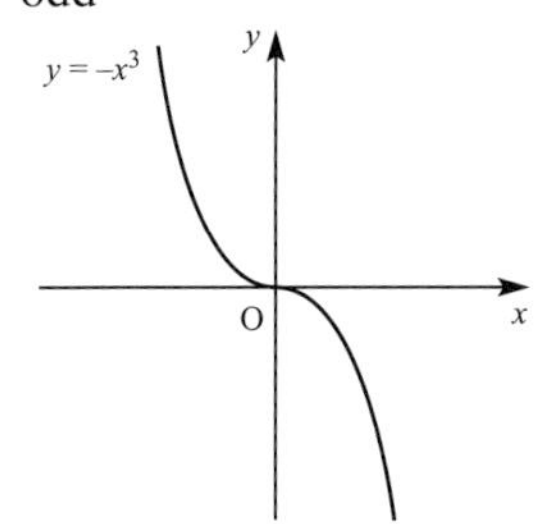

(c) neither

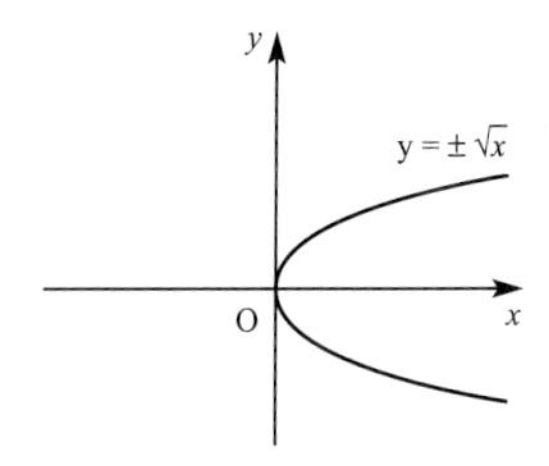

EXERCISE 2E (Page 36)

1

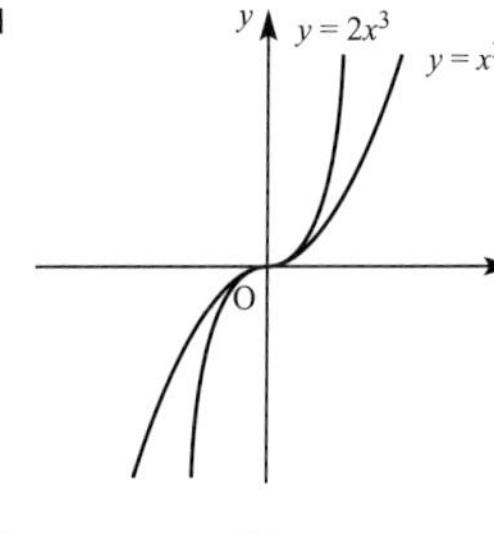

2

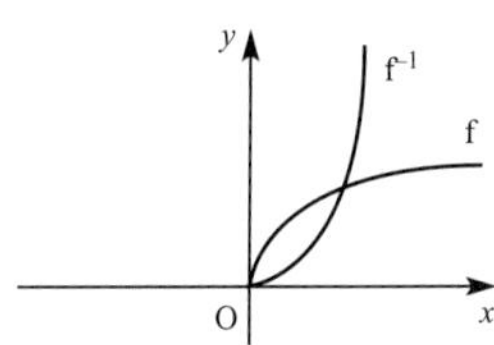

3

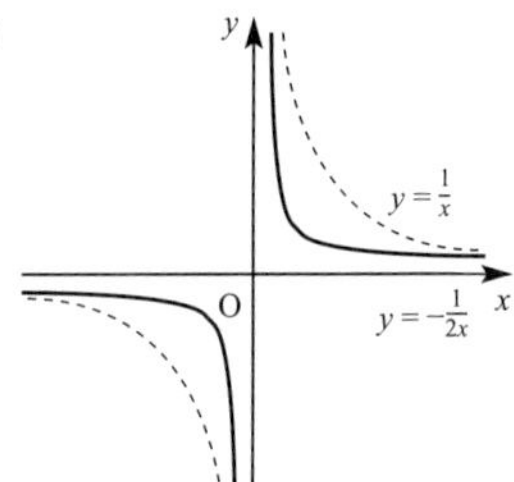

4

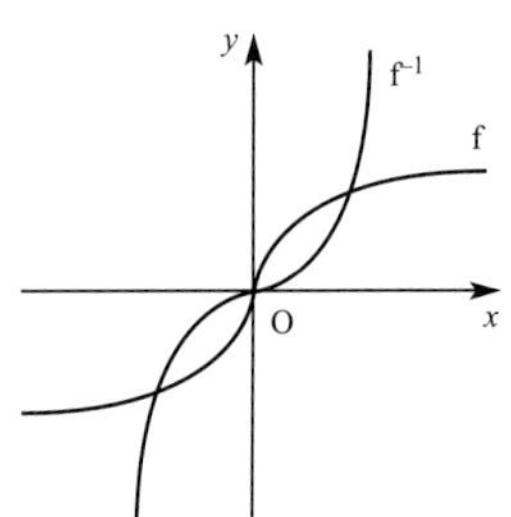

5

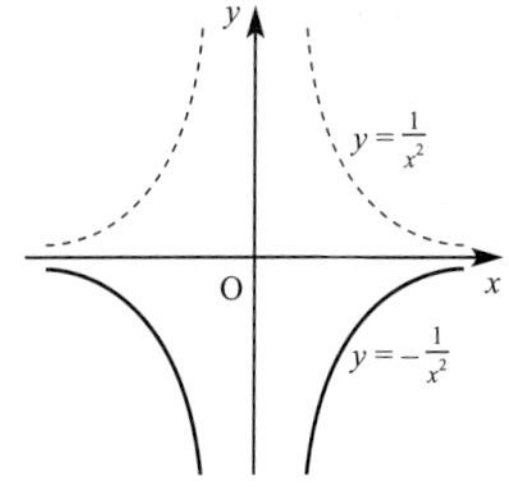

6

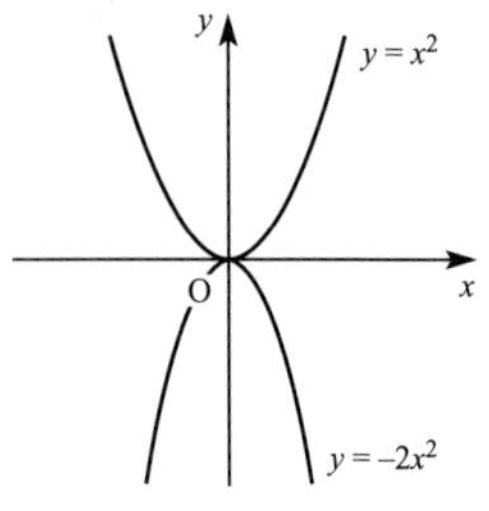

7

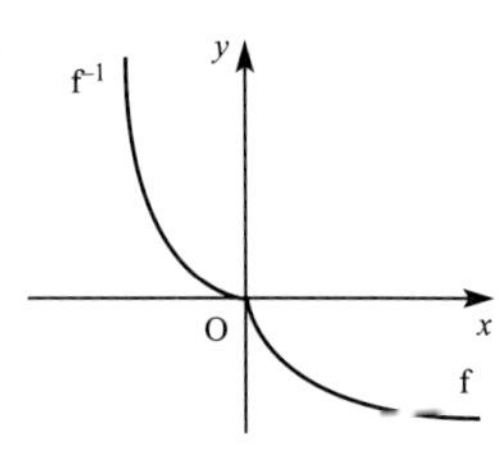

8

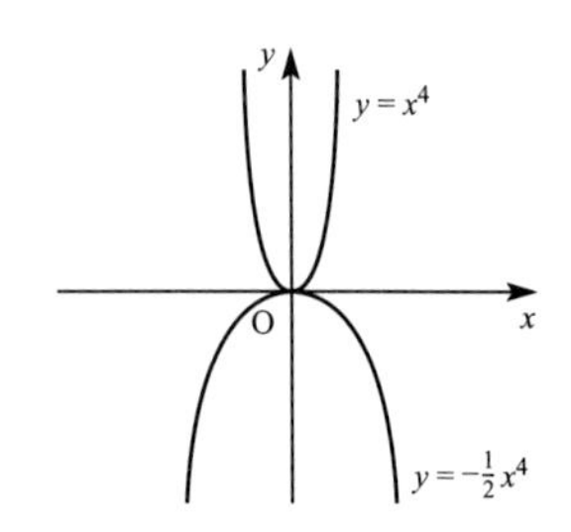

9

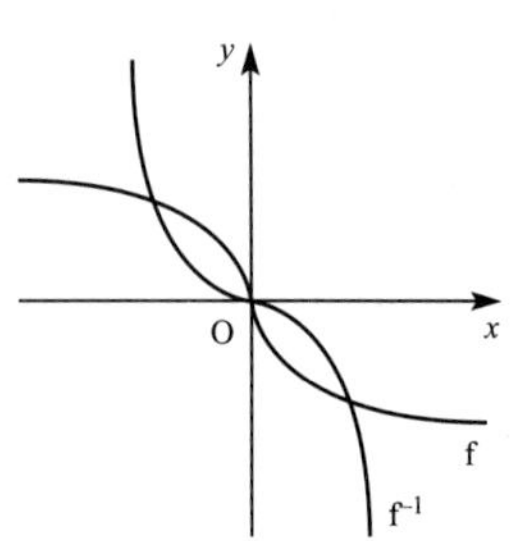

10

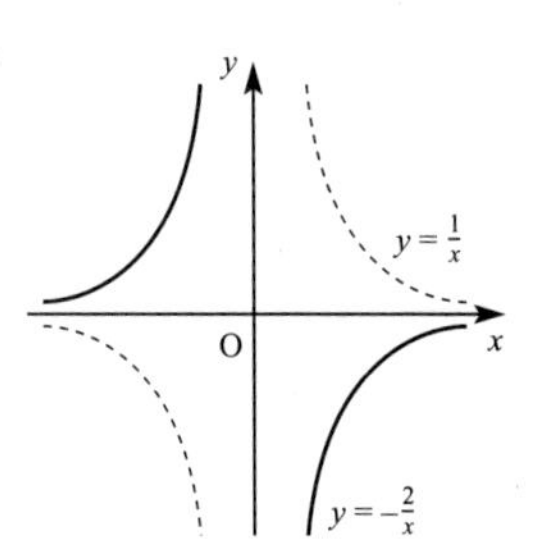

EXERCISE 2F (Page 40)

1 (a)

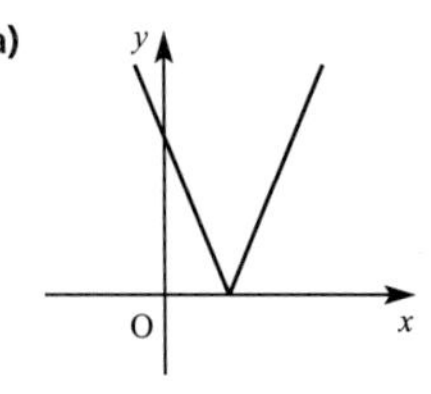

(b)

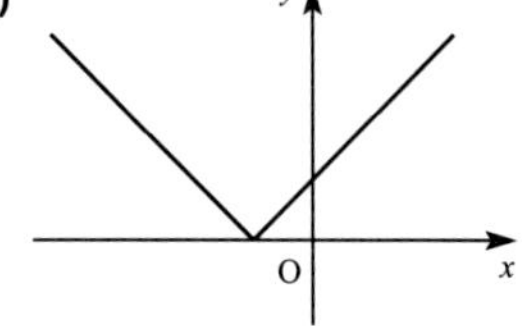

(c)

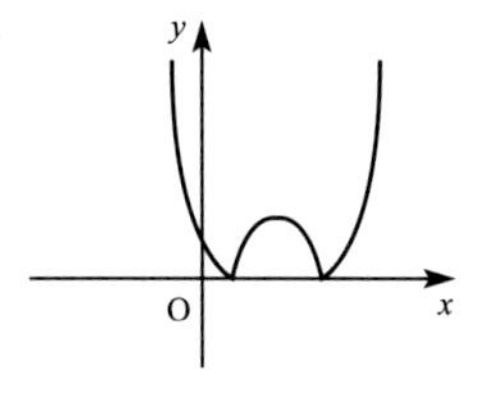

(d)

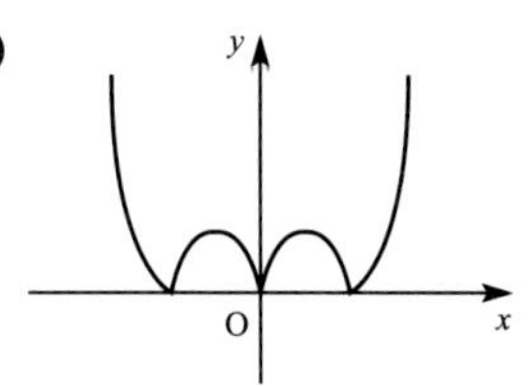

(e)

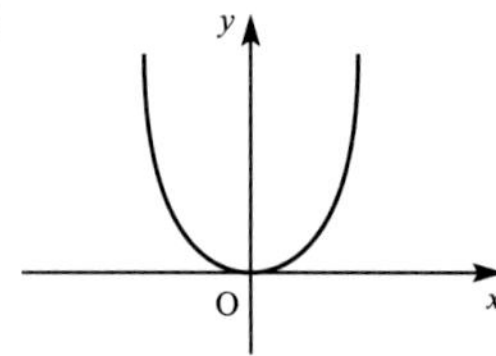

(f)

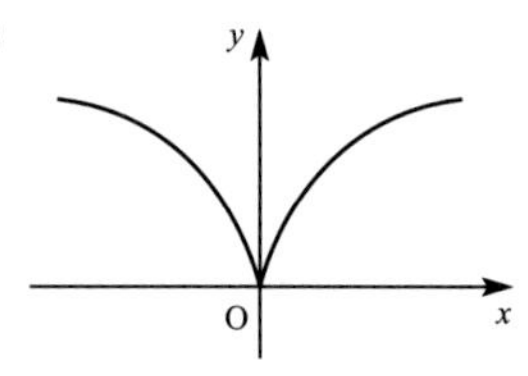

2 (a)

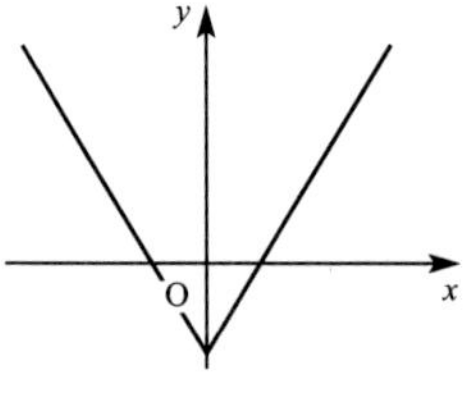

(b)

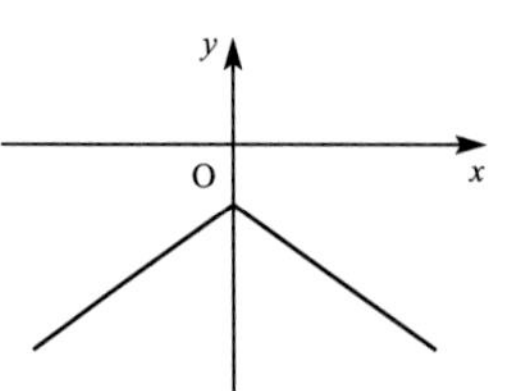

(c)

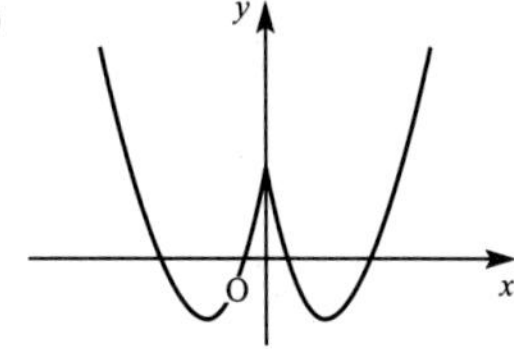

(d)

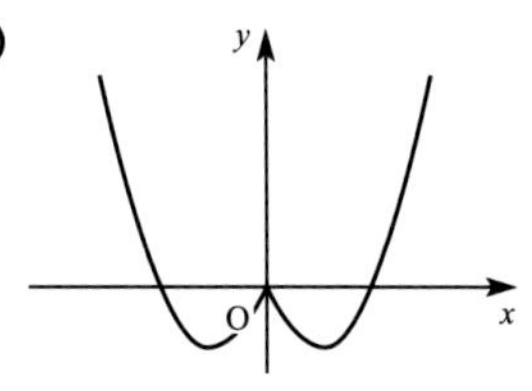

(e)

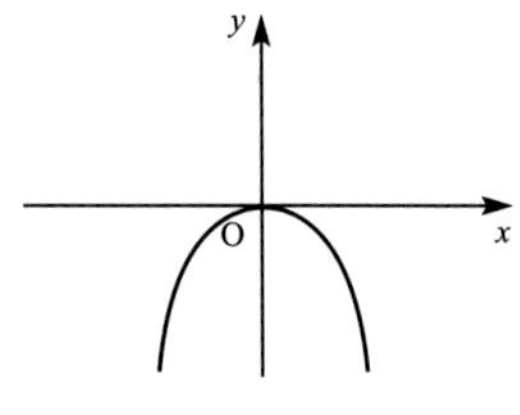

(f)

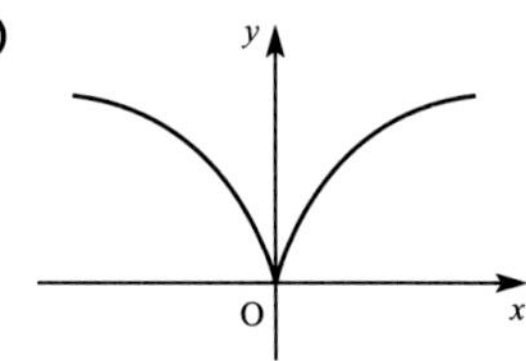

3 (a) $y = |x + 2|$

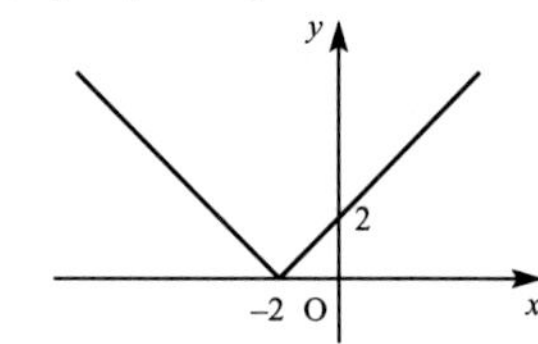

(b) $y = |3x - 2|$

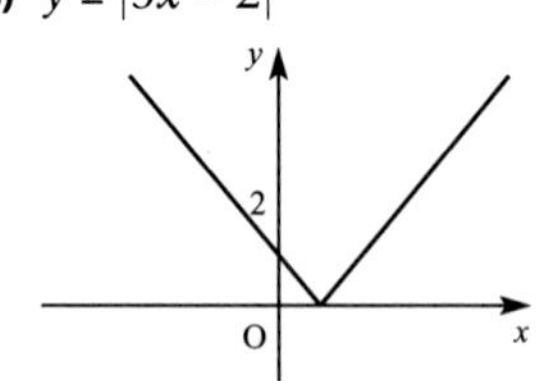

(c) $y = |x| + 2$

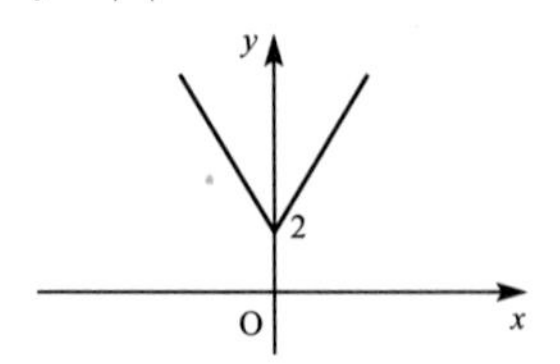

(d) $y = |x^2 - 2|$

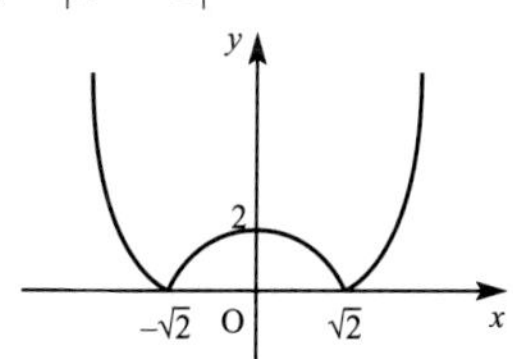

(e) $y = |x^3|$

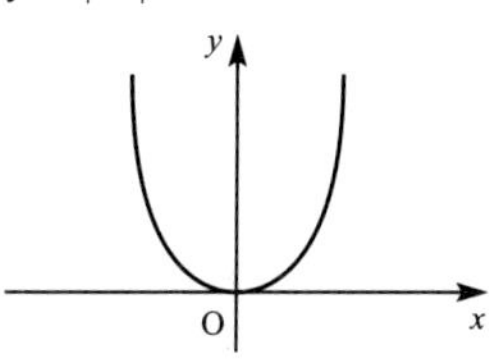

(f) $y = |x^3| - 1$

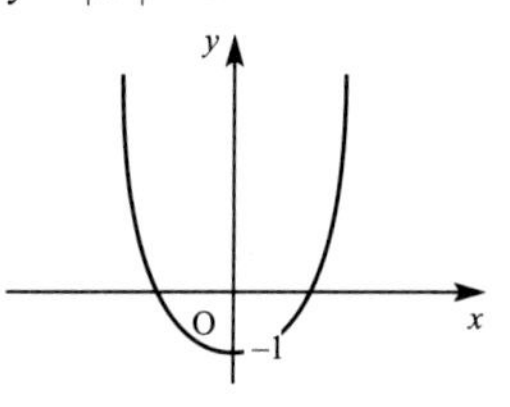

(g) $y = \left|\frac{1}{x}\right|$

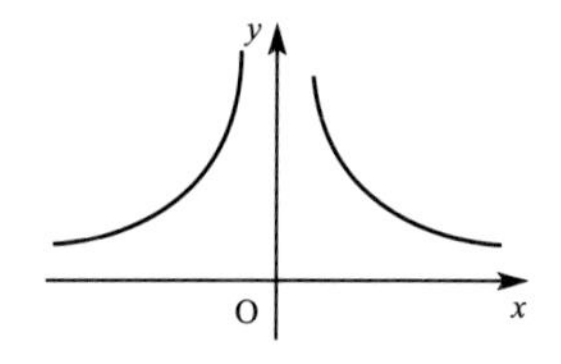

(h) $y = |3 - x|$

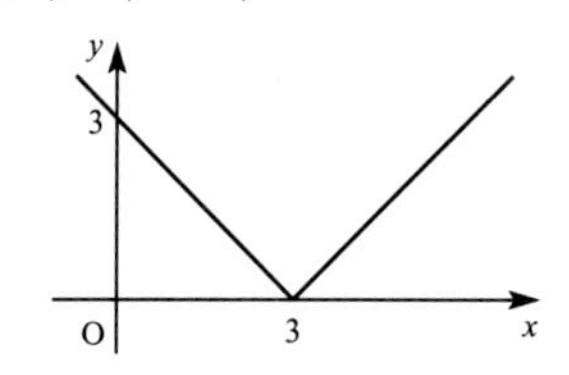

(i) $y = |9 - x^2|$

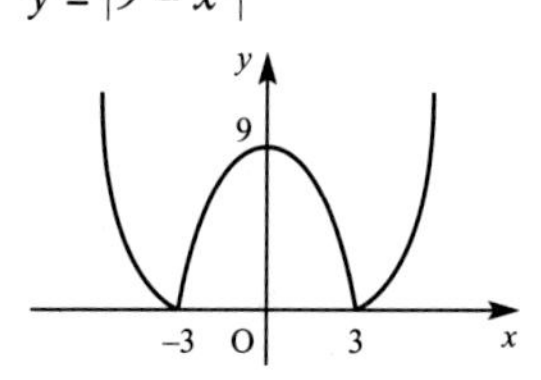

(j) $y = |\sin x|$

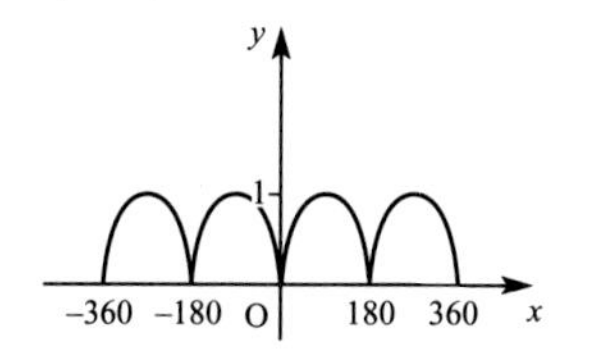

4 (a)

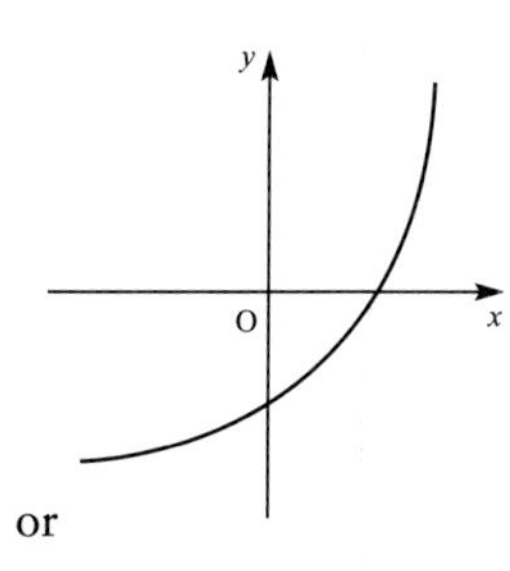

or

(b)

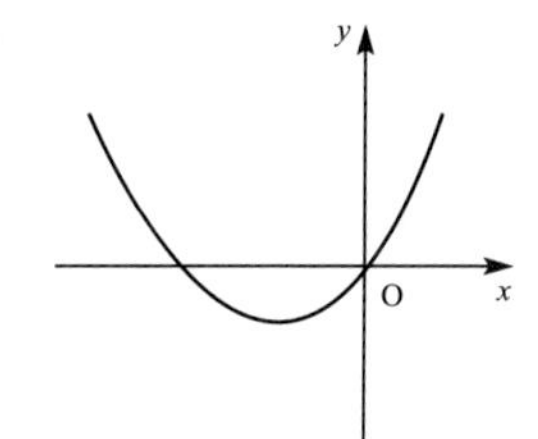

or

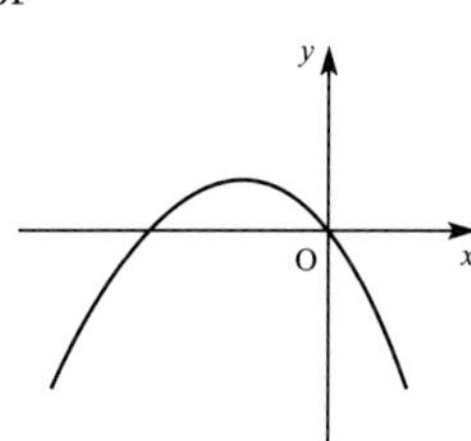

5 (a)

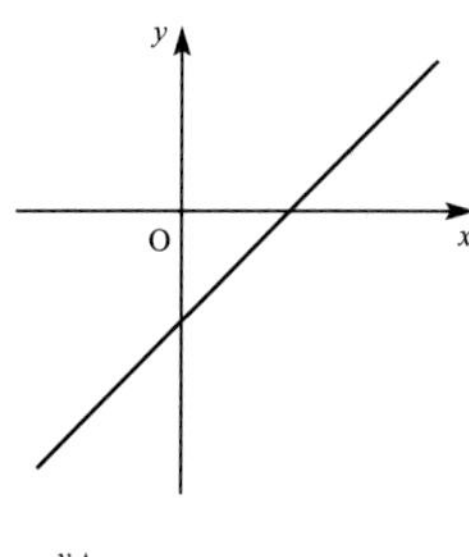

(b)

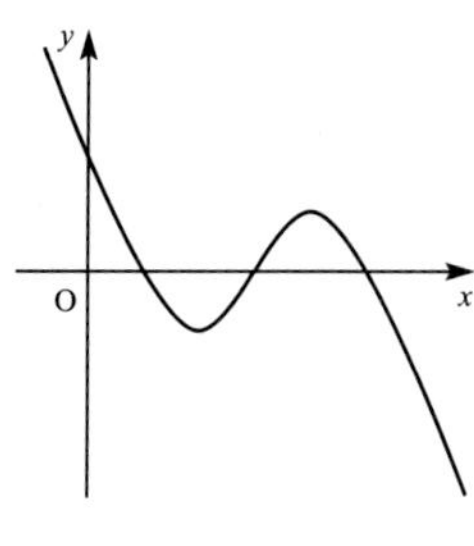

EXERCISE 2G (Page 44)

1 (a) $x = \pm 1$
(b) $x = \pm 2$

2 (a) $-0.6, 1.6$
(c) $-1, 3$
(d) $-1 < x < 3$

3 (c) $-3, 1$

4 $1 < x < 3$

5 $-2.9, -0.7, 0.5$

6 $-3.1, 0.5, 2.7$

7 $0.1, 2.9$

8 $-2 < x < 2$

9 (a) even
(b) $x = \pm 1.7$

10 $x = 1.5$

EXERCISE 2H (Page 53)

1 (a) Translation $\begin{pmatrix} 0 \\ -2 \end{pmatrix}$; $x = 0$
(b) Translation $\begin{pmatrix} -4 \\ 0 \end{pmatrix}$; $x = -4$
(c) Stretch parallel to y axis of s.f. 4, or stretch parallel to x axis of s.f. $\frac{1}{2}$; $x = 0$
(d) Stretch parallel to y axis of s.f. $\frac{1}{3}$, or stretch parallel to x axis of s.f. $\sqrt{3}$; $x = 0$;
(e) Translation $\begin{pmatrix} 3 \\ -5 \end{pmatrix}$; $x = 3$
(f) $y = (x - 1)^2 - 1$: translation $\begin{pmatrix} 1 \\ -1 \end{pmatrix}$; $x = 1$
(g) $y = (x - 2)^2 - 1$: translation $\begin{pmatrix} 2 \\ -1 \end{pmatrix}$; $x = 2$
(h) $y = 2\left[(x + 1)^2 - 1\frac{1}{2}\right]$: translation $\begin{pmatrix} -1 \\ -1\frac{1}{2} \end{pmatrix}$; then stretch parallel to y axis of s.f. 2; $x = -1$
(i) $y = 3\left[(x - 1)^2 - \frac{5}{3}\right]$: translation $\begin{pmatrix} +1 \\ -\frac{5}{3} \end{pmatrix}$; then stretch parallel to y axis of s.f. 3; $x = 1$

2 (a) Translation $\begin{pmatrix} 90° \\ 0 \end{pmatrix}$
(b) Stretch parallel to x axis of s.f. $\frac{1}{3}$

(c) Stretch parallel to y axis of s.f. $\frac{1}{2}$

(d) Stretch parallel to x axis of s.f. 2

(e) Stretch parallel to x axis of s.f. $\frac{1}{3}$ and translation $\binom{0}{2}$, in either order

3 (a) Translation $\binom{-60°}{0}$

(b) Stretch parallel to y axis of s.f. $\frac{1}{3}$

(c) Translation $\binom{0}{1}$

(d) Translation $\binom{-90°}{0}$, then stretch parallel to x axis of s.f. $\frac{1}{2}$

4 (i) (a)

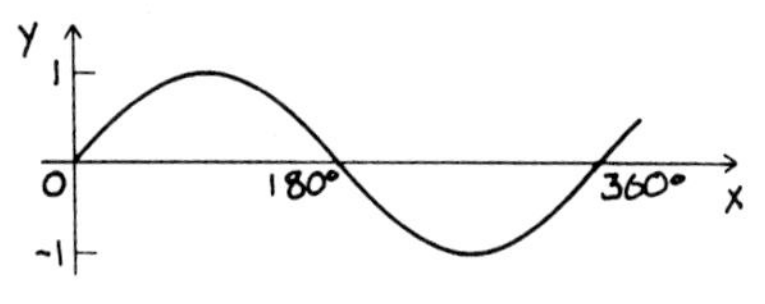

(b) $y = \sin x$

(ii) (a)

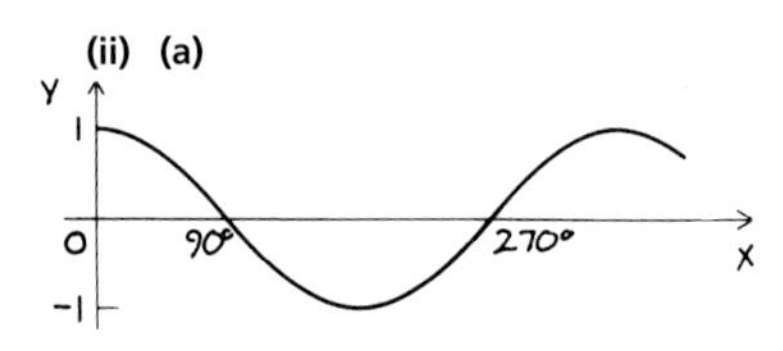

(b) $y = \cos x$

(iii) (a)

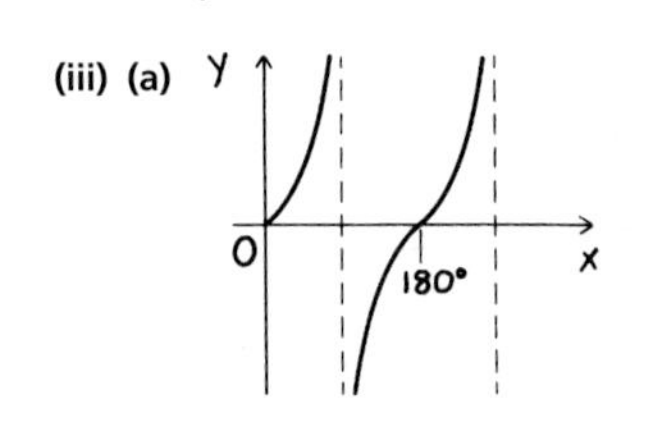

(b) $y = \tan x$

(iv) (a)

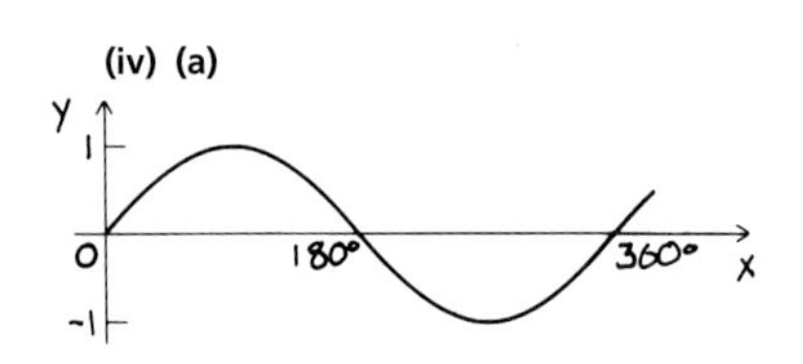

(b) $y = \sin x$

(v) (a)

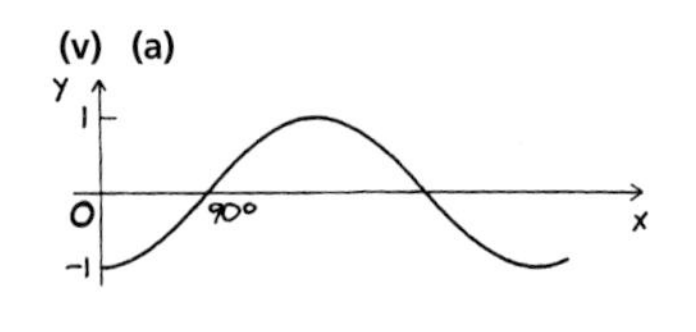

(b) $y = \cos x$

5 (a) $a = -4$

(b)

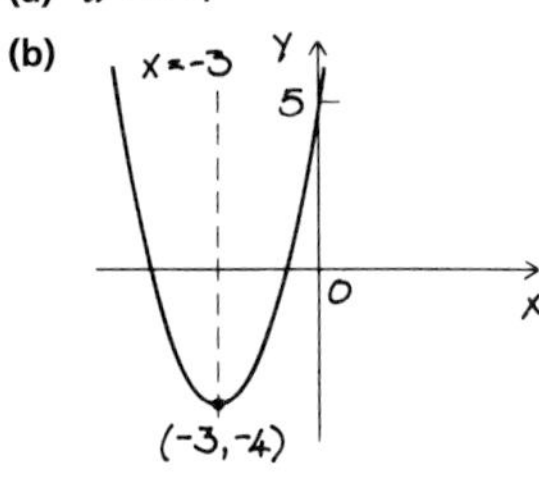

6 $p = 3$, $q = 2$

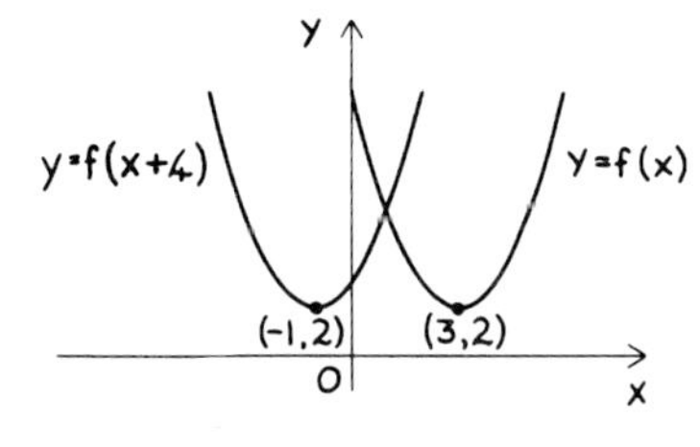

7 (a)

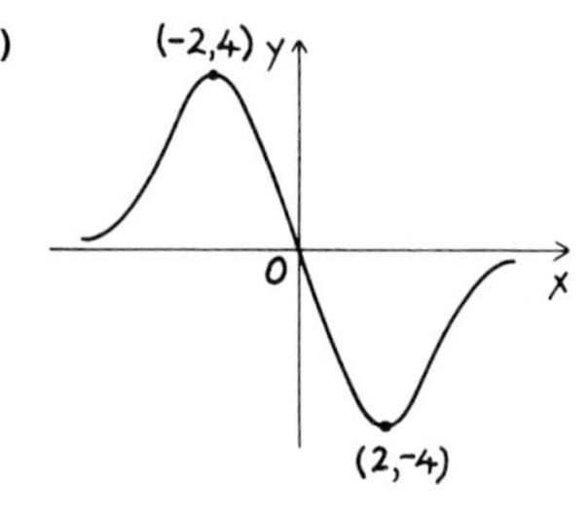

(b)

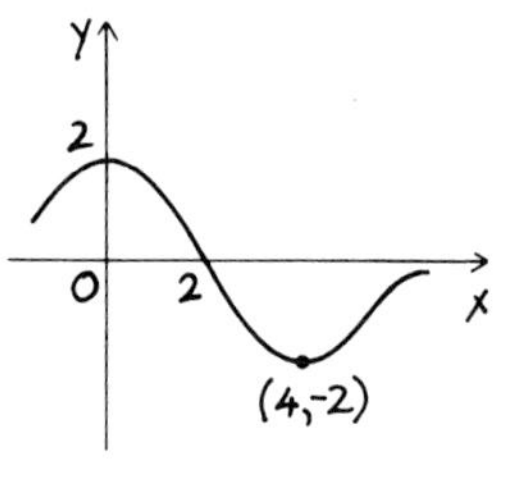

(c)

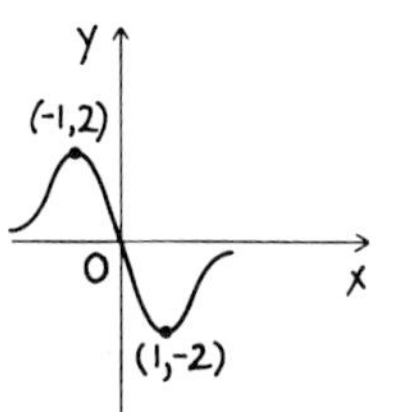

(d)

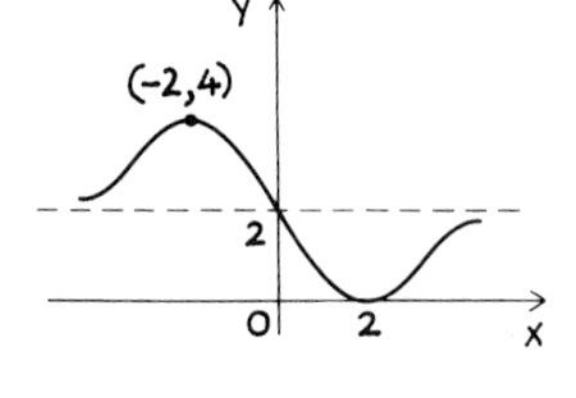

(e)

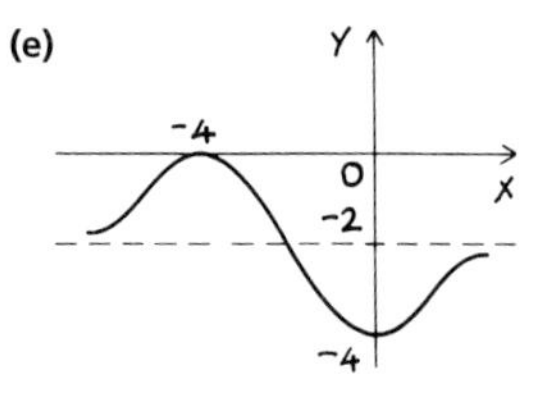

(f)

(-4,4) y O x (4,-4)

8 (a)

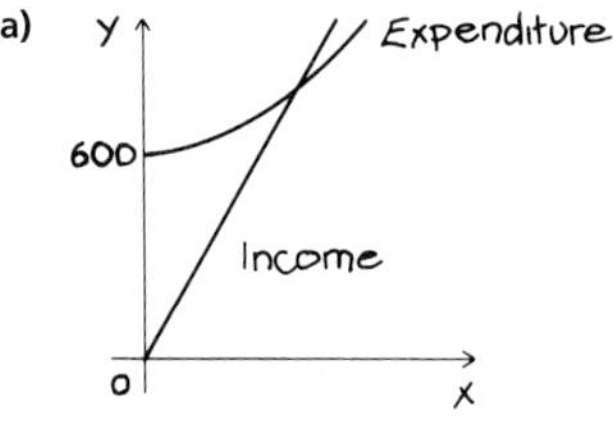

(b) 21 machines

9

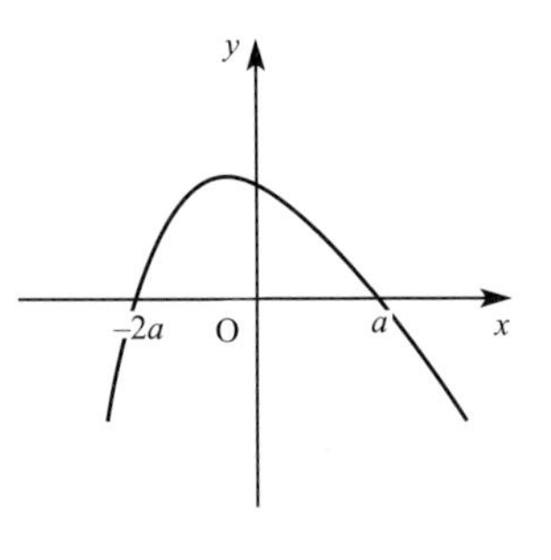

y a $-\frac{1}{2}a$ O a x

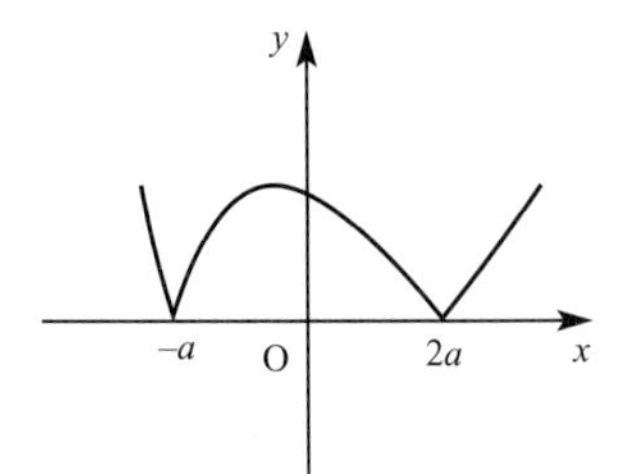

10

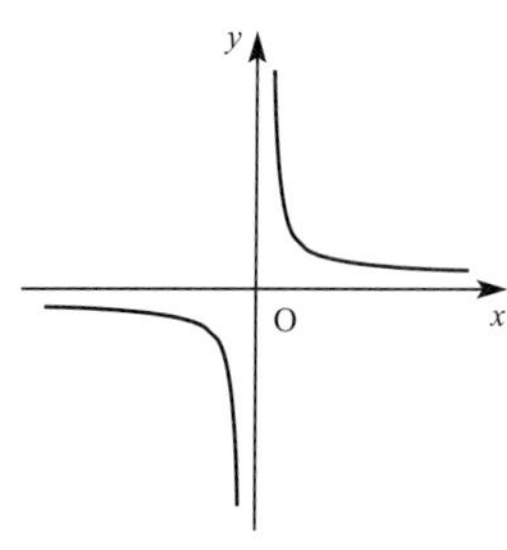

(a)

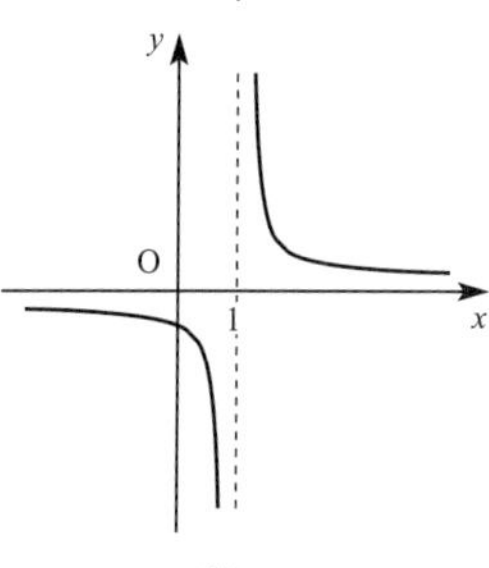

(b)

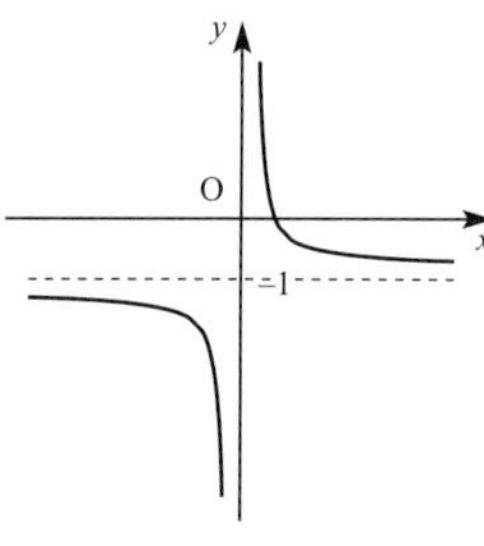

(c)

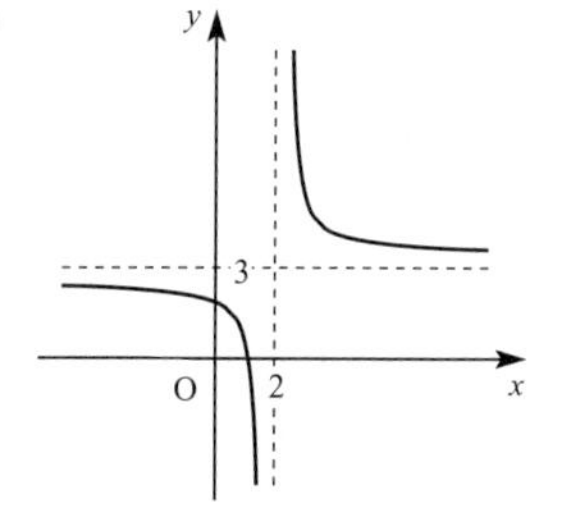

EXERCISE 2I (Page 59)

1 (a) Stretch parallel to y axis of s.f. 2, then reflection in x axis, either order; $x = 0$

(b) Reflection in x axis then translation $\binom{0}{4}$; $x = 0$

(c) $y = -(x - 1)^2$: translation $\binom{1}{0}$, and reflection in x axis, either order; $x = 1$

2 (i) (a)

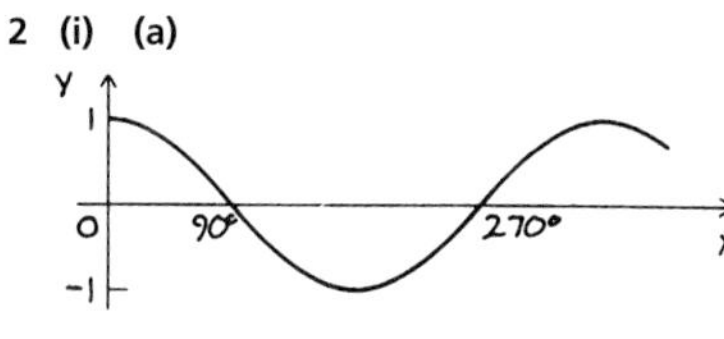

(b) $y = \cos x$

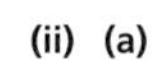

(ii) (a)

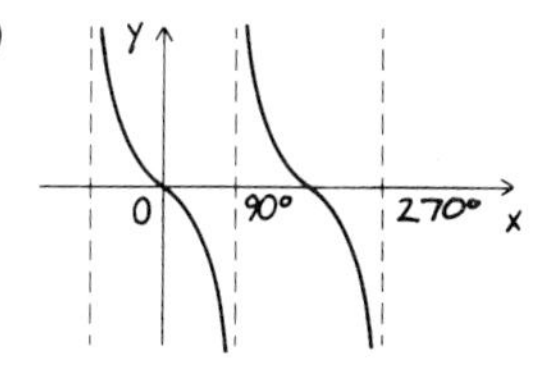

(b) $y = -\tan x$

(iii) (a)

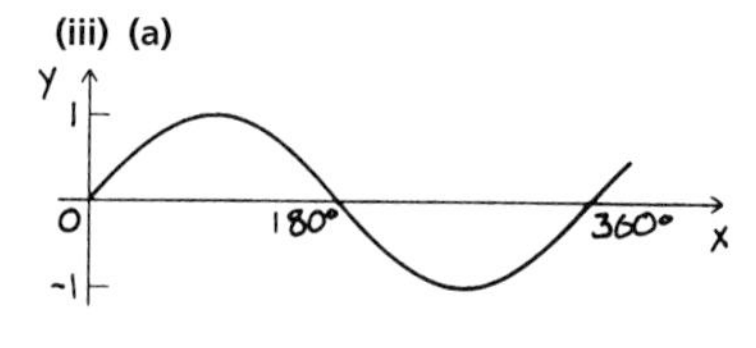

(b) $y = \sin x$

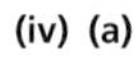

(iv) (a)

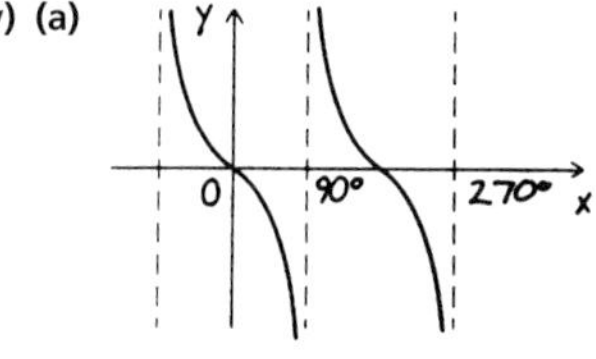

(b) $y = -\tan x$

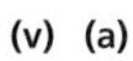

(v) (a)

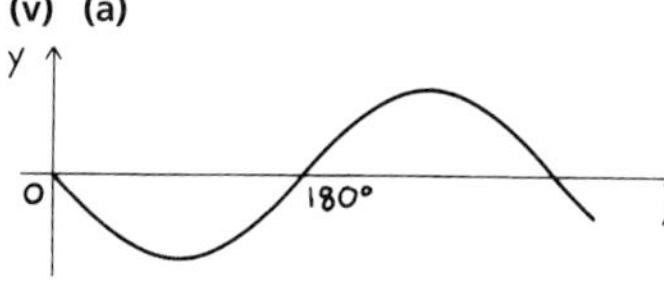

(b) $y = -\sin x$

3 (a) $a = 3$, $b = 5$

(b) Translation $\binom{3}{5}$

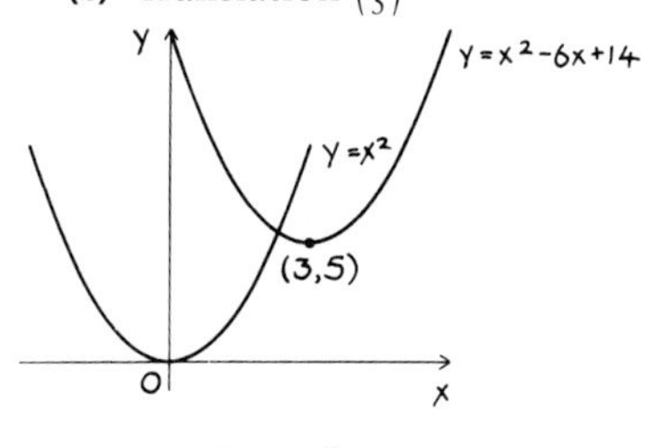

(c) $y = 6x - x^2 - 14$

4 (a)

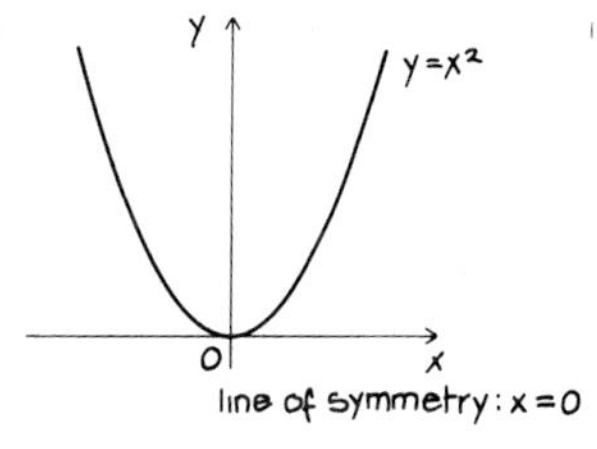

(i)

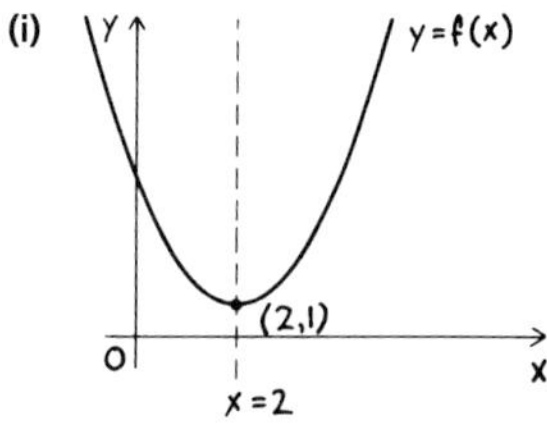

(ii)

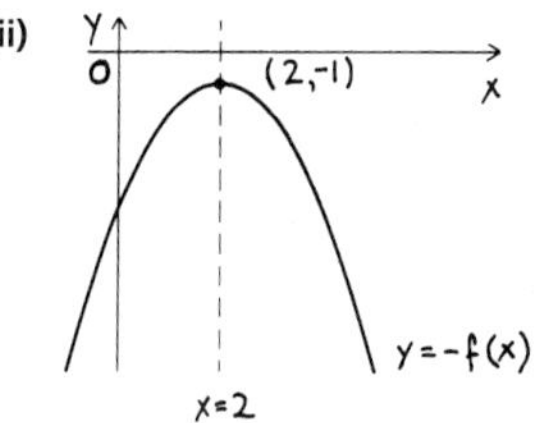

(iii)

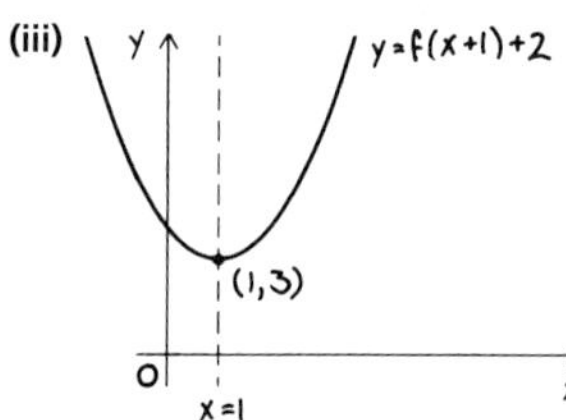

5 (a) $a = 2$, $b = 1$, $c = 3$; $(-1, 3)$

6 (a)

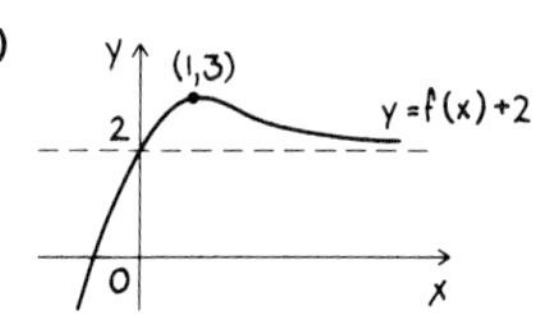

(b)

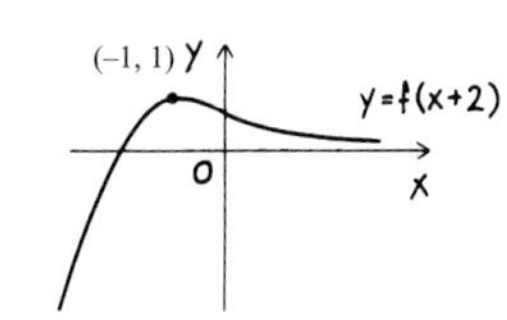

(c)

Y ↑ $(\frac{1}{2}, 1)$ y = f(2x) O x

7 (a) $\dfrac{x^2}{9} + \dfrac{y^2}{4} = 1$

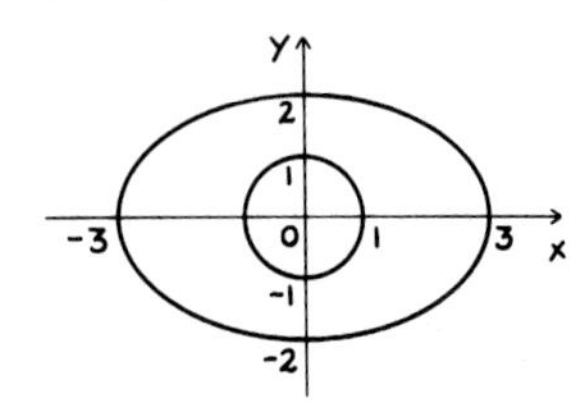

8 (a) $y = \text{f}(x + 2)$
(b) $y = -\text{f}(x)$
(c) $y = \text{f}\left(\frac{x}{2}\right)$
(d) $y = \text{f}(x) - 3$
(e) $y = \text{f}(-x)$, (or $y = 2 - \text{f}(x)$)
(f) $y = \frac{3}{2}\text{f}(x)$

9 (a) $(-1, 5)$; $y = 2$
(b) $a = -3$, $b = 5$
(d) $y = 3x^2 + 6x + 2$; $(-1, -1)$

Exercise 2J (Page 61)

1 (a) $y \geqslant 0$
(b) $-2, 8$
(c)

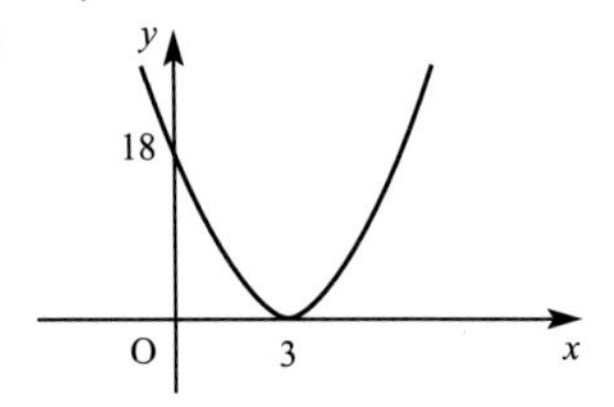

(d)

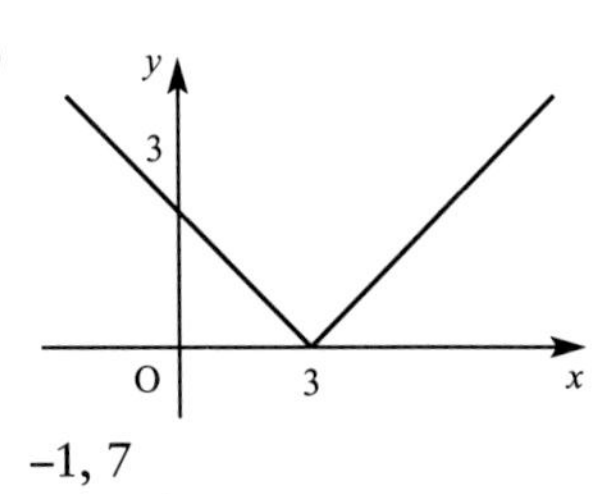

$-1, 7$

2 (a) $y \geqslant 0$
(b) $\text{f}^{-1}: x \mapsto \dfrac{x-1}{2}$ $\text{g}^{-1}: x \mapsto x^2$
(c)

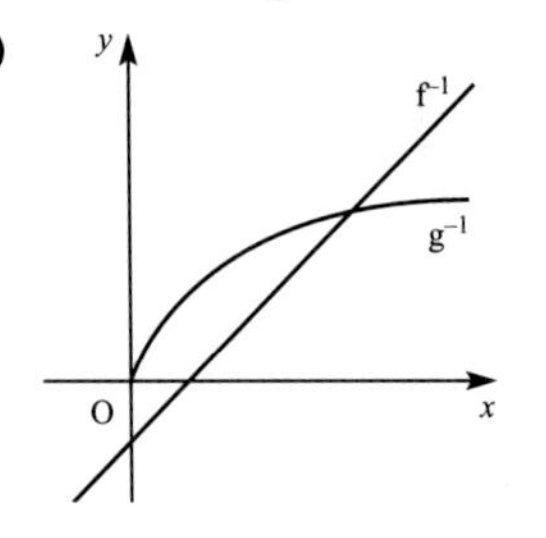

(d) one solution
(e) 3.317

3 (a)

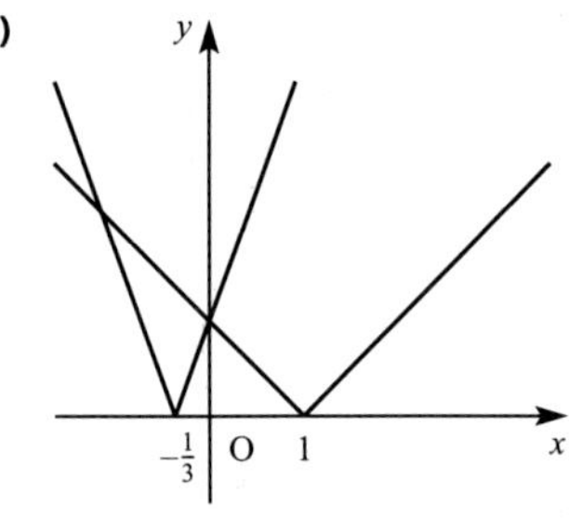

(b) $-1, 0$

4 (a) $a = 2$ $b = 1$
(b)

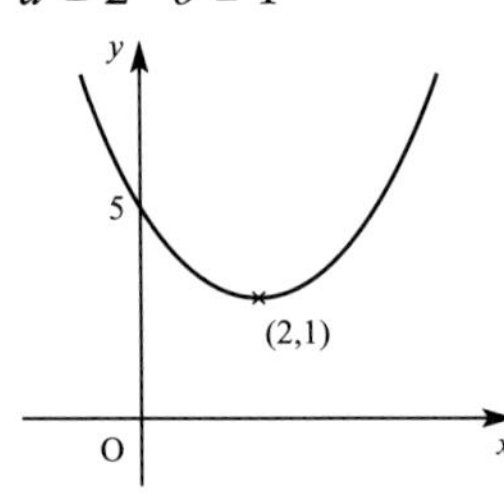

(c) $x \geqslant 2$
(d) $\text{f}^{-1}: x \mapsto \sqrt{x-1} + 2$
(e)

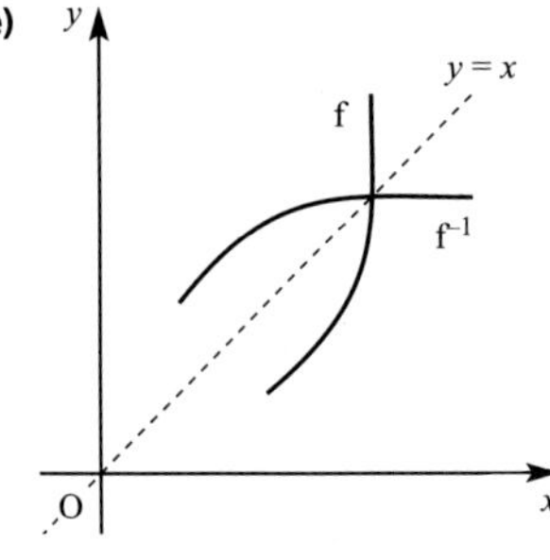

(f) 3.6

5 (a) $\text{g}^{-1}: x \mapsto \frac{1}{x} - 1 \quad x \neq 0$
(b) $\text{fg}(x) = \dfrac{1-x}{1+x}$
(c)

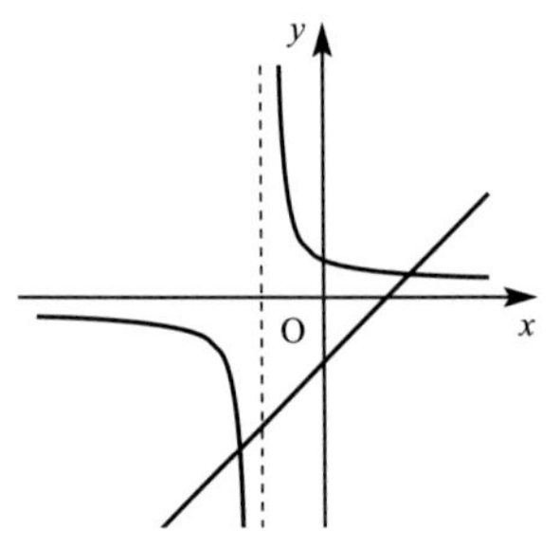

(d) $-1.3, 0.8$
(e) $-1.28, 0.781$

6 (a) $\text{Y}\left(\frac{\pi}{2}, 2\right)$ $\text{Z}\left(\frac{3\pi}{4}, 0\right)$
$\text{A} = 2$ $\text{B} = 2$
(b) $y = \text{f}\left(x + \frac{\pi}{2}\right)$

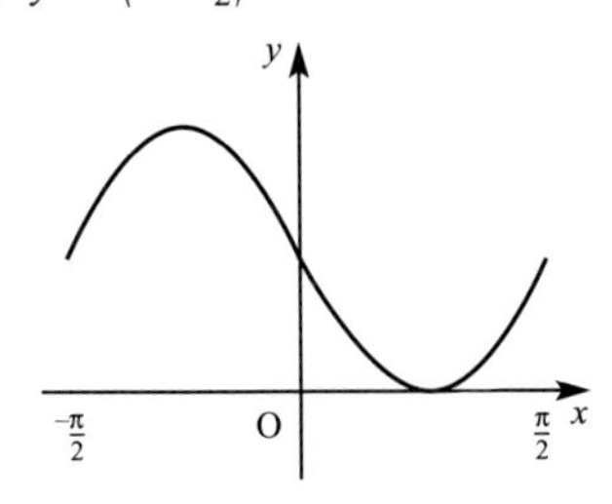

$y = |\text{f}(x) - 2|$

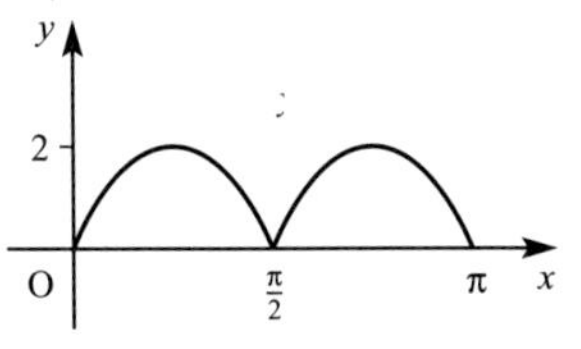

7 (a) $y = \text{f}(-x)$

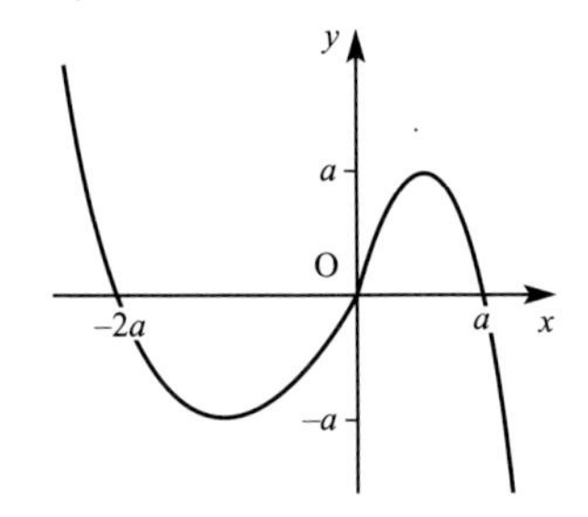

(b) $y = -\text{f}(x)$

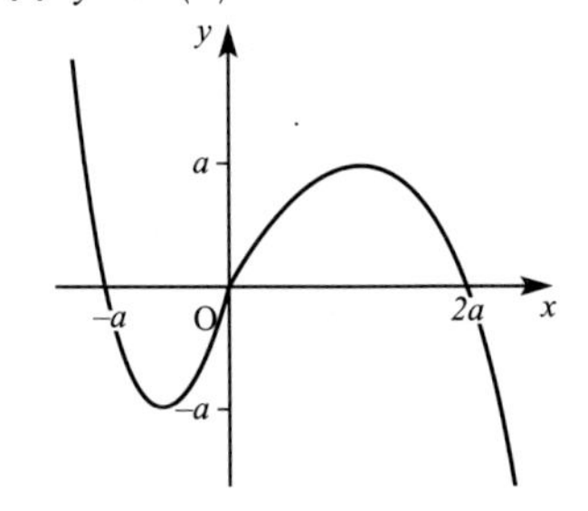

(c) $y = \text{f}(x - a)$

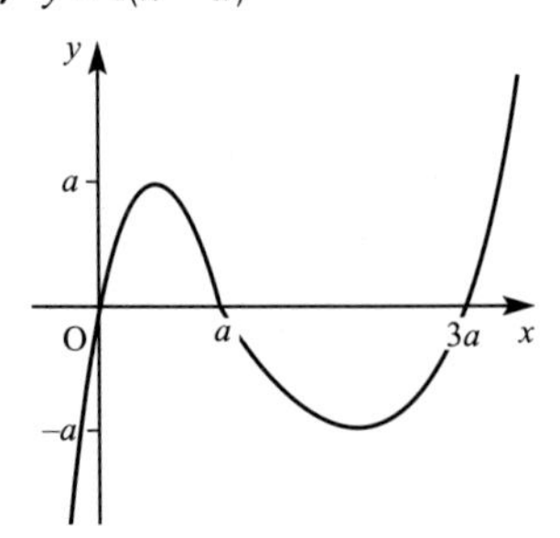

(d) $y = \mathrm{f}(x) - a$

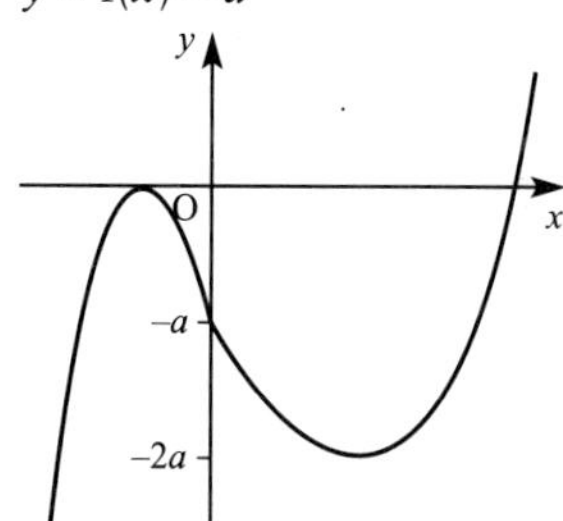

(e) $y = |\mathrm{f}(x)|$

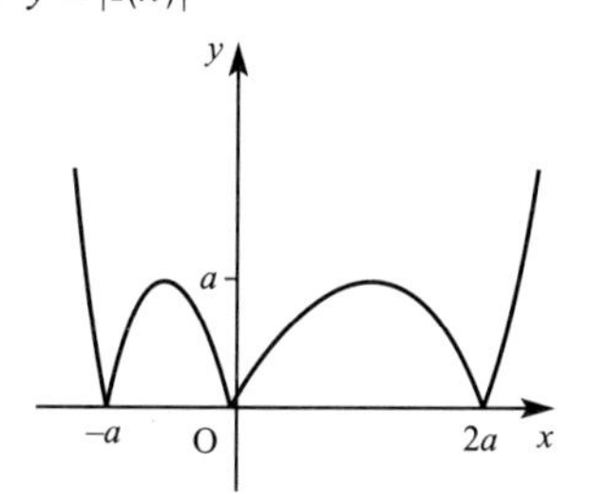

$-a < x < 0 \qquad x > 2a$

8 (a) $\mathrm{f}^{-1}: x \mapsto \frac{x+1}{4}, x \in \mathbb{R}$

(b) $\mathrm{gf}: x \mapsto \frac{3}{8x-3}, x \neq \frac{3}{8}$

(c) $-0.076, 0.826$

9 (a)

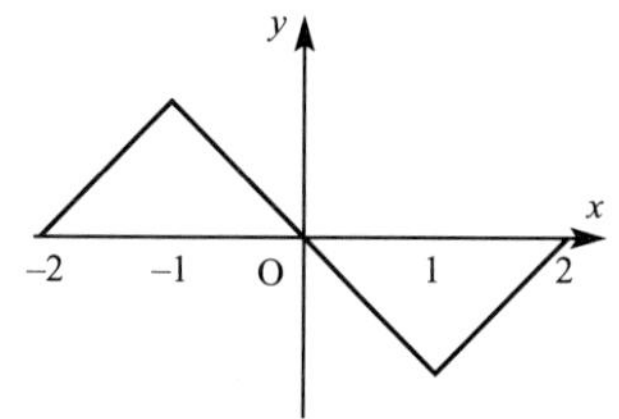

(b) $-\frac{3}{2}, -\frac{1}{2}$

0 (a)

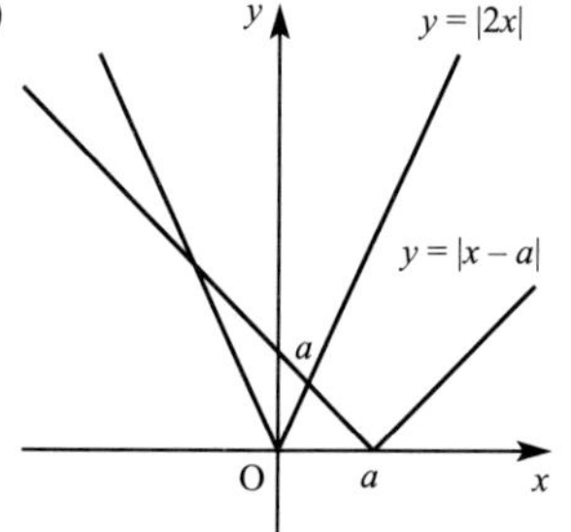

(b) $(a, 0)\ \ (0, a)$

(d) $(\frac{1}{3}a, \frac{2}{3}a)$

CHAPTER 3

EXERCISE 3A (Page 66)

1 1, 3, 6, 10, 15

2 2, 3, 5, 9, 17

3 5, 8, 11, 14, 17

4 4, 6, 4, 6, 4

5 16, 8, 4, 2, 1

6 $2, 1\frac{1}{2}, 1\frac{1}{3}, 1\frac{1}{4}, 1\frac{1}{5}$

7 $1, \frac{1}{2}, \frac{2}{3}, \frac{3}{5}, \frac{5}{8}$

8 $\sqrt{2}, \sqrt{3}, 2, \sqrt{5}, \sqrt{6},$

9 $3, 3\frac{1}{2}, 4, 4\frac{1}{2}, 5$

10 $1, 3, 4, 4\frac{1}{2}, 4\frac{3}{4}$

EXERCISE 3B (Page 68)

1 $\frac{1}{2}, \frac{1}{8}, \frac{1}{32}, \frac{1}{128}, 0$

2 2.5690, 2.5630, 2.5618, 2.5616

3 $0.\dot{3}$, 0.375, 0.37209, 0.37229

4 (a) $u_1 = 0 \Rightarrow$ 0 fixed point
$u_1 = 1 \Rightarrow$ 1 fixed point
$u_1 = 2 \Rightarrow$ diverges

(b) $u_2 = u_1^{\,2}, u_3 = u_1^{\,4}$

(c) $u_n = u_1^{\,2^{n-1}}$

5 3969

6 (a) 4, 6, 8, 10, 12

(b) 8, 12, 16, 20, 24

(c) $u_n = 4(n+1), u_{100} = 404$

7 (a) $\frac{1}{2}, \frac{1}{4}, \frac{1}{8}, \frac{1}{16}, \frac{1}{32}$

(b) 2

8 (a) $x_0 = 1$ −5, 19, 355, 126 019
$x_0 = 2$ −2, −2, −2, −2
$x_0 = 3$ 3, 3, 3, 3

(b) $x^2 - x - 6 = 0, -2, 3$

(c) Solutions to quadratic = fixed points.

9 (a) 1.192 57

(c) Answer to (a) is one of the solutions.

10 (a) $\frac{1}{3}u_1, \frac{1}{9}u_1, \frac{1}{27}u_1, \frac{1}{81}u_1$

(b) $\frac{1}{3^{n-1}}u_1$

(c) 0, 1.5

EXERCISE 3C (Page 72)

1 (a) $x^4 + 12x^3 + 54x^2 + 108x^3 + 81$

(b) $x^5 - 10x^4 + 40x^3 - 80x^2 + 80x - 32$

(c) $1 + 8x + 24x^2 + 32x^3 + 16x^4$

(d) $8 - 36x + 54x^2 - 27x^3$

(e) $64 + 192x + 240x^2 + 160x^3 + 60x^4 + 12x^5 + x^6$

(f) $16x^4 - 96x^3 + 216x^2 - 216x + 81$

2 (a) $1 + 7x + 21x^2 + 35x^3$

(b) $x^8 - 4x^7 + 7x^6 - 7x^5$

(c) $64 + 96x + 60x^2 + 20x^3$

3 (a) 6

(b) 3

(c) 70

4 (a) $1 + 5x + 10x^2 + 10x^3 + \ldots$
$1 - 5x + 10x^2 - 10x^3 + \ldots$

(b) $1 - 5x^2$

(c) Same because $[(1-x)(1+x)]^5 = (1-x^2)^5$

5 $1 + 3x + 6x^2 + 7x^3 + 6x^4 + 3x^5 + x^6$

EXERCISE 3D (Page 78)

1 (a) $x^4 + 4x^3 + 6x^2 + 4x + 1$

(b) $1 + 7x + 21x^2 + 35x^3 + 35x^4 + 21x^5 + 7x^6 + x^7$

(c) $x^5 + 10x^4 + 40x^3 + 80x^2 + 80x + 32$

(d) $64x^6 + 192x^5 + 240x^4 + 160x^3 + 60x^2 + 12x + 1$

(e) $32x^5 - 240x^4 + 720x^3 - 1080x^2 + 810x - 243$

(f) $8x^3 + 36x^2y + 54xy^2 + 27y^3$

2 (a) 6
(b) 15
(c) 20
(d) 15
(e) 1
(f) 220
(g) 220
(h) 1365
(i) 1

3 (a) 56
(b) 210
(c) 673 596
(d) −823 680
(e) 70

4 (a) $1 + 4x + 6x^2 + 4x^3 + x^4$
(b) 1.008
(c) 0.0024%

5 (a) $6x + 2x^3$

6 (a) $32 - 80x + 80x^2 - 40x^3 + 10x^4 - x^5$
(b) 31.208
(c) 0.000 13%

7 (a) $x^6 + 6x^4 + 15x^2 + 20 + \frac{15}{x^2} + \frac{6}{x^4} + \frac{1}{x^6}$
(b) $16x^4 - 16x^2 + 6 - \frac{1}{x^2} + \frac{1}{16x^4}$
(c) $1 + \frac{10}{x} + \frac{40}{x^2} + \frac{80}{x^3} + \frac{80}{x^4} + \frac{32}{x^5}$

8 (a) £2570
(b) True value = £2593.74
(c) 0.92%

9 (b) $x = 0, -1$ and -2

10 (a) $1 + 24x + 264x^2 + 1760x^3$
(b) 1.268 16
(c) 0.0064%

Exercise 3E (Page 79)

1 2.1520, 2.1125, 2.1423, 2.1197

2 (a) fixed point of −1
(b) fixed point of 2
(c) $x^2 - x - 2 = 0, -1, 2$
(d) $u_3 = u_1^4 - 4u_1^2 + 2$

3 (a) 35
(b) 4, 9, 16, 25
(c) $(n + 1)^2$

4 (a) $1, \frac{1}{3}, -\frac{1}{9}$
(b) 5.986

5 (a) $15 - 4k$
(b) $-8k^2 + 30k - 30$
(c) $-\frac{1}{4}, 4$

6 $81x^4 - 36x^2 + 6 - \frac{4}{9x^2} + \frac{1}{81x^4}$

7 4096, 24 576, 67 584, 112 640

8 (a) $1 - 44x + 880x^2 - 10\,560x^3$
(b) 0.637 44

9 (a) $81 + 216x + 216x^2 + 96x^3 + 16x^4$
(b) $81 - 216x + 216x^2 - 96x^3 + 16x^4$
(c) 1154

10 (a) 8
(b) $4\frac{3}{8}$

Chapter 4

Exercise 4A (Page 86)

1 (a) 30°, 150°
(b) 78.5°, 281.5°
(c) $\frac{\pi}{4}, \frac{5\pi}{4}$
(d) $-\frac{\pi}{3}, -\frac{2\pi}{3}, \frac{4\pi}{3}, \frac{5\pi}{3}$
(e) 63.4°, 243.4°

2 (a) 30°, 150°, 210°, 330°
(b) $-\frac{3\pi}{2}, -\frac{5\pi}{6}, -\frac{\pi}{6}, \frac{\pi}{2}, \frac{7\pi}{6}, \frac{11\pi}{6}$
(c) −148.3°, −58.3°, 31.7°, 121.7°
(d) $-\frac{11\pi}{12}, -\frac{5\pi}{6}, -\frac{5\pi}{12}, -\frac{\pi}{3}, \frac{\pi}{12}, \frac{\pi}{6}, \frac{7\pi}{12}, \frac{2\pi}{3}$
(e) $-\frac{5\pi}{6}, -\frac{\pi}{3}, \frac{\pi}{6}, \frac{2\pi}{3}, \frac{7\pi}{6}, \frac{5\pi}{3}, \frac{13\pi}{6}, \frac{8\pi}{3}$

3 (a) 0°, 60°, 300°, 360°
(b) 19.5°, 160.5°, 270°
(c) 63.4°, 45°, 243.4°, 225°
(d) $\frac{\pi}{2}, \frac{3\pi}{2}$
(e) 6.1°, 23.9°, 66.1°, 83.9°, 126.1°, 143.9°

4 (a)

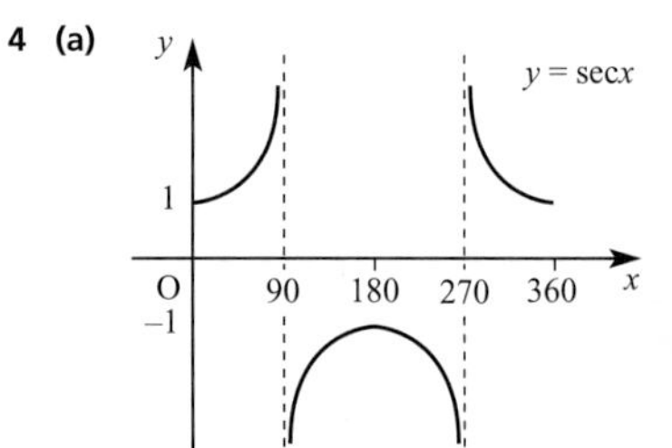

(b)

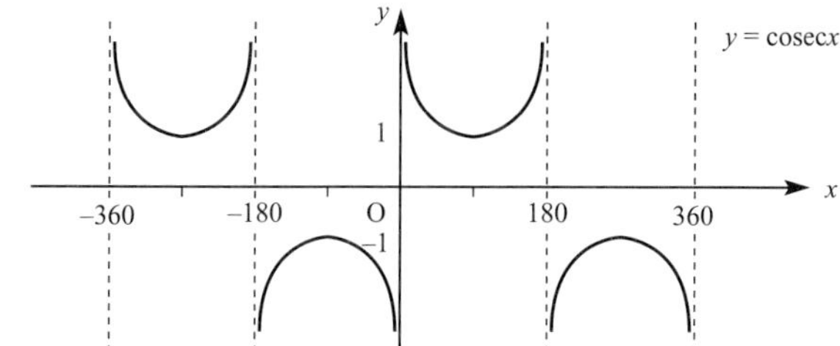

(c)

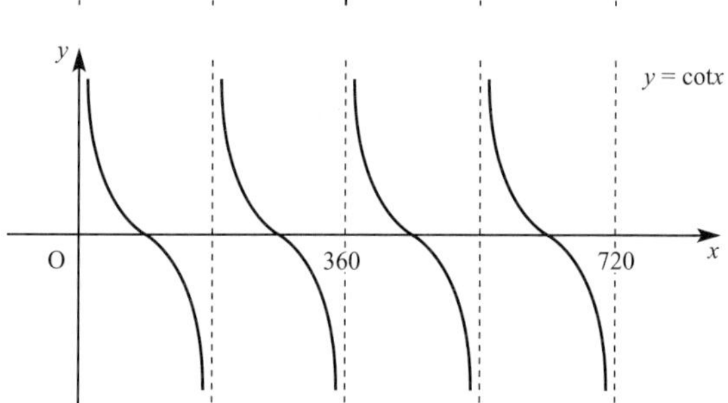

(d)

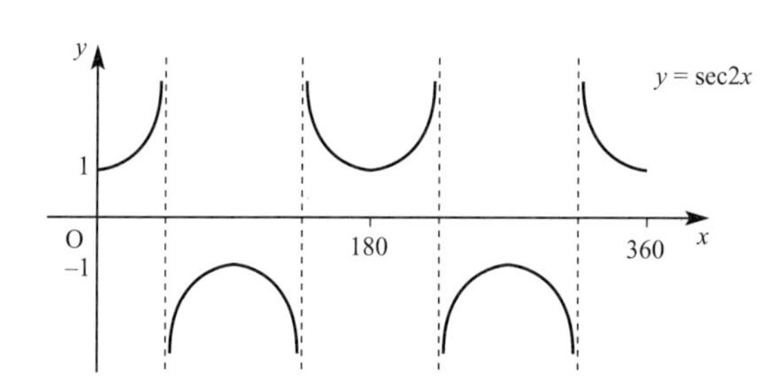

(e)

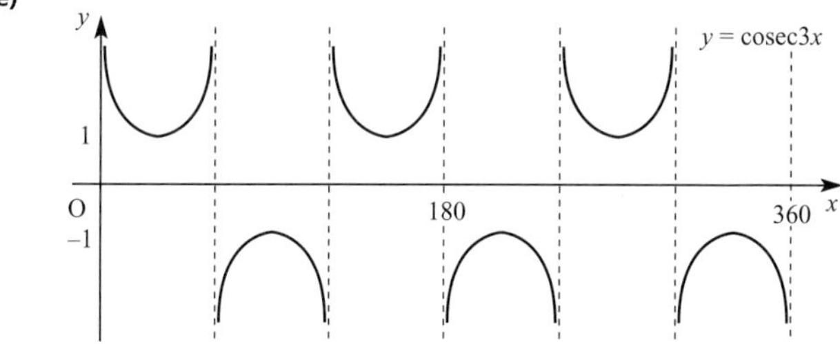

5 (a) -2
(b) -2
(c) $\sqrt{3}$
(d) 2
(e) -2

EXERCISE 4B (Page 92)

1 (a) 90°, 210°, 330°
(b) 45°, 63.4°
(c) 48.2°, 60°, 300°, 311.8°
(d) 26.6°, 45°, 206.6°, 225°
(e) 26.6°, 135°, 206.6°, 315°
(f) 60°, 70.5°
(g) 33.7°, 45°
(h) $\frac{\pi}{6}, \frac{5\pi}{6}, \frac{3\pi}{2}$
(i) $0, \frac{2\pi}{3}, \frac{4\pi}{3}, 2\pi$
(j) 63.4°, 123.7°, 243.4°, 303.7°

3 (a) $B = 60°$ $C = 30°$
(b) $\sqrt{3}$

4 (a) $L = N = 45°$
(b) $\sqrt{2}, \sqrt{2}, 1$

5 $\frac{1}{2}, -\frac{3}{4}$
30°, 150°, 228.6°, 311.4°
$8\sin^2\theta + 2\sin\theta - 3 = 0$
$\theta = 30°$

EXERCISE 4C (Page 96)

1 (a) $\frac{\sqrt{3}}{2\sqrt{2}} + \frac{1}{2\sqrt{2}}$
(b) $-\frac{1}{\sqrt{2}}$
(c) $\frac{\sqrt{3}}{2\sqrt{2}} + \frac{1}{2\sqrt{2}}$
(d) $\frac{\sqrt{3} - 1}{\sqrt{3} + 1}$
(e) $\frac{\sqrt{3} + 1}{\sqrt{3} - 1}$

2 (a) $\frac{1}{\sqrt{2}}(\sin\theta + \cos\theta)$
(b) $\frac{1}{2}(\sqrt{3}\cos\theta + \sin\theta)$
(c) $\frac{1}{2}(\sqrt{3}\cos\theta - \sin\theta)$
(d) $\frac{1}{\sqrt{2}}(\cos2\theta - \sin2\theta)$
(e) $\frac{\tan\theta + 1}{1 - \tan\theta}$
(f) $\frac{\tan\theta - 1}{1 + \tan\theta}$

3 (a) $\sin\theta$
(b) $\cos4\phi$
(c) 0
(d) $\cos2\theta$

4 (a) 15°
(b) 157.5°
(c) 0° or 180°
(d) 111.7°
(e) 165°

5 (a) $\frac{\pi}{8}$
(b) 2.79 radians

EXERCISE 4D (Page 102)

1 (a) 14.5°, 90°, 165.5°, 270°
(b) 0°, 35.3°, 144.7°, 180°, 215.3°, 324.7°, 360°
(c) 90°, 210°, 330°
(d) 30°, 150°, 210°, 330°
(e) 0°, 138.6°, 221.4°, 360°

2 (a) $-\pi, 0, \pi$
(b) $-\pi, 0, \pi$
(c) $\frac{-2\pi}{3}, 0, \frac{2\pi}{3}$
(d) $\frac{-3\pi}{4}, \frac{-\pi}{4}, \frac{\pi}{4}, \frac{3\pi}{4}$
(e) $\frac{-11\pi}{12}, \frac{-3\pi}{4}, \frac{-7\pi}{12}, \frac{-\pi}{4}, \frac{\pi}{12}, \frac{\pi}{4}, \frac{5\pi}{12}, \frac{3\pi}{4}$

3 $3\sin\theta - 4\sin^3\theta$,
$\theta = 0, \frac{\pi}{4}, \frac{3\pi}{4}, \pi, \frac{5\pi}{4}, \frac{7\pi}{4}, 2\pi$

4 51°, 309°

5 $\cot\theta$

6 $\dfrac{\tan\theta(3 - \tan^2\theta)}{1 - 3\tan^2\theta}$

8 **(b)** 63.4°

9 **(a)** 180°
(b) 126.8°, 180°
(c) 0°, 53.1°, 306.9°, 360°
(d) 97.2°, 262.8°

11 **(a)** 90°, 323.1°
(b) 0°, 67.4°, 360°
(c) 143.1°, 306.9°
(d) 90°, 202.6°

12 $\dfrac{(1 - t)^2}{1 + t^2}$

Exercise 4E (Page 106)

1 **(a)** $\sqrt{2}\cos(\theta - 45°)$
(b) $5\cos(\theta - 53.1°)$
(c) $2\cos(\theta - 60°)$
(d) $3\cos(\theta - 41.8°)$

2 **(a)** $\sqrt{2}\cos\left(\theta + \frac{\pi}{4}\right)$
(b) $2\cos\left(\theta + \frac{\pi}{6}\right)$

3 **(a)** $\sqrt{5}\sin(\theta + 63.4°)$
(b) $5\sin(\theta + 53.1°)$

4 **(a)** $\sqrt{2}\sin\left(\theta - \frac{\pi}{4}\right)$
(b) $2\sin\left(\theta - \frac{\pi}{6}\right)$

5 **(a)** $2\cos(\theta - (-60°))$
(b) $4\cos(\theta - (-45°))$
(c) $2\cos(\theta - 30°)$
(d) $13\cos(\theta - 22.6°)$
(e) $2\cos(\theta - (-150°))$
(f) $2\cos(\theta - (-135°))$

6 **(a)** $13\cos(\theta + 67.4°)$
(b) Max 13, min – 13
(c)

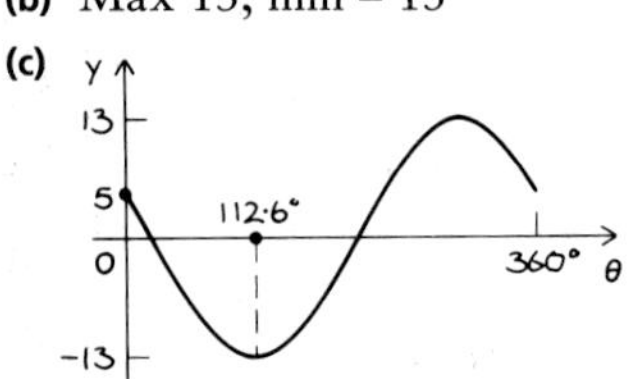

(d) 4.7°, 220.5°

7 **(a)** $2\sqrt{3}\sin\left(\theta - \frac{\pi}{6}\right)$
(b) Max $2\sqrt{3}$, $\theta = \frac{2\pi}{3}$; min $- 2\sqrt{3}$, $\theta = \frac{5\pi}{3}$
(c)

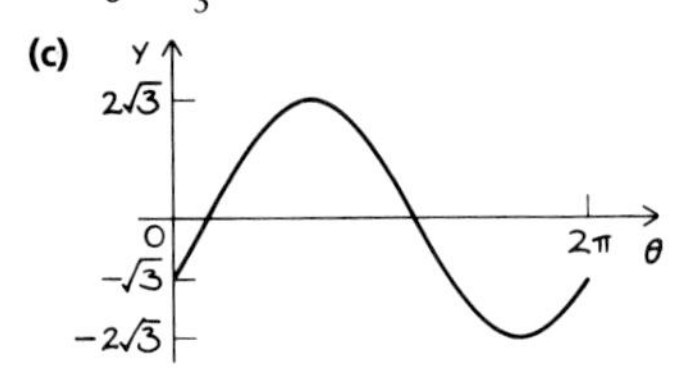

(d) $\frac{\pi}{3}$, π

8 **(a)** $\sqrt{13}\sin(2\theta + 56.3°)$
(b) Max $\sqrt{13}$, $0 = 16.8°$;
min $-\sqrt{13}$, $\theta = 106.8°$
(c)

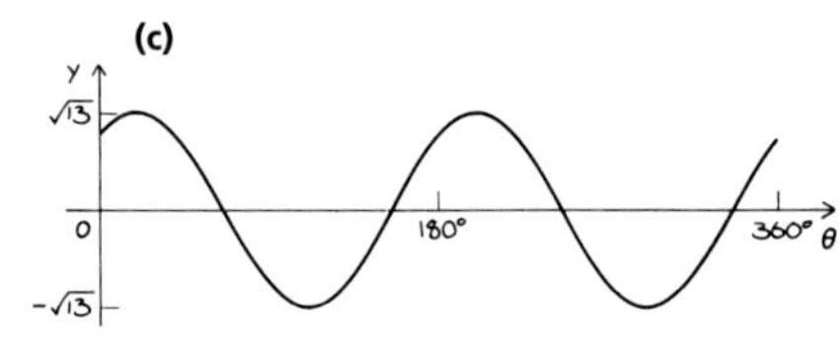

(d) 53.8°, 159.9°, 233.8°, 339.9°

9 **(a)** $\sqrt{3}\cos(\theta - 54.7°)$
(b) Max $\sqrt{3}$, $\theta = 54.7°$;
min $-\sqrt{3}$, $\theta = 234.7°$
(c)

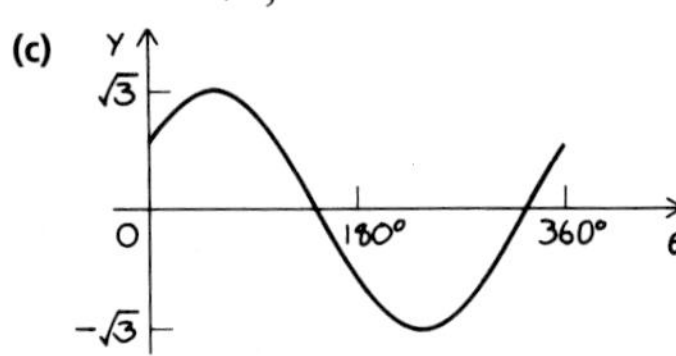

(d) Max $\dfrac{1}{3 - \sqrt{3}}$, $\theta = 234.7°$;
min $\dfrac{1}{3 + \sqrt{3}}$, $\theta = 54.7°$

10 **(b)** 30.6° or 82.0°

Exercise 4F (Page 110)

1 **(a)** $2\cos3\theta\sin\theta$
(b) $2\cos3\theta\cos2\theta$
(c) $-2\sin5\theta\sin2\theta$
(d) $\cos\theta$
(e) $\sqrt{2}\sin3\theta$

2 20°, 90°, 100°, 140°

3 $\dfrac{\tan4\theta}{\tan\theta}$

4 $0, \frac{\pi}{4}, \frac{3\pi}{4}, \pi, \frac{5\pi}{4}, \frac{7\pi}{4}, 2\pi$

5 $\cos(\theta + 43°)$

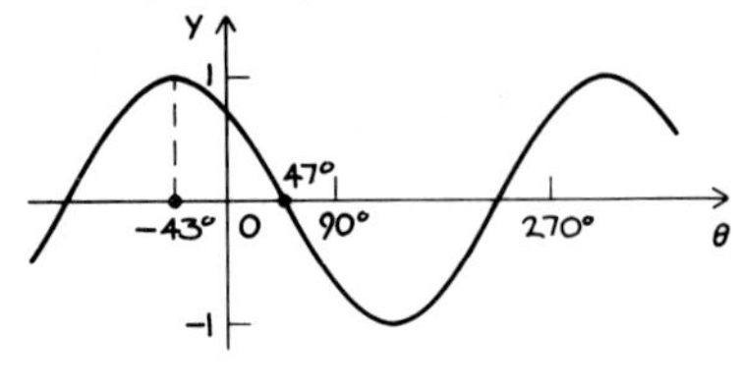

8 **(a)** 0°, 45°, 60°, 120°, 135°, 180°, 225°, 240°, 300°, 315°, 360°,
(b) 0°, 60°, 120°, 180°, 240°, 300°, 360°
(c) 0°, 45°, 135°, 180°, 225°, 315°, 360°
(d) 30°, 60°, 90°, 120°, 150°, 210°, 240°, 270°, 300°, 330°,

9 **(a)** $\cos4\theta + \cos2\theta$
(b) $\cos2\theta - \cos8\theta$
(c) $2(\sin10\theta - \sin2\theta)$
(d) $\frac{1}{2}(\sin4\theta + \sin2\theta)$

10 **(a)** $\dfrac{\sqrt{3} + 1}{2}$
(b) $\dfrac{1}{2\sqrt{2}}$
(c) $\dfrac{\sqrt{3} + 1}{4}$
(d) $\dfrac{1}{4}$

Exercise 4G (Page 111)

1 **(a)** $R = 25$, $\alpha = 73.7°$
(b) –36.9°, –110.6°
(c) 25 when $x = -73.7°$

2 **(a)** $\cos2x = 2\cos^2x - 1$
(b) $0, \frac{2\pi}{3}, \frac{4\pi}{3}, 2\pi$
(c) $0, \frac{4\pi}{3}$

3 **(a)** $\dfrac{1 - \sqrt{3}}{4}$
(b) 22.5°, 67.5°, 90°, 112.5°, 157.5°, 202.5°, 247.5°, 270° 292.5°, 337.5°

4 (a)

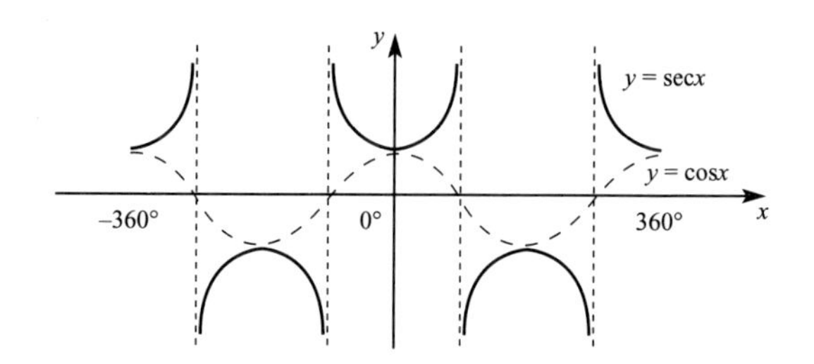

(b) 18.4°, 71.6°, 198.4°, 251.6°

5 (a) $R = 10$, $\alpha = 0.93$

(b) –10, –0.64

(c) 1.34

6 $\frac{7\pi}{6}, \frac{11\pi}{6}, \frac{\pi}{2}$

7 (b) 0.85, 2.29, 4.71

(c) 1.70, 4.59

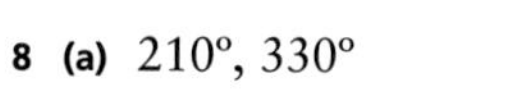

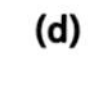

8 (a) 210°, 330°

(b) 2.01

9 (a) $R = 25$, $\alpha = 73.7°$

(b) 20.6°, 126.8°

(c) $\frac{1}{630} \leqslant f(\theta) \leqslant \frac{1}{5}$

10 (a) $R = 15$ $= 53.1°$

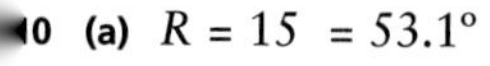

(b) 156.9°, 276.9°

(c) $\frac{\pi}{3}, \frac{4\pi}{3}$

CHAPTER 5

EXERCISE 5A (Page 122)

1 (a)

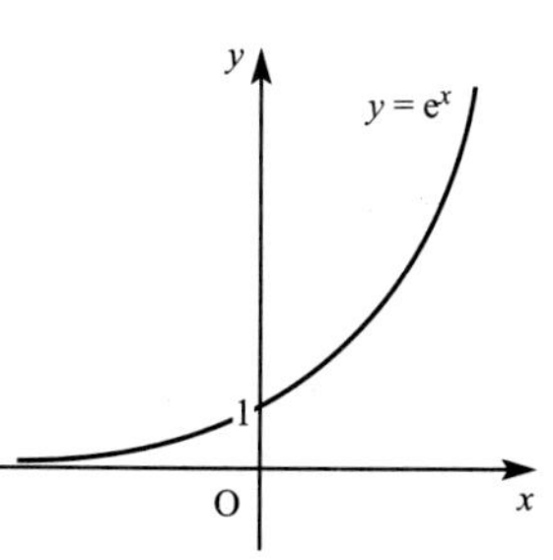

(b)

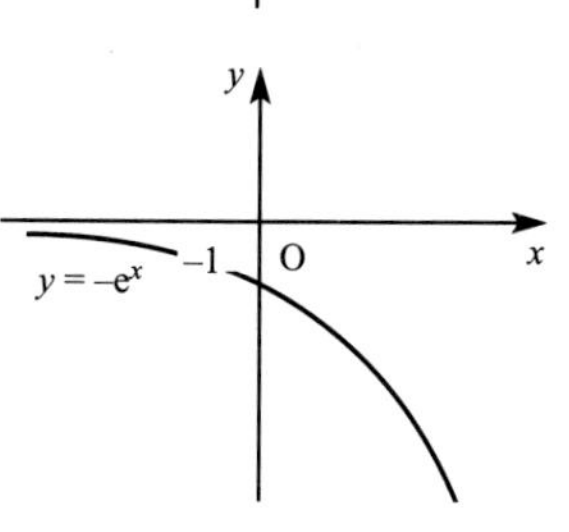

(c)

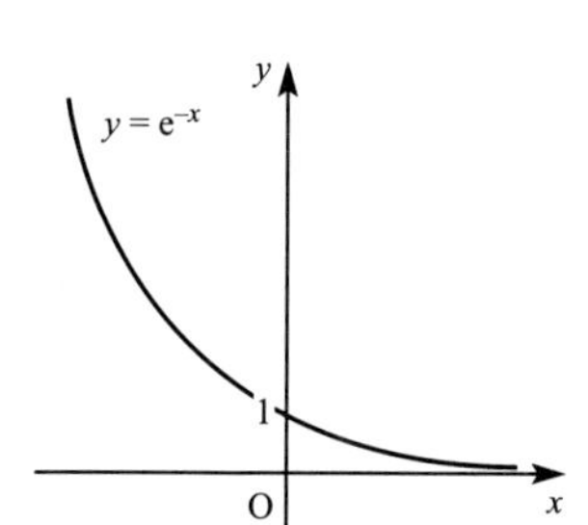

(d)

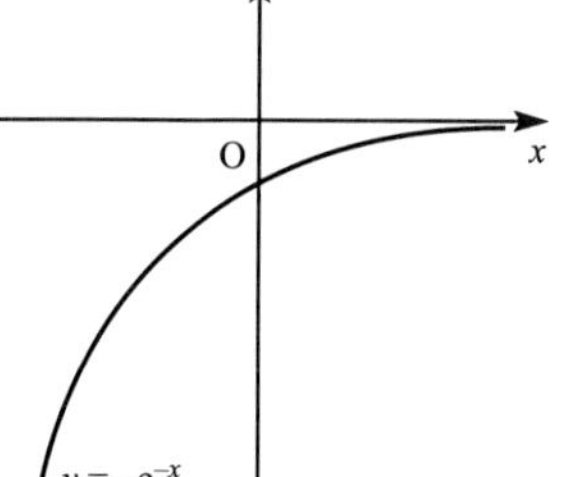
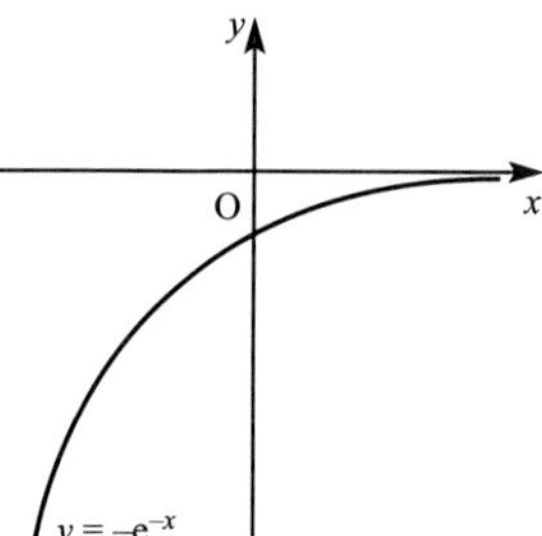

2 (a)

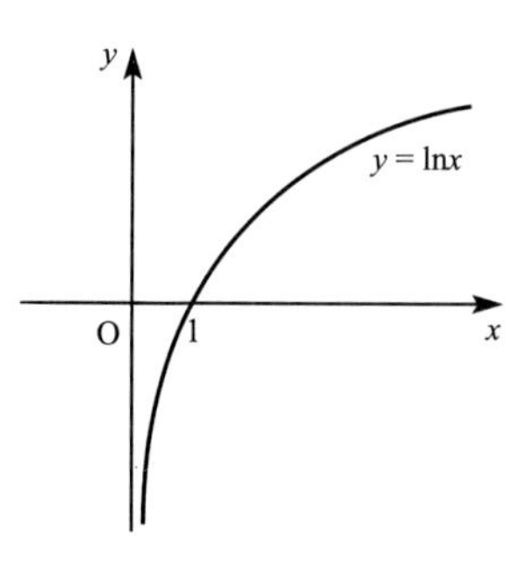

(b)

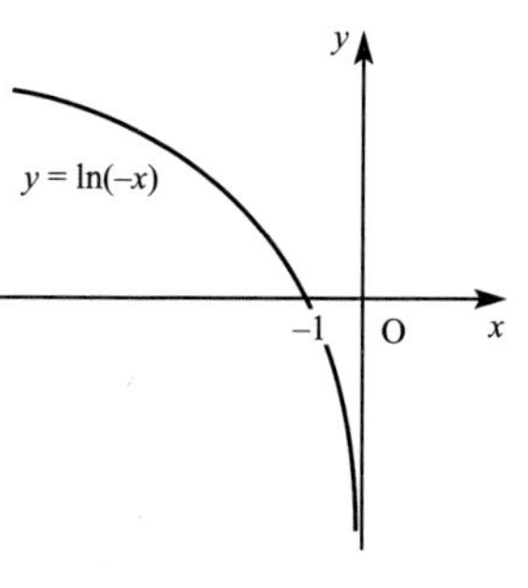

(c)

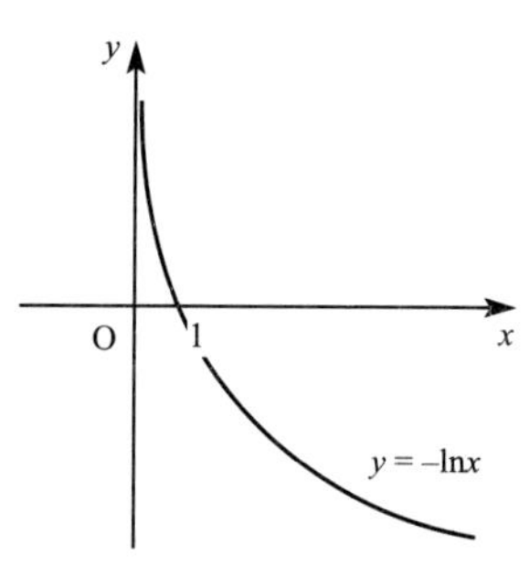

(d)

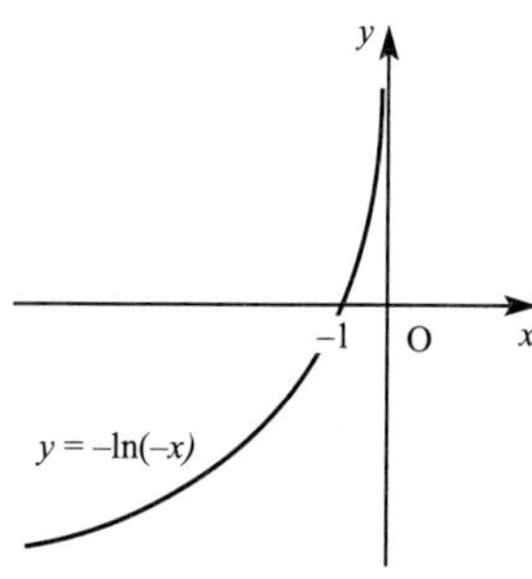

3 (a)

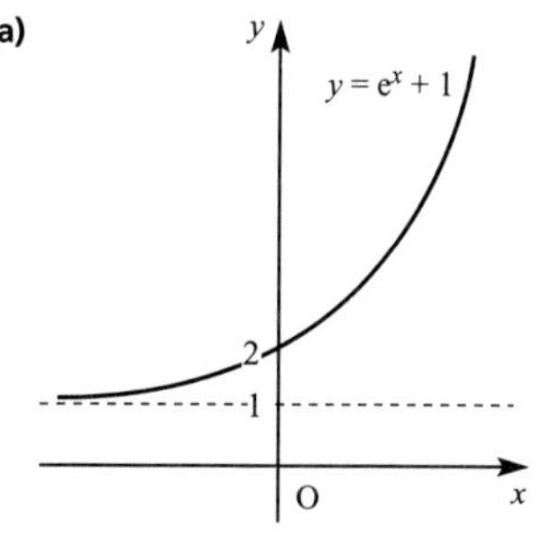

(b)

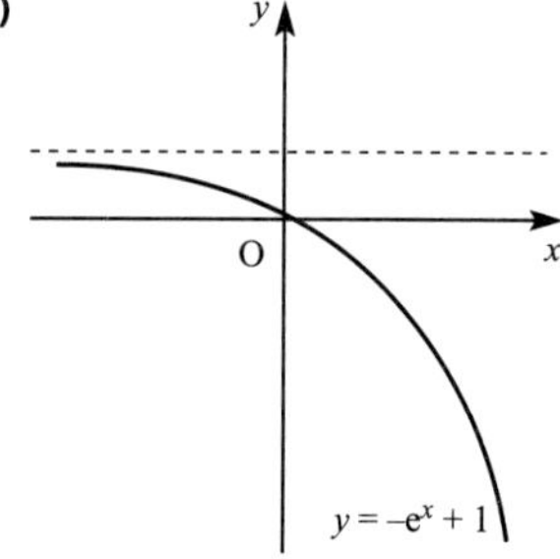

(c)

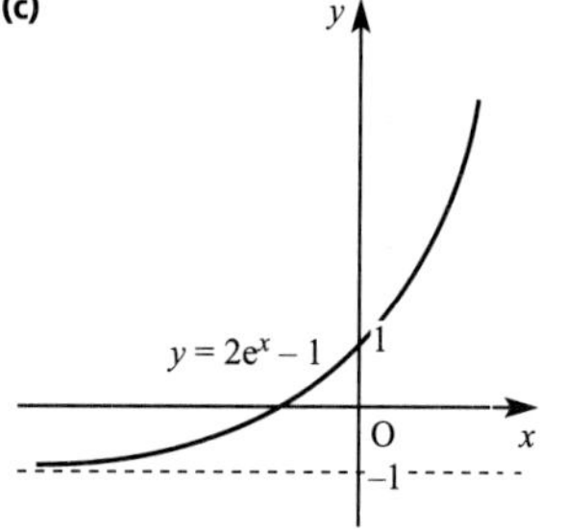

(d)

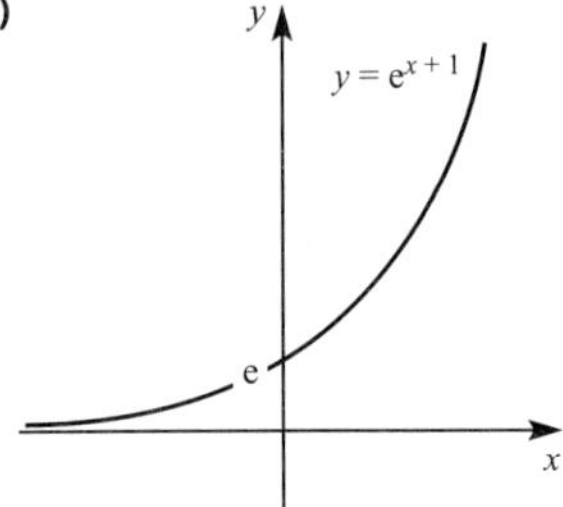

4 (a)

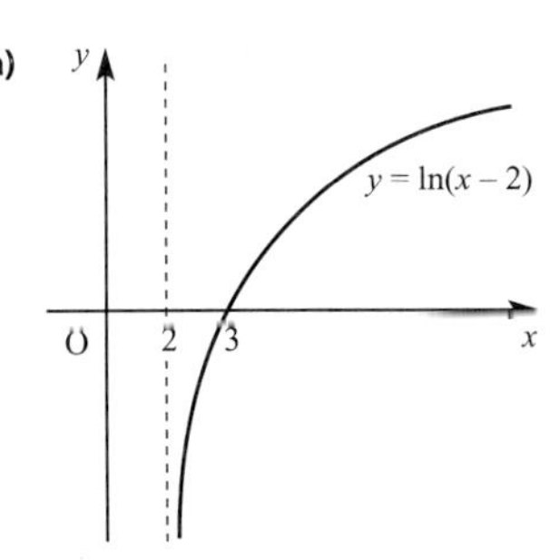

(b)

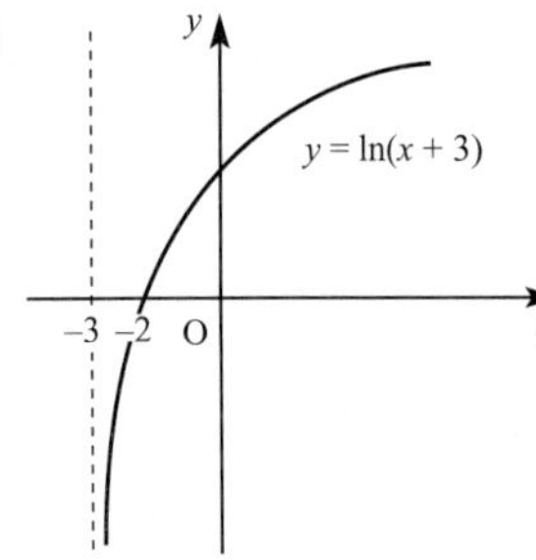

(c)

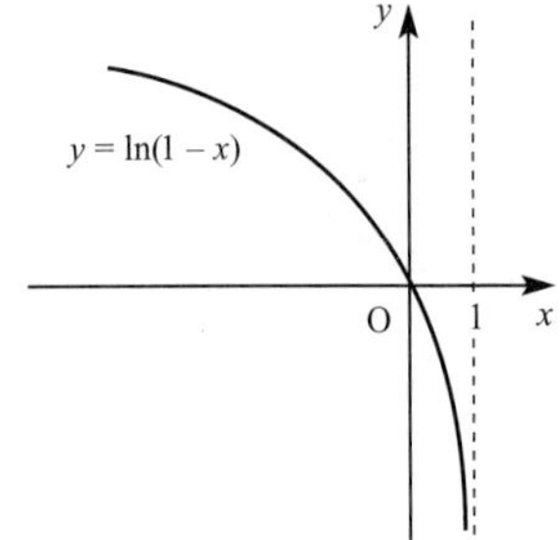

(d)

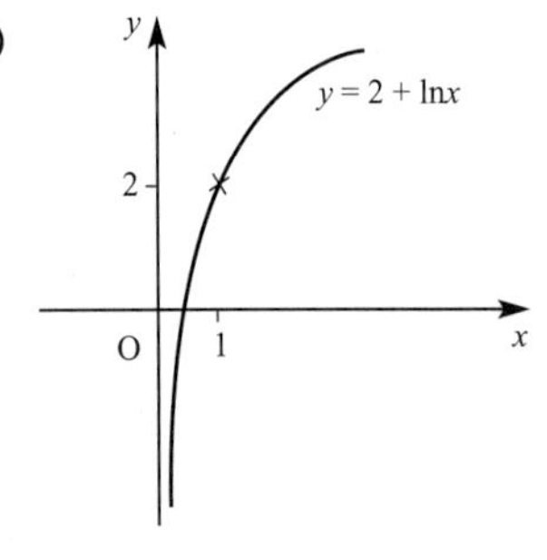

5 (a)

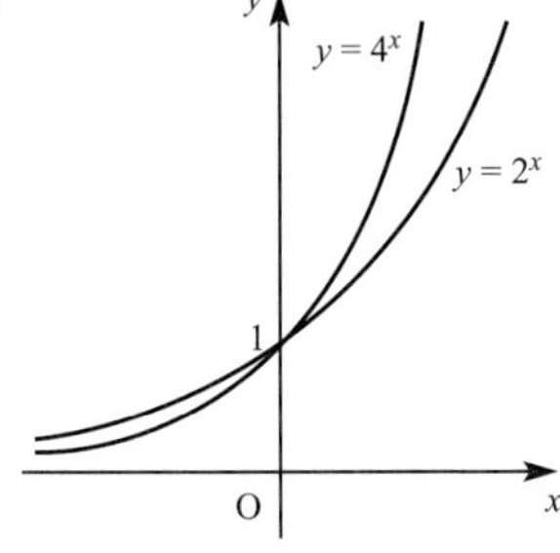

(b) $4^x = (2^2)^x = 2^{2x}$
stretch scale factor $\frac{1}{2}$ parallel to x axis, y axis invariant

6 (a)

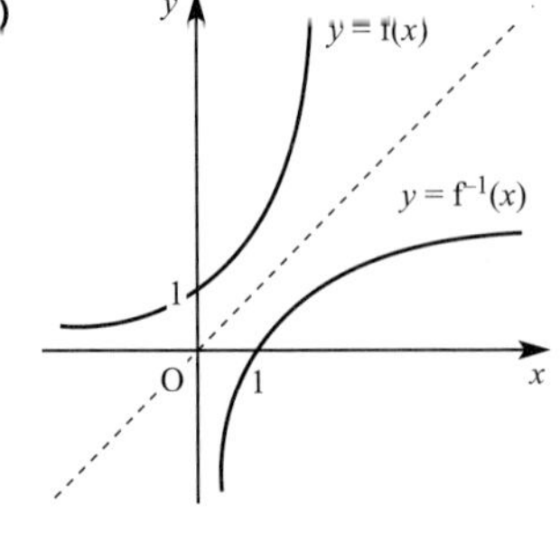

(b) Reflection in $y = x$

(c) $f^{-1}(x) = \log_{10}x$

7

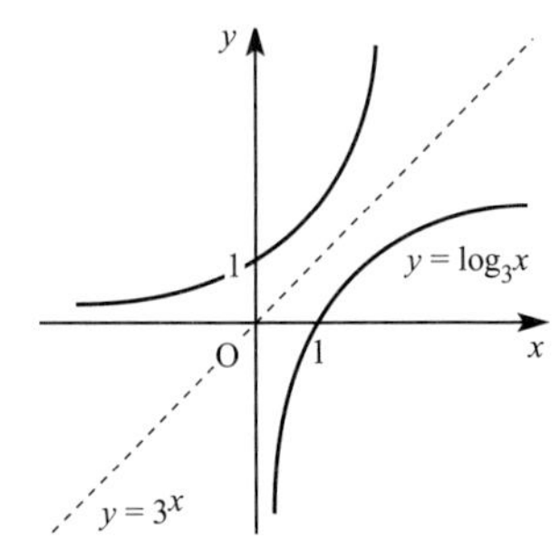

$y = 3^x$

8 (a)

(b) $f^{-1}(x) = e^x + 3$

(c) range $y > 3$

EXERCISE 5B (Page 126)

1 (a) 0.693
(b) 3.08
(c) 1.61
(d) –3.91
(e) 9.90
(f) –3.00
(g) 2.69
(h) 1.26
(i) 21.2
(j) 0.231

2 (a) 7.39
(b) 4.35
(c) 143
(d) 2.74
(e) 1.04
(f) –0.105
(g) 1.38
(h) 1.45
(i) 52.4
(j) 0.273

3 $p = 25e^{-0.02t}$

4 $x = \ln\left(\frac{y-5}{y_0-5}\right)$

5 (a)

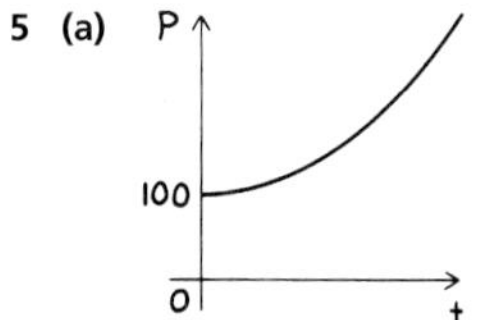

(b) 100
(c) 1218
(d) 185 years

6 (a)

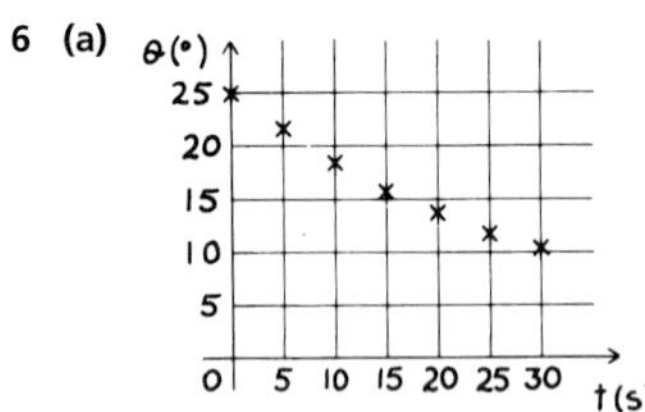

(b) 25°
(c) 4.1°
(d) 22

7 **(a)**

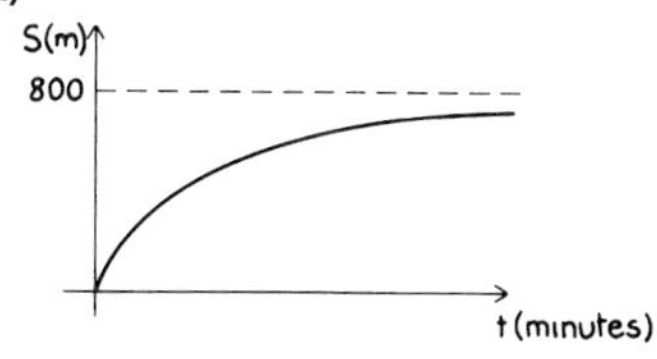

(b) 621.5 m

(c) 8.07 am (to nearest minute)

(d) Never

8 **(a)**

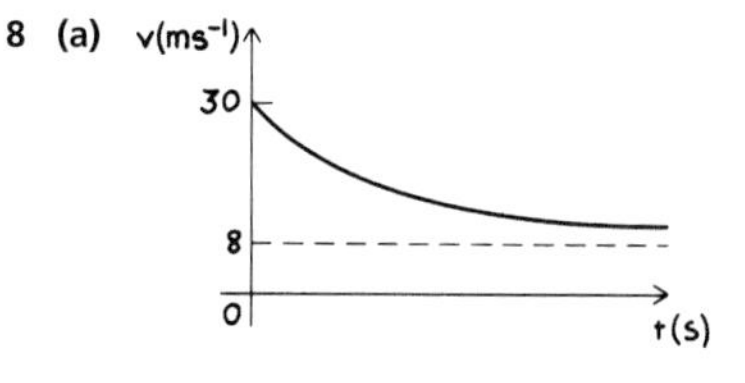

(b) 30 ms^{-1}, 8 ms^{-1}

(c) 8.33 ms^{-1}

(d) 8.7 s

9 **(a)** 250

(b) 6.9

(c)

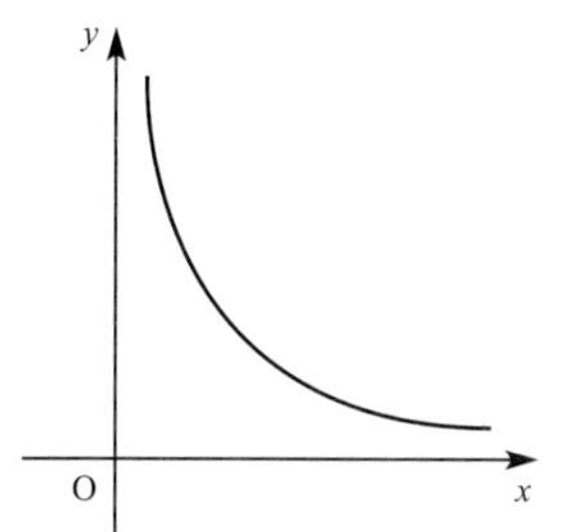

10 **(a)** Yes

(b) 1115

(c) 26 days

EXERCISE 5C (Page 131)

1 **(a)** $\log_3 81 = 4$

(b) $\log_2 256 = 8$

(c) $\log_a l = b$

(d) $\log_y x = z$

2 **(a)** $10\,000 = 10^4$

(b) $216 = 6^3$

(c) $b = c^q$

(d) $y = n^x$

3 **(a)** $\log p + \log q - \log r$

(b) $\log p - \log q - \log r$

(c) $2\log p + 3\log q$

(d) $\log p + \frac{1}{3}\log r - \frac{1}{2}\log q$

4 **(a)** $\log 10$

(b) $\log 2$

(c) $\log 36$

(d) $\log \frac{1}{7}$

(e) $\log 3$

(f) $\log 4$

(g) $\log 4$

(h) $\log \frac{1}{3}$

(i) $\log \frac{1}{2}$

(j) $\log 12$

5 **(a)** 3

(b) –4

(c) $\frac{1}{2}$

(d) 0

(e) 4

(f) –4

(g) $\frac{3}{2}$

(h) $\frac{1}{4}$

(i) $\frac{1}{2}$

(j) –3

6 **(a)** 19.93

(b) –9.97

(c) 9.01

(d) 48.32

(e) 1375

(f) 4.64

(g) 6.11

(h) 10.48

(i) 8.58

(j) –5

7 **(a)** –2, 1.58

(b) 2

8 $\log_3 \dfrac{x+1}{2x}$

$\frac{1}{17}$

9 **(a)**

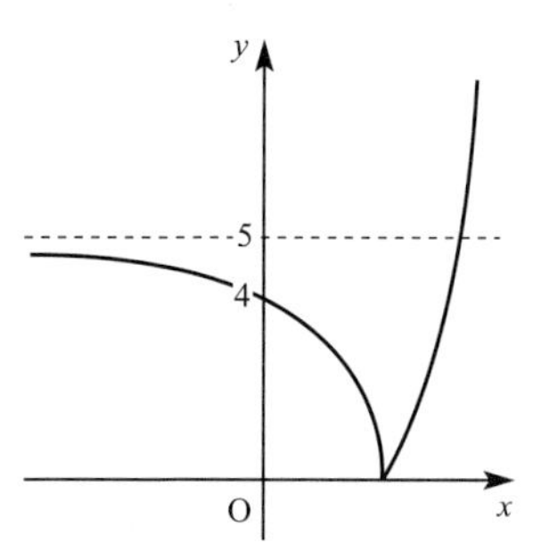

(b) $0.613 < x < 1.086$

10 **(a)** $P = P_0 e^{\frac{-t}{30}}$

(b) $P_0 = 1000$

EXERCISE 5D (Page 132)

1 **(a)** $\log_2\left(\dfrac{x+4}{x-1}\right)$

(b) $\frac{4}{3}$

(c) $\frac{13}{8}$

2 **(a)** 1.43

(b) 1.43 or 1.68

3 **(a)** $S = 6000$

(b) $t = 110$

$P = P_0 \times 2^{\frac{t}{15}}$

99.7 hours

4 **(a)**

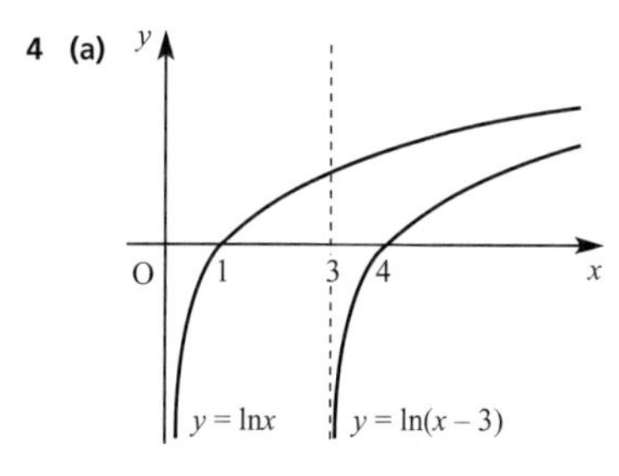

translation $\binom{3}{0}$

(b) $y = e^{x+2}$

stretch scale factor e^2 parallel to y axis, x axis invariant.

5 **(a)** Levels off

(b) 164

(c) 8.4 weeks

(d) 260

6 **(c)** $y^2 - 1000.1y + 100 = 0$

(d) $x = -1$ or $x = 3$

7 **(a)** range $\geqslant k$

(b) 2k

(c) $\ln(x-k)$, domain $x > k$

(d)

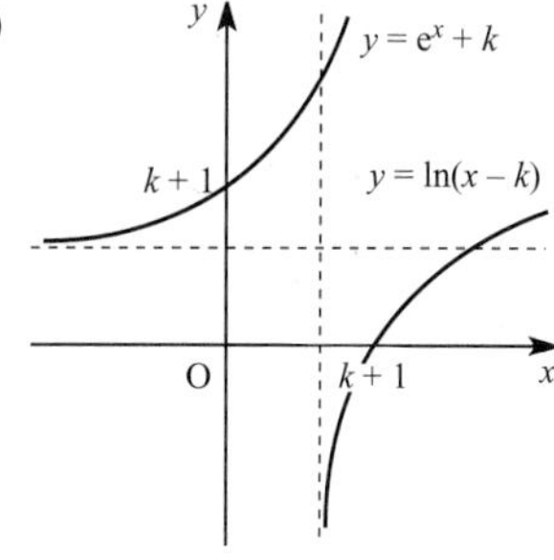

8 **(a)** $y\ln 2 = x$
(c) 2.43

9 **(a)** 330
(b) 317 to 345

10 **(a)** $a = b^p$
(d) 1.46

CHAPTER 6

EXERCISE 6A (Page 139)

1 **(a)** e^x
(b) $2e^x$
(c) $-3e^x$
(d) $2 + 8e^x$
(e) $0.1e^x - x$
(f) $\frac{1}{2\sqrt{x}} - \frac{e^x}{2}$

2 **(a)** $\frac{1}{x}$
(b) $\frac{2}{x}$
(c) $\frac{1}{x}$
(d) $\frac{4}{x}$
(e) $\frac{4}{x}$
(f) $-\frac{1}{x}$
(g) $-\frac{2}{x}$
(h) $-\frac{6}{x}$
(i) $1 - \frac{1}{x}$
(j) $-\frac{2}{x^2} + \frac{4}{x}$

3 **(a)** $e + 1$
(b) $\frac{3}{2} - 4e^2$
(c) $4 + 2e$
(d) $2e^4 - \frac{5}{8}$
(e) $\frac{-5}{4} - 5e^2$
(f) 3

5 **(a)** 1030 agents
(b) 1000 agents/year

EXERCISE 6B (Page 142)

1 **(a)** $2y = 8x - 15$
(b) $(-\frac{1}{8}, -8)$
(c)

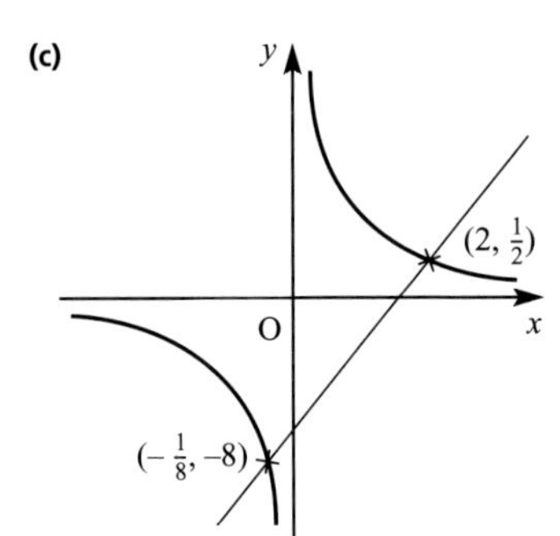

2 $y = 4x - 2$

3 **(a)** $x = \ln 4$
(b) 4
(c) $y = 4x - 4\ln 4$

4 **(a)**

translation $\binom{2}{0}$
(b) Using (a) gradient of $\ln x$ is $\frac{1}{x}$ so gradient of $\ln(x - 2)$ must be $\frac{1}{x - 2}$
(d) $3y = x - 3 - 6\ln 3$

7 **(b)** $16y = 3x + 28$

9 $x + (e - 1)y = 2(e - 1)\ln 5 + e^2 - 4e + 4$

EXERCISE 6C (Page 143)

1 **(a)** $h = \ln 4$
(b) $y = 4x + 3 - 4\ln 4$

2 **(b)** $15y = 4x - 121$

3 **(b)** $y = 2x + 3$

4 Maximum at $x = 2$

5 Tangent: $y = (e + 2)x + 1$
Normal: $x + (e + 2)y = e^2 + 5e + 7$

6 **(a)** $h = 3$
(c) $3x + 2y = 9 + 2\ln 18$

7 **(a)** $\frac{dy}{dx} = 2e^x - 4$
(b) $x = \ln 2$
(c) $y = 4 - 4\ln 2$
(d) minimum

8 **(a)** $\frac{dy}{dx} = -\frac{1}{x} + \frac{2}{x^3}$
(c) $4y = 16x - 33$

9 **(b)** $y = (e^2 - 4)x - (e^2 - 4)$

10 **(b)** $\frac{5}{4} + \frac{1}{2}\ln 2$

CHAPTER 7

EXERCISE 7A (Page 148)

1 **(a)** $e^x + c$
(b) $2e^x + c$
(c) $3e^x + c$
(d) $e^x + 2x + c$
(e) $20x + 15e^x + c$
(f) $4x - e^x + c$
(g) $x^2 + 3e^x + c$
(h) $\frac{1}{2}e^x + \frac{2}{3}x^{\frac{3}{2}} + c$
(i) $4e^x - \frac{1}{x^2} + c$
(j) $e^{x+2} + c$

2 **(a)** 1.72
(b) 12.8
(c) 1.63
(d) −47.2
(e) −0.198
(f) 1
(g) 32.8
(h) −255
(i) 6.39
(j) 33.1

4 **(a)**

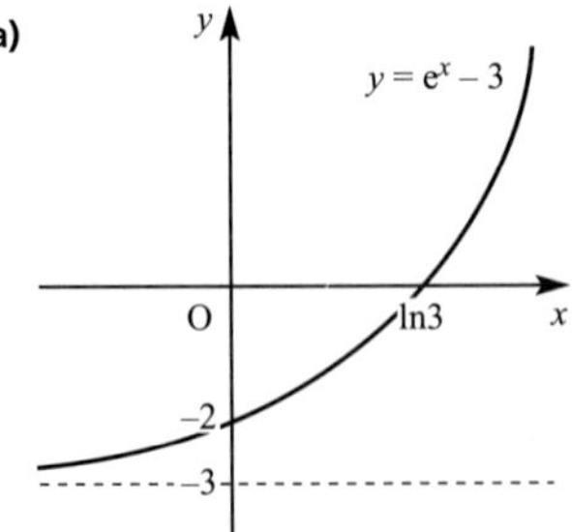

EXERCISE 7B (Page 151)

1 **(a)** $2\ln x + c$
(b) $\frac{1}{2}\ln x + c$
(c) $\frac{3}{4}\ln x + 5x + c$
(d) $\ln x + e^x + c$
(e) $\frac{1}{2}\ln x + \sqrt{x} + c$
(f) $-\frac{1}{x} - \ln x + c$
(g) $4\ln x + x + c$
(h) $5\ln x + \frac{1}{10}x^2 + c$
(i) $\frac{1}{2}x^2 + 2x + \ln x + c$
(j) $\frac{1}{2}x^2 - \ln x + c$

2 **(a)** 0.693
(b) 4.16
(c) −3.97
(d) −1.10
(e) 2.81
(f) 0.189
(g) 10.7
(h) 10.5
(i) 3.64
(j) −3.36

3 **(a)**

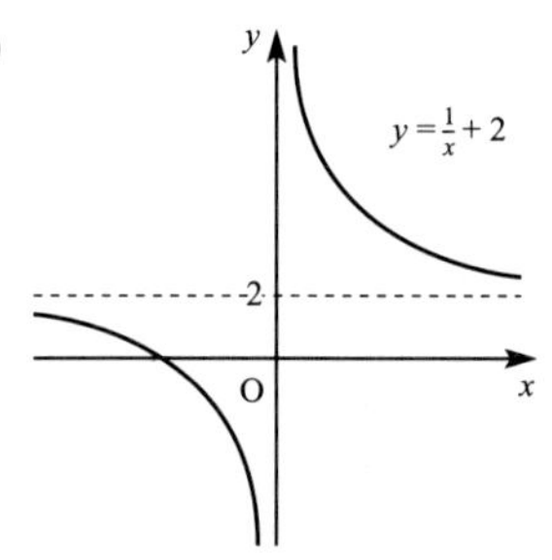

(b) $2 + \ln\left(\frac{3}{2}\right)$

5 −1.616

EXERCISE 7C (Page 156)

1 **(a)**

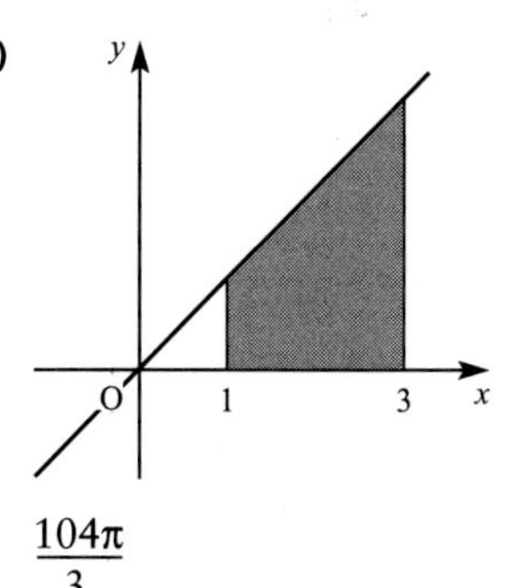

$\frac{104\pi}{3}$

(b)

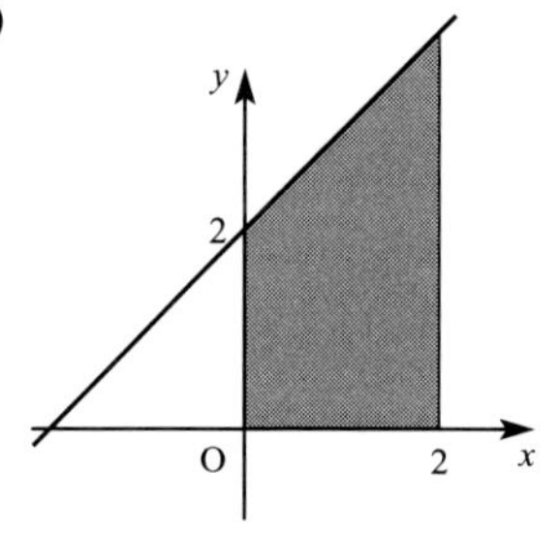

$\frac{56\pi}{3}$

(c)

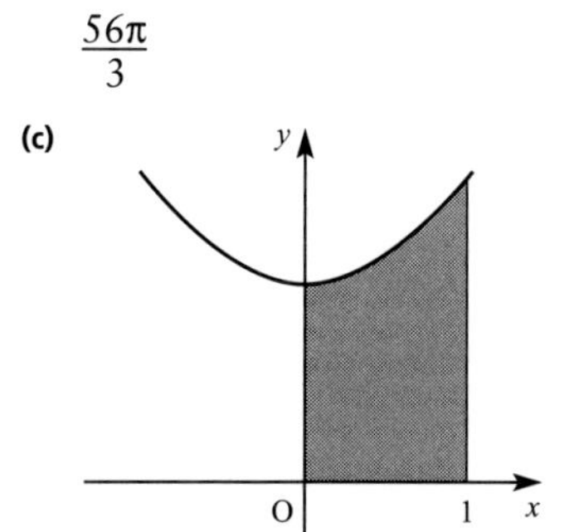

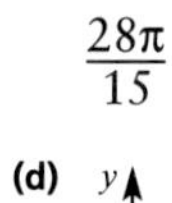

$\frac{28\pi}{15}$

(d)

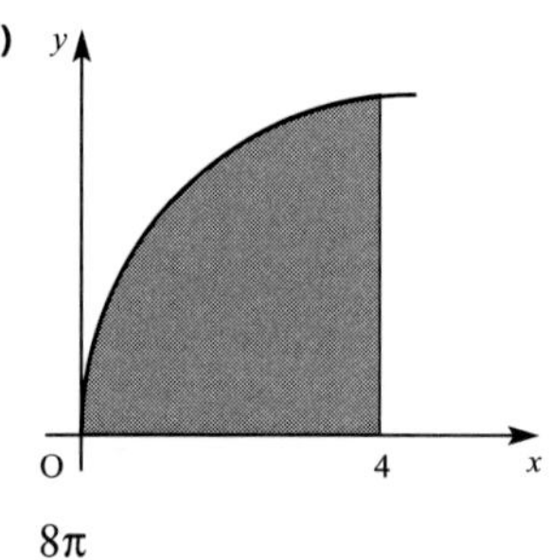

8π

(e)

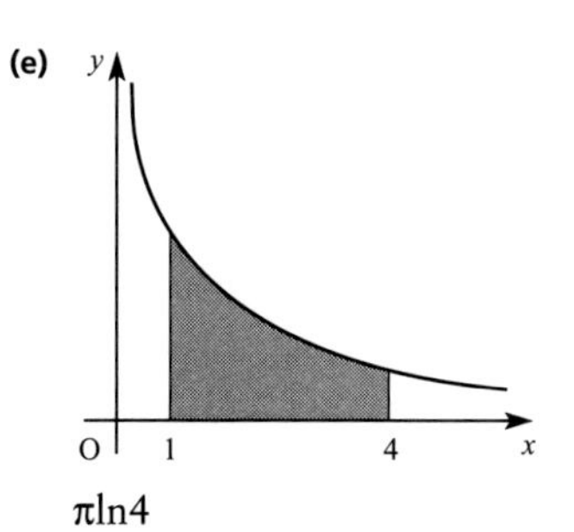

$\pi\ln 4$

(f)

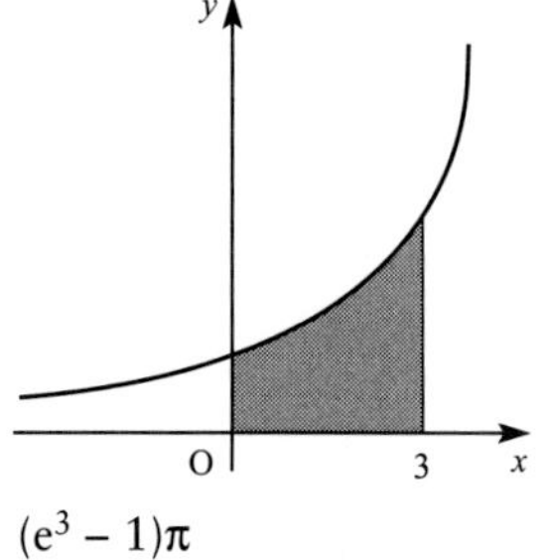

$(e^3 - 1)\pi$

2 **(a)**

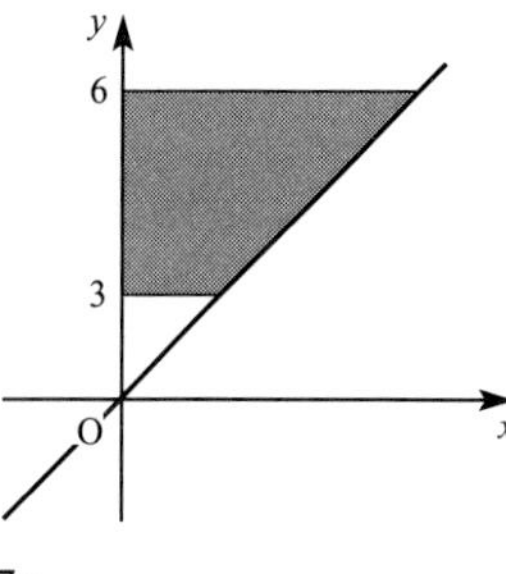

7π

(b)

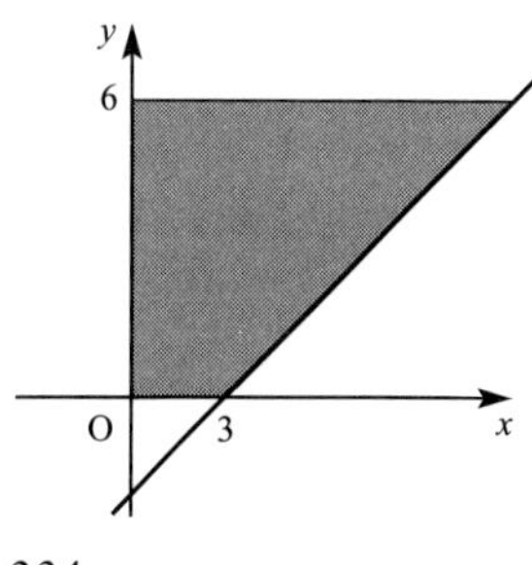

234π

(c)

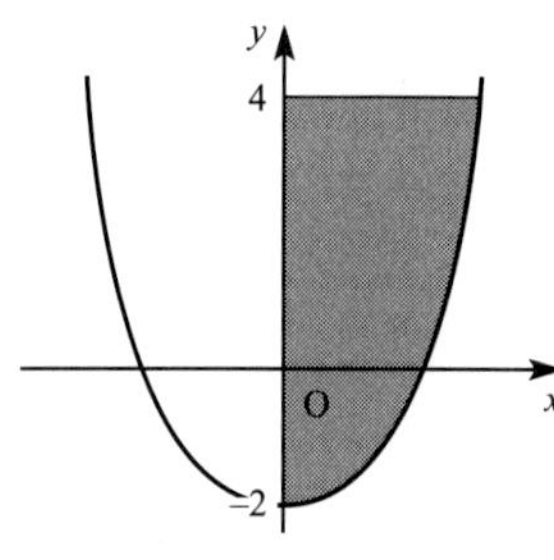

18π

(d)

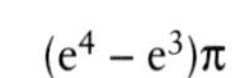

$(e^4 - e^3)\pi$

(e)

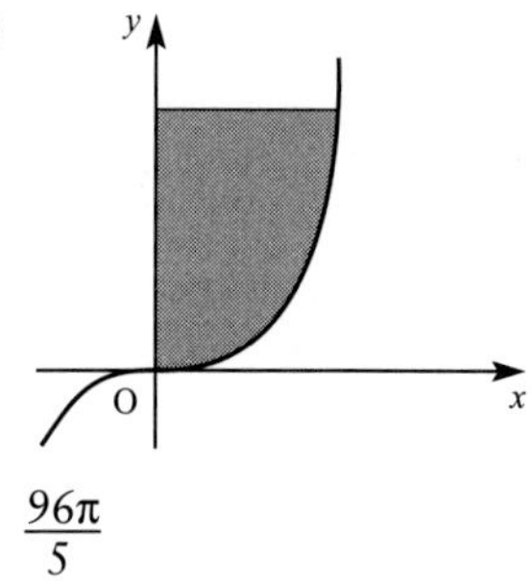

$\frac{96\pi}{5}$

(f)

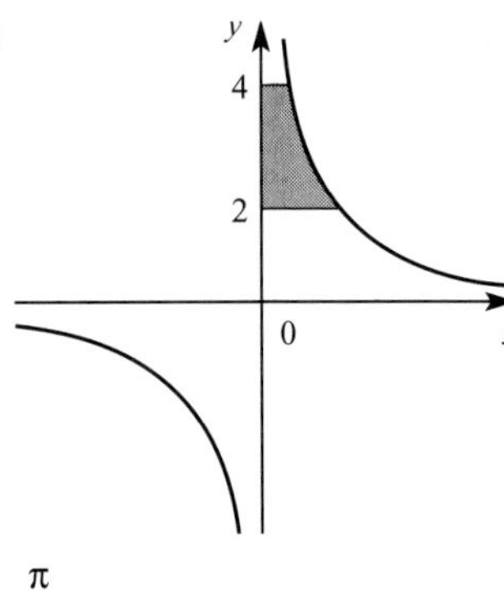

π

3 (a) A(–3, 4) B(3, 4)
(b) 36π

4 (a)

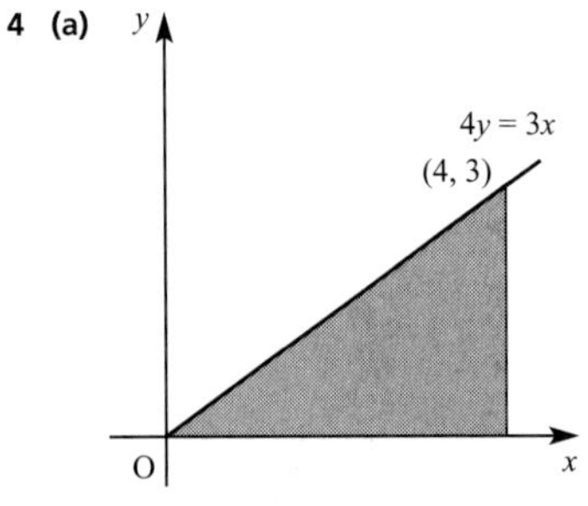

(b) 12π

5 (a) $\frac{333\pi}{250} = 1.332\pi$
(b) $2\pi\ln 10 = 4.61\pi$

6 (a) 2094 cm^3
(b) 1038 cm^3

7 (a)

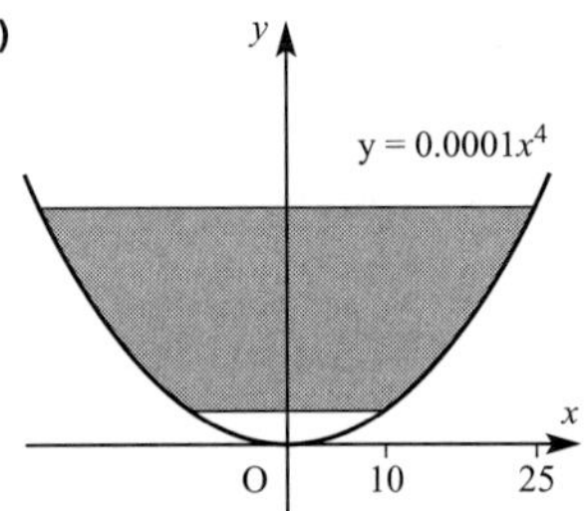

(b) 43.2 litres

8 (a)

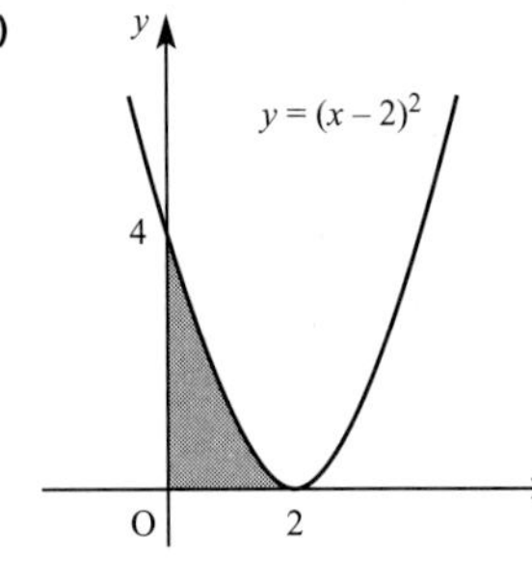

(b) $2\frac{2}{3}$
(d) $\frac{32\pi}{5}$

9 (a) $\frac{726\pi}{5}$
(b) $x = 1, x = 3$

EXERCISE 7D (Page 159)

1 (a) $\frac{1}{2}(e^2 + 7)$
(b) 94.9

2 (b) $p = \frac{62}{5}, q = 3, r = 2$ (or $q = \frac{3}{2}$, $r = 4$)

4 $y = \frac{2}{3}x^{\frac{3}{2}} + \frac{1}{4}\ln x + c$; $20 + \frac{1}{2}\ln 3$

5

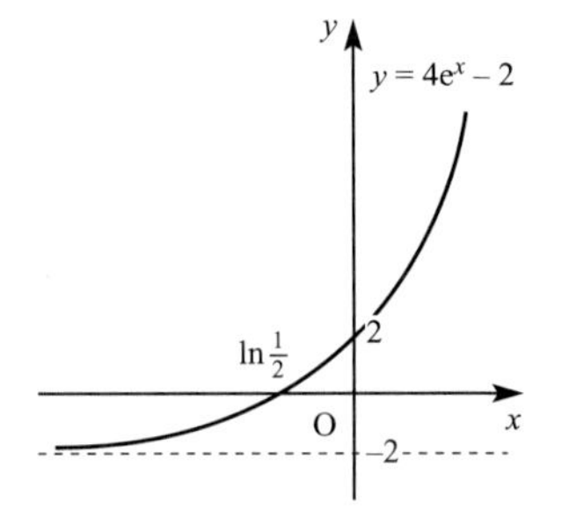

6 $a = 8$

7 $3e^2 - 5$

8 (b) 105

9 (a) $p = 5, q = -2$

10 $\frac{7\pi}{2}$

CHAPTER 8

EXERCISE 8A (Page 168)

1 –1.62, 1.28

2 (a) [–2, –1]; [1, 2]; [4, 5]
(b)

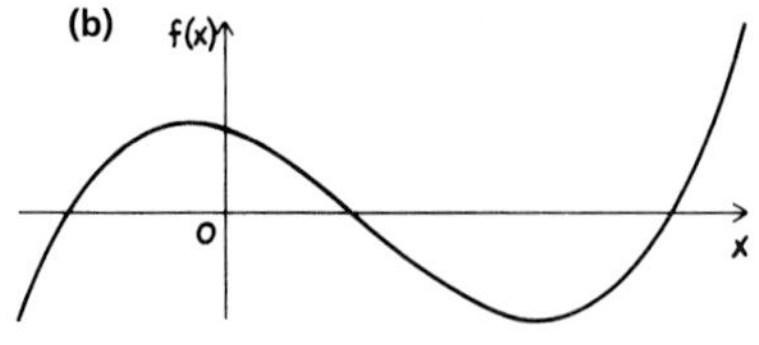

(c) –1.51, 1.24, 4.26
(d) $a = -1.511\,718\,75$ $n = -8$
$a = 1.244\,386\,172$ $n = -12$
$a = 4.262\,695\,313$ $n = -10$

3 (a) [1, 2]; [4, 5]
(b) 1.857, 4.536

4 (b)

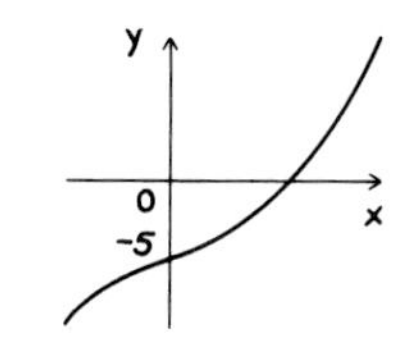

(c) 1.154

5 (a) 2
(b) [0, 1]; [1, 2]
(c) 0.62, 1.51

6 (a)

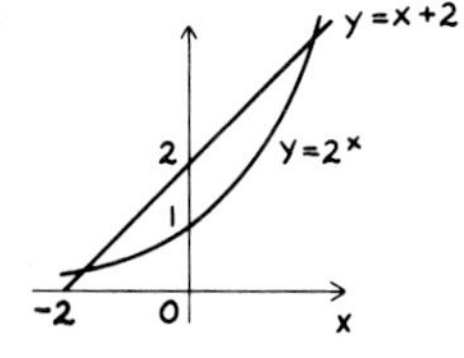

(b) 2 roots
(c) 2, –1.690

7 –1.88, 0.35, 1.53

8 (a) (ii) No root
(iii) Convergence to a non-existent root
(b) (ii) $x = 0$
(iii) Success
(c) (ii) $x = 0$
(iii) Failure to find root

9 $a = 4$
0.8

10 –3, 0.5, 2;
0.53

EXERCISE 8B (Page 176)

1 (c) 1.521

2 (c) 2.120

3 (c) $x = \sqrt[3]{3 - x}$
(d) 1.2134

4 (b) 1.503

5 (a)

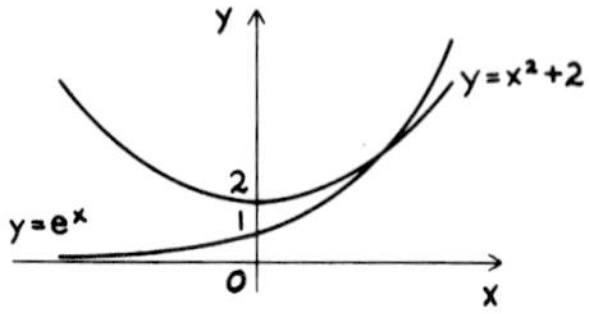

(b) Only one point of intersection

(d) 1.319 (to 3 dp.)

6 (a)

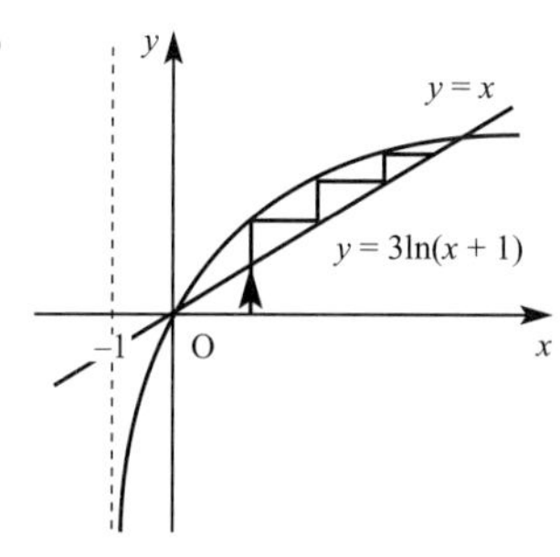

(b) 5.711

7 (a)

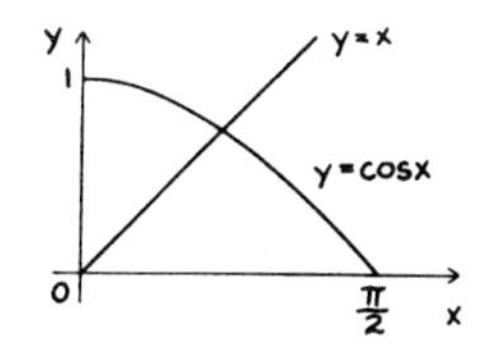

(b) 0.739 09

8 (a) $x^2 - 3x + 1 = 0$

(b) 0.382

(c)

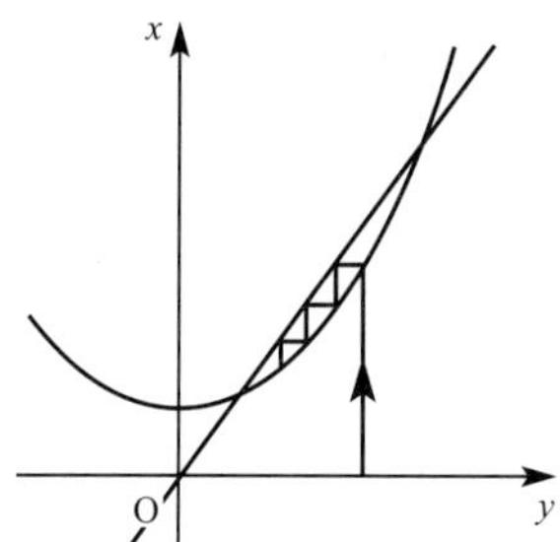

(d) 2.62

EXERCISE 8C (Page 182)

1 2.6286, 2.5643, 2.5475; overestimate

2 5.1667, 5.1167, 5.1032; overestimate

3 1.2622, 0.9436, 0.8395; overestimate

4 6.1061

5 (a)

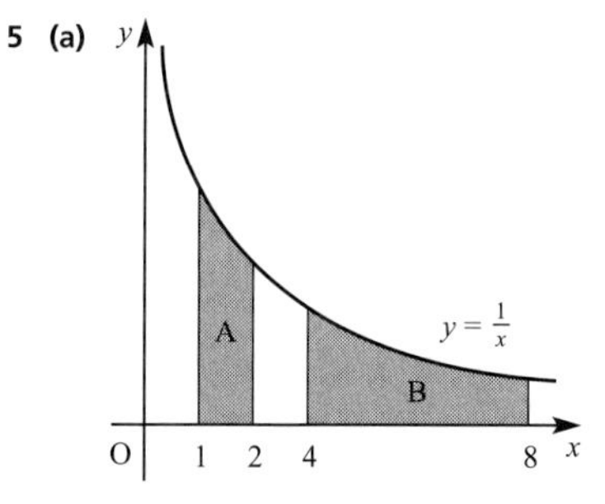

(b) for A and B

(i) 0.7083

(ii) 0.6970

(iii) 0.6941

(c) A and B are of equal area

6 (a) 458 m

(b) A curve is approximated by a straight line. The speeds are only given to 1 dp.

7 (a) 3.1311

(b) 3.1389, 3.1409

(c) 3.14

8 (a) 7.32775

(b) overestimate

9 (b) 12.6597, too small

(c) $2\frac{1}{3}$

(d) 0.055%

10 (a) 1.05101, 1.53862, 0.53565

(b) $1 + 5x^2 + 10x^4 + 10x^6 + 5x^8 + x^{10}$, 0.52715

(c) (i) Too high – trapezia above graph.

(ii) Too low – remaining terms would increase the value.

EXERCISE 8D (Page 185)

1 (b) $N = 23$

(c) $[-3, -2]$

2 (b) $N = -3$

(c) 1

3 (b)

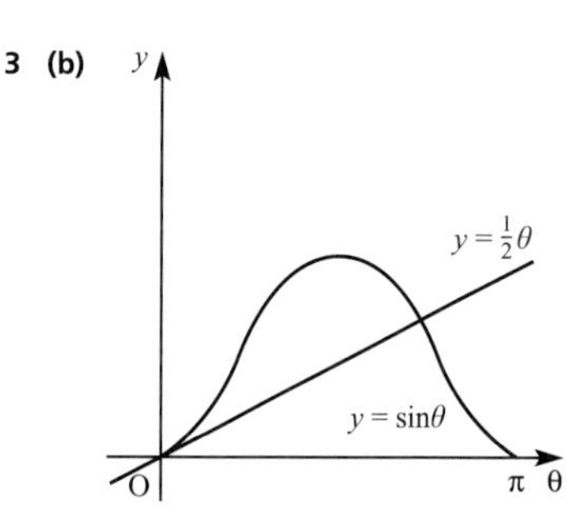

(c) [1.89, 1.90]

4 (a) $a = 2$, $b = 5$, 3.449

(b) –1.449

5 $p = 5$, $q = 6$, $r = 5$
1.708

6 (a)

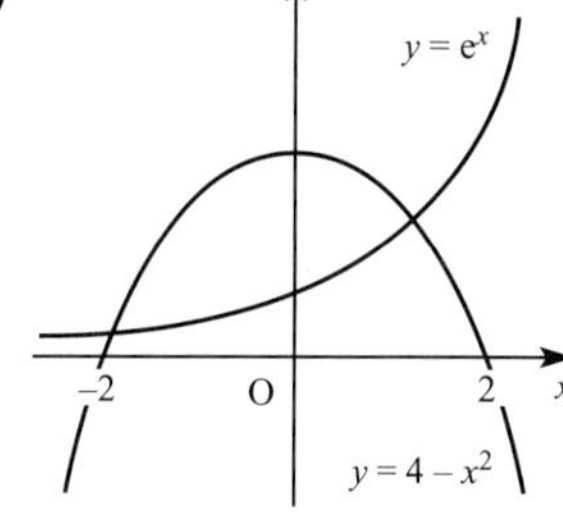

(b) –1.9646

(c) No real solution

7 (a) $a = 6$, $b = 2$

(b) 2.60

8 6.27
overestimate

9 0.5242
$\pi = 3.145$

10 (a) 5.382

(b) $e^2 - e + \ln 2$

(c) 0.336%

APPENDIX

EXERCISE AA (Page 191)

2 If $b^2 - 4ac < 0$ then there are no real roots

3 Not true when $n = 4$

5 If $A = 30$ and $B = 60$ then
LHS $= \infty$ and RHS $= \sqrt{3} + 2$.

7 Line 3

8 $n = 11 \Rightarrow 2047 = 23 \times 89$

10 Let $n = 2k + 1$
then $n^2 = 4k^2 + 4k + 1 \Rightarrow$ odd

11 $\cos 60^\circ + \cos 30^\circ = \dfrac{1 + \sqrt{3}}{2}$

$\cos 90^\circ = 0$

14

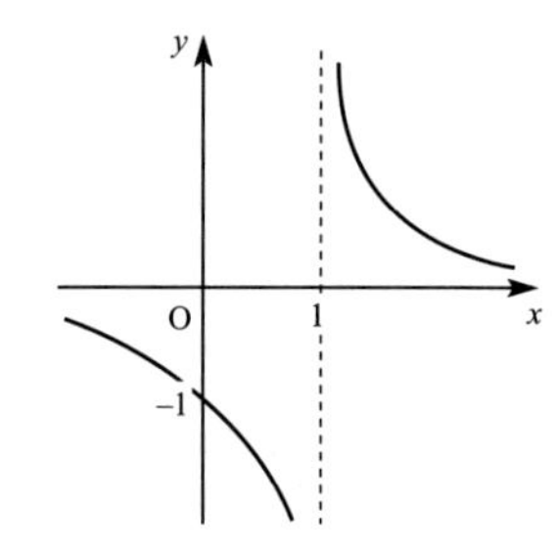

16

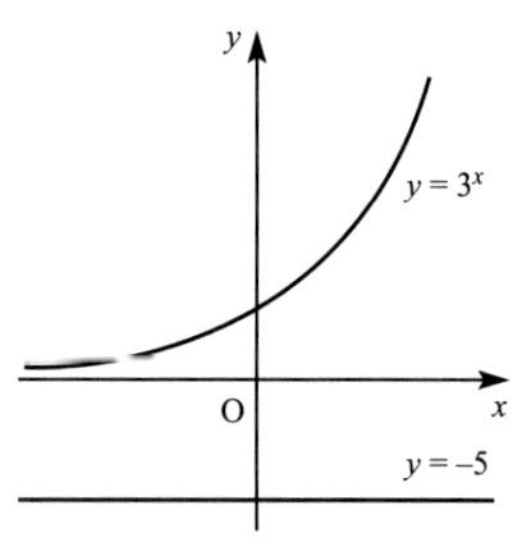

No point of intersection

17 Cannot integrate numerator and denominator separately.
$1 + \frac{1}{2}\ln 2$

20 Let p = largest prime.
Let $q = 2 \times 3 \times 5 \times 7 \times 11 \times \ldots \times p$
consider $q + 1$.

INDEX